The Landscape of Place-Names

Hope Dale, Shropshire, from Halford at its south end. Wenlock Edge centre, Aymestrey Limestone Escarpment on the right. Lower Dinchope is just visible, with its stream flowing west through Wenlock Edge. The other *hop* settlements, listed on page 136, are at springs from which streams flow through the eastern escarpment. On the left of the picture is a belt of territory which the Anglo-Saxons called *Langanfeld*, 'long open space'. The village of Cheney Longville, bottom left, derives its name from this, and so does Longville-in-the-Dale, nine miles north-east. Also in this strip is Felhampton *'feld* settlement'. The 'long open space' is bounded on the west by the Church Stretton hills. The Wrekin can be seen on the skyline (photograph: Arnold Baker).

THE LANDSCAPE OF PLACE-NAMES

by

MARGARET GELLING

and

ANN COLE

SHAUN TYAS
STAMFORD
2000

© Margaret Gelling and Ann Cole, 2000

Word-processed by the publisher

Published by

SHAUN TYAS
(an imprint of 'Paul Watkins')
18 Adelaide Street
Stamford
Lincolnshire
PE9 2EN

ISBN

1 900289 25 3 (hardback)

1 900289 26 1 (paperback)

Printed in the United Kingdom by the Alden Group, Oxford

CONTENTS

List of Illustrations	vi
Map of the pre-1974 County Boundaries	viii
List of Abbreviations	ix
Introduction	xii
Chapter 1: Rivers and Springs, Ponds and Lakes	1
Chapter 2: Marsh, Moor and Floodplain	36
Chapter 3: Roads and Tracks: River Crossings and Landing Places	65
Chapter 4: Valleys, Hollows and Remote Places	97
Chapter 5: Hills, Slopes and Ridges	143
Chapter 6: Woods and Clearings	220
Chapter 7: Ploughland, Meadow and Pasture	262
Appendix: the Chilterns: a Case Study, *Ann Cole*	288
Bibliography	317
Index	324

ILLUSTRATIONS

Figures
1. Map of pools at Baulking (p. 5)
2. Distribution map of *brōc* and *burna* in south-central England (p. 8)
3. *ēa-tūn* map (p. 15)
4. *flēot* map (p. 17)
5. Map of *flēot* names near The Wash (p. 17)
6. Map of Latin words in English place-names (p. 19)
7. *mere* sketches (p. 23)
8. *mere* map (p. 24)
9. *mere-tūn* map (p. 25)
10. Map of *ēg* names in the Ock Valley (p. 41)
11. *fenn* map (p. 45)
12. Map of *hamm* names in the Rother Levels (p. 51)
13. *hȳth* map (p. 84)
14. Map of Bleadney and environs (p. 85)
15. Map of *hȳth* names in The Fens (p. 86)
16. Hudspeth sketch (p. 89)
17. *botm* map (p. 99)
18. *corf* sketches (p. 103)
19. *cumb* sketches (p. 104)
20. *cumb* map (p. 105)
21. Map of Chilterns showing *cumb* and *denu* (p. 115)
22. *denu* sketches (p. 116)
23. *denu* map (p. 122)
24. *halh* sketches (p. 124)
25. *halh* map (p. 126)
26. *hop* sketches (p. 134)
27. *hop* map (p. 138)
28. (a) and (b) *beorg* sketches (pp. 146–7)
29. *clif* sketches (p. 154)
30. *copp* sketches (p. 158)
31. *crūg* sketches (p. 160)
32. *dūn* country (p. 164)
33. *dūn* sketches (p. 166)
34. *hlinc* sketches (p. 181)

35. *hlith* sketches (p. 183)
36. *hōh* sketches (p. 187)
37. *ofer* sketches (p. 201)
38. *ōra/ofer* map (p. 204)
39. Droitwich saltways (p. 205)
40. *ōra* sketches (p. 207)
41. Escrick/Wheldrake moraine (p. 215)
42. *scelf* sketches (p. 217)
43. *bearu*, 'grove' and **holt** map (p. 222)
44. *tūn* and *lēah* map (p. 238)
45. *sceaga* map (p. 246)
46. *æcer* map (p. 263)
47. *ersc* map (p. 268)
48. *feld* map (p. 273)
49. *land* map (p. 280)

Appendix figures
50. Chilterns: location (p. 288)
51. Pre-historic (p. 291)
52. Roman (p. 292)
53. *halh* (p. 294)
54. *hrycg* (p. 297)
55. *hōh* (p. 298)
56. *ōra* and *yfre* (p. 299)
57. *well*, *funta*, etc. (p. 300)
58. *burna*, *brōc*, etc. (p. 301)
59. *mere* (p. 302)
60. *ēg*, *hamm* (p. 304)
61. *fenn*, *mēos*, etc. (p. 305)
62. *weg*, *hȳth*, etc. (p. 306)
63. *ford* (p. 307)
64. *feld* and *falod* (p. 310)
65. *lēah* (p. 311)
66. *wudu* (p. 313)
67. *grāf*, *holt* (p. 314)
68. *fyrhth*, etc. (p. 315)

Preliminary illustrations
Frontispiece: Hope Dale, Shropshire.
Map of the pre-1974 county boundaries in England, Wales and southern Scotland (p. viii).

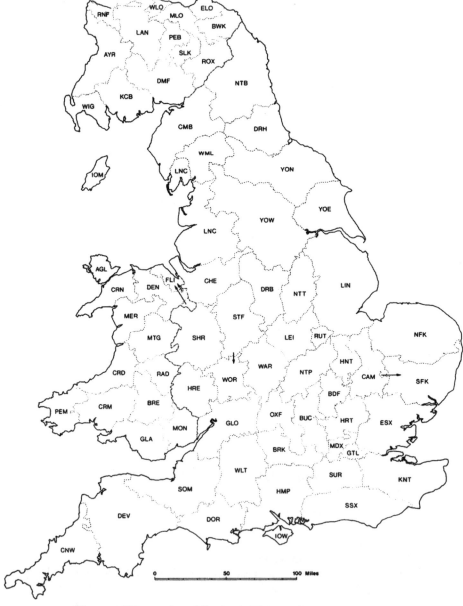

The pre-1974 counties of England, Wales and southern Scotland

viii

ABBREVIATIONS AND NOTES

COUNTY ABBREVIATIONS (the counties referred to are those which preceded the reorganisation of 1974).

ABD	Aberdeenshire	FLI	Flintshire
AGL	Anglesey	GLA	Glamorgan
ANG	Angus	GLO	Gloucestershire
ARG	Argyllshire	GTL	Greater London
AYR	Ayrshire	HMP	Hampshire
BDF	Bedfordshire	HNT	Huntingdonshire
BNF	Banffshire	HRE	Herefordshire
BRE	Brecknockshire	HRT	Hertfordshire
BRK	Berkshire	INV	Inverness-shire
BTE	Bute	IOM	Isle of Man
BUC	Buckinghamshire	IOW	Isle of Wight
BWK	Berwickshire	KCB	Kirkcudbrightshire
CAI	Caithness	KCD	Kincardineshire
CAM	Cambridgeshire	KNR	Kinross-shire
CHE	Cheshire	KNT	Kent
CLA	Clackmannanshire	LAN	Lanarkshire
CMB	Cumberland	LEI	Leicestershire
CNW	Cornwall	LIN	Lincolnshire
CRD	Cardiganshire	LNC	Lancashire
CRM	Carmarthenshire	MDX	Middlesex
CRN	Caernarvonshire	MER	Merionethshire
DEN	Denbighshire	MLO	Midlothian
DEV	Devon	MON	Monmouthshire
DMF	Dumfriesshire	MOR	Morayshire
DNB	Dunbartonshire	MTG	Montgomeryshire
DOR	Dorset	NAI	Nairnshire
DRB	Derbyshire	NFK	Norfolk
DRH	Durham	NTB	Northumberland
ELO	East Lothian	NTP	Northamptonshire
ESX	Essex	NTT	Nottinghamshire
FIF	Fife	ORK	Orkney

OXF	Oxfordshire		SSX	Sussex
PEB	Peebleshire		STF	Staffordshire
PEM	Pembrokeshire		STL	Stirlingshire
PER	Perthshire		SUR	Surrey
RAD	Radnorshire		SUT	Sutherland
RNF	Renfrewshire		WAR	Warwickshire
ROS	Ross and Cromarty		WIG	Wigtownshire
ROX	Roxburghshire		WLO	West Lothian
RUT	Rutland		WLT	Wiltshire
SFK	Suffolk		WML	Westmorland
SHE	Shetland		WOR	Worcestershire
SHR	Shropshire		YOE	Yorkshire (East Riding)
SLK	Selkirkshire		YON	Yorkshire (North Riding)
SOM	Somersetshire		YOW	Yorkshire (West Riding)

For the purpose of this book, county areas are those which obtained before the reorganisation of 1974. Except for the recognition of Greater London, this conforms to the policy pursued by the English Place-Name Society, and by other ongoing series such as the Victoria County Histories. The map (p. viii) shows these counties. To have used the post-1974 divisions would have made reference to EPNS volumes and to *The Concise Oxford Dictionary of English Place-Names* unacceptably complicated, and the 1974 arrangements have in any case been in a state of flux while the book has been under compilation.

The county surveys published by the English Place-Name Society are referred to as 'EPNS BRK' etc. Volume numbers are given when a county survey consists of several volumes paginated individually, but page numbers only are given when a several-volume survey is paginated consecutively.

OTHER ABBREVIATIONS

a.	*ante* ('earlier than')
acc.	accusative
c.	*circa* ('approximately')
dat.	dative
DB	Domesday Book
EPNS	English Place-Name Society
fem.	feminine
Fm	Farm
freq	*frequenter* ('frequently')
gen.	genitive
GR	grid reference
masc.	masculine

ME	Middle English
NED	*A New English Dictionary*, ed. J. A. H. Murray and others, 1888–1933 (reissued as *The Oxford English Dictionary*)
nom.	nominative
OE	Old English
OFr	Old French
ON	Old Norse
O.S.	Ordnance Survey
pers. n.	personal name
pl.	plural
PrCorn	Primitive Cornish
PrCumb	Primitive Cumbric
PrW	Primitive Welsh
sg.	singular
s.v.	*sub voce* ('under the word')
t.	*tempore* ('in the time of')

* An asterisk before a word or personal name indicates that the item is not on record except for its presumed occurrence in place-names.

Words and personal names which are cited in etymologies are OE except where otherwise indicated (i.e., by ON, PrW etc.).

Words which occur with their modern form and sense as qualifying elements in place-names (like *sea* in Seaford, *bright* in Brightwell, *rush* in Rushbrooke) are for the most part not translated: it can be assumed that when no translation is given the meaning is the same as in the modern language.

Throughout the book some familiarity is assumed with the commoner OE and ON habitative words, such as *hām, tūn, wīc, worth, bý* and *thorp*.

Place-name elements which are headwords in the book are printed in bold when they are referred to under other headings.

INTRODUCTION

The main topic to be addressed in this Introduction is the relationship of the present study to my earlier book, *Place-Names in the Landscape*, published in 1984 and reprinted in 1993. This was a pioneering work. It was the first study to be made of the type of settlement-name which has been labelled 'topographical'. These names, which define settlements by describing their physical surroundings, contrast with the other main type, labelled 'habitative', which has as the main component (the 'generic') a word for a farm, manor-house, village or town. The two categories overlap. The majority of English place-names are compounds of two elements, and in many of those which are classified as habitative because the generic is a settlement-term like *tūn*, *hām*, *cot*, *wīc*, *worth* the first element (the 'qualifier') is a landscape-term. Here may be instanced many common *tūn* names, like Compton, Denton, Fenton, Haughton, Marston, Merton, and such compounds as Eyecote GLO, Fordham CAM, ESX, NFK, Welwick YOE. The reverse phenomenon, in which a topographical term is qualified by a habitative one, is much rarer, but some instances occur: at Bottesford LEI, LIN, e.g., there was a building by the ford, and Berden ESX and Burden DRH were valleys in which there were byres.

In spite of this overlapping there is a clear distinction between the two ways of defining a settlement: in the topographical names the main emphasis is on geographical features which were felt to be of particular significance, rather than on buildings which had been constructed there.

For much of the 75-year-long history of organised English place-name study topographical names were held in low esteem by scholars. They were considered to have little to offer to the historian desirous of using place-name evidence in the reconstruction of the history of the post-Roman period. In fact the greatest of these historians, Sir Frank Stenton, said that they were 'trivial' and 'accidental'. Attention was concentrated on the habitative names and on some other categories, particularly the Hastings/Reading type which refer to groups of people and the small category which are believed to refer to pagan religious practices. Topographical settlement-names began to seem more important in the second half of the 1960s, when a number of studies disputed earlier beliefs in the chronological sequence of place-name types. This major upheaval in place-name studies is described in my book *Signposts to the Past*, published in 1978 and updated in editions of 1988 and 1997. Here it is sufficient to say that, contrary to earlier opinions, the topographical type of

INTRODUCTION

settlement-name emerged from these studies as the one most likely to have been coined by English-speaking immigrants in the 5th and 6th centuries. Once attention had been focused on these names it became apparent that they had much more to offer than had previously been appreciated, both to the mainstream and to the local historian. It also became apparent that they had hitherto unsuspected depths of subtlety and significance.

Failure to appreciate the nature of topographical settlement-names was due in part to the deficiencies of the modern English language. All languages are rich in words for things which are of major concern to the speakers, be it snow for Eskimos or camels for Arabs. Characteristics of the land are the primary concern of subsistence farmers, and the vocabulary of the Anglo-Saxons was extremely rich in that area. This is not the case in modern English. The quantity of words for hills and valleys discussed in Chapters 4 and 5 of the present book outnumbers the words in common use today, and this is far from being a complete tally, as only words which are comparatively frequent in settlement-names are included. This diversity of vocabulary is evidenced also in Old English terms for watercourses, wet areas, roads, woods and land used for various types of farming activity. The shrinking of the topographical vocabulary causes translations into modern English to be much less meaningful than the original name would have been to an Anglo-Saxon. Grendon and Longdon are merely 'green hill' and 'long hill' in modern English, and this gives no indication of the significance which attached to the word **dūn** when the names were first coined. In the 4th edition of *The Concise Oxford Dictionary of English Place-Names*, the standard reference book, Lynch is translated 'hill', Lythe is 'slope' and Hoo is 'spur of land'. Investigations reported in this book make it possible to enlarge considerably on such definitions, but discussion is required as well as translation. The inadequacy of modern English for conveying the full meaning of Old English place-name terms is perhaps most clearly instanced by the word **wæsse**, discussed in Chapter 2, which before 1984 was misleadingly accorded the translation 'swamp'.

Study of topographical names in relation to the actual landscape has made it clear that groups of words which can be translated by a single modern English word such as 'hill' or 'valley' do not contain synonyms. Each of the terms is used for a different type of hill, valley or whatever, and many of the words have connotations which are not simply geomorphological. Valleys called **cumb** offered totally different prospects as settlement-sites from those called **denu**; a hill called **dūn** was likely to be the site of a large village, while one called **beorg** might have a single farm or be the site of a church. Many topographical words would convey not just an image of the place but also a wealth of information about the likely size, status and pattern of farming practised by the community living there.

THE LANDSCAPE OF PLACE-NAMES

The key to Anglo-Saxon topographical naming lies in the precise choice of one of the many available words for streams, marshes, roads, valleys, hills, woods and farmland, and the concept which underlies both the 1984 book and the present one is that much of the precise meaning of these terms can be discerned by study of the names in relation to the existing landscape. Except where destroyed by motorways or new towns the contours are still there. Names relating to wet land and woodland present greater problems than those which refer to hills and valleys, but even here a great deal is still discernible. The principle of systematic field-work and map-work in the investigation of topographical place-names seems simple and obvious, but it was nevertheless overlooked until I embarked upon it in the late 1970s.

The 1984 book demonstrated the general validity of the concept, but it was based more on map-work than on field-work, and it was not illustrated, which was a fundamental defect since the subject is essentially visual. Also, since most of the statements were being made for the first time, each had to be defended in great detail. Much of what was argued at length there can be taken as established now, and this makes it possible to write more smoothly. Inside *Place-Names in the Landscape* was a more reader-friendly book crying to be let out, and it is hoped that *The Landscape of Place-Names* is that book. The articles on many of the elements in the earlier book contained long lists of names which some readers felt were an interruption to the argument. These lists have now been placed in separate reference sections, and etymologies can be more conveniently found there than in the glossarial index at the end of the earlier book. The division of the material into seven categories which was initiated in 1984 has continued to appear satisfactory, and the same or similar chapter headings are used in the present study.

A turning point in the study came in 1981, when Ann Cole attended a weekend school which I conducted at Rewley House (now the Kellogg Centre for Continuing Education) in Oxford. This encounter led to the partnership which has resulted in the present book. Ann is a geographer, while I am primarily a philologist: the benefits of this combination are obvious. Also, Ann is a skilled draughtswoman, and this has made it possible to produce an illustrated book. The illustrations comprise distribution maps, which are a vital element in all branches of place-name study, and line drawings which are made from photographs. Together over the years we have built up an archive of coloured slides which are used in lectures all over Britain, and the sketches printed here are traced from projections of these. The 1984 project was over-ambitious for one person with no specialised knowledge of geography or geology and no artistic or cartographic skills. The present study is much richer and we hope it conveys something of the pleasures and excitement of our shared addiction. The Appendix on The Chilterns is entirely Ann Cole's work. Otherwise the writing is mine while the maps (with two exceptions) and the line drawings are hers. Shared responsibility for checking and proof

INTRODUCTION

reading should have resulted in a higher standard of accuracy than was achieved in the 1984 book.

The years of field-work since 1984 have amply confirmed the major discovery, which is that these names represent a system which operated over most of England, from Kent to Northumberland and from the east coast to Offa's Dyke. It is not so apparent in the south-west peninsula, where English speech arrived several centuries later, and this accords with many other indications that the full glory of the topographical naming system belongs to the early part of the Anglo-Saxon period. In order to establish the country-wide currency of topographical terms it was necessary to amalgamate the information in the English Place-Name Society's county surveys and to unscramble *The Concise Oxford Dictionary of English Place-Names*, in which the words to be studied feature mostly as the second elements in compounds. Only a country-wide study in which the names were not treated in alphabetical order could have revealed the overall significance of topographical names.

The possibility of a systematic use of specialised terms for identical land-formations in Kent and Northumberland has been viewed with understandable scepticism by some critics, but we have an archive of photographs which demonstrates that this is so. The **ofer** which gives name to Shotover near Oxford corresponds closely to the **ofer** which overlooks Wooler in Northumberland. The **hōh** of Ivinghoe Beacon in the Chilterns, with its 'heel' and instep, is replicated by the end of the ridge on which stands the Northumberland village of Ingoe, and by many others in the country between them. The small, man-made ponds vital to farmers in dry areas are designated **mere** in the hinterland of Brighton and on the Yorkshire Wolds. Valleys called **cumb** and **denu** are clearly differentiated from each other in all areas except Devon, and everywhere a **beorg** is a small rounded hill and a **dūn** is a larger eminence which affords a particularly good settlement-site.

How could such a system arise in the circumstances of post-Roman Britain? Here one encounters philosophical and linguistic problems which I am reluctant to pronounce upon. Some guesses should probably be hazarded, however, though I should prefer to confine my role to that of demonstrating that the phenomenon exists.

The nature of the English landscape may be of crucial relevance. It is remarkably varied, and variety often occurs within small spaces. In this it contrasts with Continental landscapes, where everything is on a much larger scale. It is likely that immigrants accustomed to the vast coastal marshes and the great plains and forests of northern Europe were impressed by this variety and found it a linguistic challenge. They would share the same inherited vocabulary and they may, when faced with the same visual challenges in Kent and Northumberland, have responded with the same items in that vocabulary.

THE LANDSCAPE OF PLACE-NAMES

Another factor in the establishment of a country-wide system may have been the perceptions of travellers. This was not suggested in the 1984 book, but Ann Cole's researches over the last decade have steadily made it seem more important. We are not succumbing to a modern version of the Old-Straight-Track lunacy, but we are asserting that, as demonstrated by a number of maps in this volume, some landform terms show a more than accidental relationship to known ancient routes, like the Icknield Way, and to the major Roman roads which retained their importance throughout the middle ages. The perceptions of seafarers would also be important, as would the prominence of some riverine features when seen from boats. The discussions of **ofer** and **ōra** in Chapter 5 give details of meaningful relationships to views from ancient routes and from rivers and the sea.

Travellers' perceptions would probably not initiate a naming system, but they could have played a major part in its development and stabilisation. If a settlement was to acquire a topographical name there would often be more than one possible choice of feature to be referred to in the generic, and it is possible that travellers' perceptions of the place would influence the eventual selection. A country-wide system of topographical naming would certainly facilitate the giving of directions to travellers, and would also make it easy to know when a destination was in sight. A traveller to the Gloucestershire village of Bromsberrow, for instance, would recognise it by its rounded **beorg**, anyone approaching the Shropshire settlement of Wentnor would recognise the **ofer** as he traversed the Long Mynd, a person travelling to a place with a **dūn** name would expect a large, prosperous settlement in open country, and so on. I hope that readers will note which of the names discussed in this book are on routes which they travel and will look for the landmarks. Motorways (many of which run parallel with ancient roads) are excellent for this purpose. The close resemblance of the hills at Crookbarrow WOR and Creechbarrow SOM can be seen from the M5, for instance, and there is a good view of Crouch Hill overlooking Banbury from the M40. The view of the Gordano Valley, south of Bristol, from the M5 gives pleasure to the driver who knows that the name means 'open valley with a triangular shape'. It is, of course, important to remember that the lorry driver behind is not aware that one has this hobby!

The system of topographical naming which we claim to have decoded is in the Old English language which developed in this country among the Germanic immigrants who came in the centuries following the end of Roman rule in Britain. Most of the newcomers came in the 5th and 6th centuries. They adopted a few Celtic terms, such as *crūc* and *penn*, from the descendants of the Romano-Britons, and at a later date, in the late 9th and the 10th centuries, some Old Norse terms were integrated into the vocabulary. But the majority of English place-names, of every category, are Old English.

INTRODUCTION

The wholesale replacement of earlier British place-names by Old English ones has not been satisfactorily explained. The view current in the latter part of the 19th century and in the first half of the present one was that this new naming marked a new colonisation of a land where the population had dwindled, settlements had decayed and there had been widespread reversion to swamp and woodland. The cases for and against the 'clean sweep' theory of post-Roman history have been extensively argued, and the discussions are summarised in *Signposts to the Past* with updates in the 1988 and 1997 editions. It is my belief that the Germanic immigrants found a country where much of the land was under cultivation, and that infiltration into established settlements was at least as common as colonisation. The point which needs to be stressed here is that when I speak of Old English topographical settlement-names I mean names applied to settlements which for the most part had been long-established when the speakers of English first saw them. This has not yet been understood by all commentators. In the introduction to *A Dictionary of English Place-Names*, published in 1991, A. D. Mills says (p. xxii):

> Topographical names ... consisted originally of a description of some topographical or physical feature, either natural or man-made, which was then transferred to the settlement near the feature named ... Thus names for rivers and streams, springs and lakes, fords and roads, marshes and moors, hills and valleys, woods and clearings, and various other landscape features became the names of inhabited places.

This implies that the Anglo-Saxons gave names like Faringdon and Stottesdon to hills, those like Pusey and Charney to dry patches in marshland, or those like Harpenden and Gaddesden to valleys, and later transferred these names to settlements. I would ask the reader to look at the illustrations in this book, particularly Figs 10, 12, 21, 32 and 44, and to consider whether it is likely that the 'islands' in the Ock Valley, the 'promontories' in the Rother Levels, the 'valleys' in the Chilterns, the 'hills' in dūn country and all the 'clearings' denoted by lēah were uninhabited when the Anglo-Saxons came. It seems much more likely that there were settlements at many of these sites, and that the Anglo-Saxons were naming both site and settlement. I have used the term 'quasi-habitative' for many landscape terms which occur in names of places where geography dictates that that is where people would choose to live. I am in this respect an unrepentant geographical determinist: when reviewers first applied this term to my work I felt like Molière's character who found he had been talking prose.

The probability of the 'quasi-habitative' use of many Old English topographical terms has relevance also to the much-debated subject of the origins of the English village. A view much favoured in recent years is that the type of settlement known to historical geographers as 'nucleated' was a comparatively late phenomenon, perhaps linked to the adoption of open-field farming in the 10th century. In the last two decades several landscape historians have developed the model of small, shifting settlements which only coalesced into nucleated villages at the end of the Anglo-Saxon period; and field-walking and study of pottery recovered by this means support the view that in the East Midlands and in East Anglia there were a number of small early Saxon settlements in parishes which later contained a single large village. It has not, however, been demonstrated that this model is relevant to all areas; and many of the settlements whose names are discussed in this book are in situations so closely delimited by topography that it is hardly possible that there should have been a significant change in their position since they were named by the earliest English speakers. The ēg settlements of the Ock Valley, the Somerset Levels and the eastern fenlands can hardly represent late coalescing of smaller settlements in the surrounding wetlands. The **hop** settlements of Hope Dale in Shropshire must always have been in the funnel-shaped hollows between Wenlock Edge and the Aymestrey limestone escarpment. Some less constricted sites offer scope for movement: settlements with names in **dūn** occupy large hills, and those with names in **denu** could have moved up and down their long valleys, but this would not have made the names any less applicable to their sites. The evidence adduced in this book for a precise relationship between topographically-named villages and the geographical features referred to by the generics of their names, deserves consideration by people engaged in discussions about village origins.

When the current of historical thinking shifted in the 1960s from the 'clean sweep' theory of post-Roman developments to a belief in a large measure of continuity, some extravagant assertions were made about the antiquity of most settlements in the English countryside. There was a period in which a hostile response was encountered by suggestions that any settlement might be of post-Roman origin, and this attitude was taken even to those with Old Norse names. Historical opinion swings like a pendulum from one extreme to another, and here, as in other disputes about the course of events between the end of Roman rule and the Norman Conquest, place-name evidence can perform the useful function of steering people away from the lunatic extremes. It is a reasonable assumption that there has been a more or less steady expansion of settlement and arable farming from the Neolithic period to the present day: more or less, because there will have been ebb and flow, with reoccupation of sites after periods of disuse. Norse place-names in eastern and northern England are probably to be associated with an expansion of settlement after the Viking invasions of the late 9th and 10th centuries, but not many of these names have topographical generics. New settlements before that may be indicated by the use of **æcer**, **feld** and **land**, and some of the woodland terms discussed in Chapter 6 are also relevant here.

INTRODUCTION

There will have been expansion of settlement and formation of new names throughout the Anglo-Saxon period, but it is my belief that many of the names discussed in this book date from the 5th century, and that they record perceptions of the landscape and of the situations of ancient settlements in that landscape which are those of the earliest Anglo-Saxon immigrants. In some areas where archaeology has demonstrated an early 5th-century Anglo-Saxon presence there are concentrations of topographical names referring to factors likely to have been of major concern to such immigrants. One such area is the Thames Valley from Reading to Faringdon where, particularly on the Berkshire side, clusters of name in īeg refer to dry sites in land liable to flooding, while others (including Wantage) refer to the canalised water channels of the vital drainage system, and to the fords which facilitated passage over streams and wet ground. South central Essex, the hinterland of the early Anglo-Saxon settlements near Mucking, is another. Here there are comments on the formidable Thames marshes (as in Thurrock, 'ship's bilge') and an emphasis on the dūn sites, such as Basildon, raised 50-100 feet above the wet ground. There are no relevant records from the 5th century, so in the nature of things none of these names is recorded until centuries later, and it cannot be proved that they date from the earliest coming of Germanic immigrants: but there would be no later period at which such matters were of such immediate concern to English speakers, and it seems to me perverse to regard these names as replacements for other types which earlier scholars arbitrarily designated 'early'. It is most unlikely that the low hills of Basildon, Horndon, Laindon and Ockendon were bare landscape features when the early 5th-century Anglo-Saxon settlers of Mucking first saw this country, or that the upper Thames 'island' sites plotted on Fig. 10 had no dwellings when the first Anglo-Saxons were buried in the cemeteries between Reading and Faringdon, so, as stated above, the generics dūn and ēg should be regarded as denoting both settlement and site.

A good case can be made for the prevalence of topographical settlement-names in the earliest decades of English speech, but this does not, of course, mean that all such names are 'early'. They predominate also in some areas, such as the arc of country south and east of Birmingham, which may only have been transformed from a belt of pasture land to one of arable farmland at the end of the Anglo-Saxon period. Much valuable work has been done on the chronology of place-name types since the great reversal of opinion in the late 1960s, but only general guidelines have been established, and there is need for refinement and for detailed studies of particular regions. Some broad conclusions can be drawn from Professor Barrie Cox's analysis (Cox 1976) of English place-names recorded before A.D. 731, but allowance needs to be made for the bias of the available sources. Records from this date are predominantly concerned with ecclesiastical history, which leads to the inclusion of a large number of monastic sites: about 50 in the total of 224 places. Sites of synods and episcopal seats are also prominent, and there are several battlefields. This corpus may not give a valid indication of the types of name borne by farming settlements. Another bias lies in the accidental nature of record survival, which leads to the names in question being heavily concentrated in some counties, particularly Kent, Sussex and Northumberland, while some other counties have no names at all which are recorded in surviving documents dating from between *c.*670 and *c.*730. The distribution of the items included in Cox 1976 is shown on a map in Cole 1991.

In spite of the deficiencies of this corpus of 224 early-recorded names some salient characteristics noted by Cox probably have a wider significance. There were more British names current *c.*730 than was the case later. Among the English names there is a considerable preponderance of topographical items, though (as Cox notes) not all of these are settlement-names. The poor showing of *tūn* and **lēah** may fairly be taken as an indication that these only became overwhelmingly predominant after A.D. 730. A salient feature which is particularly relevant to the present discussion is the number of 'lost' names in Cox's list. More than 60 of his total of 224 are in this category, which suggests that a great deal of name replacement happened after the early 8th century and presumably had been happening from the 5th century on.

The unstable nature of place-names in the earliest centuries of English speech is a weapon which could be used to support contrary hypotheses. When scholars were clinging to the *-ingas* theory and defending it in the teeth of archaeological evidence one of the defences advanced was that some areas, particularly the Cambridge and Oxford regions, might have been well provided with names of the Reading/Hastings type but these were replaced later by the names shown in records to have been current from the 10th century on. There is no way of disproving this, or of proving my contrary assertion that names of the Faringdon/Wantage/Oxford type are likely to have been primary. A later stage of name replacement is, however, amply documented in west Berkshire and the adjacent part of Wiltshire. Records from the mid 10th century are abundant here, and these show that at that date names like Woolstone ('Wulfric's estate') and Aughton ('Æffe's estate') were replacing topographical ones like *Æscesbyrig* and Collingbourne. Details of these developments and a fuller discussion of the Berkshire, Essex and Warwickshire evidence can be found in Chapter 5 of *Signposts to the Past*, and there is a still more detailed discussion of the Berkshire evidence in EPNS survey, pp. 812–35.

Evidence concerning the nature of the earliest English place-names is circumstantial, and the best that can be attempted is the establishment of a probability. To the statistics of the earliest recorded names, which are not conclusive owing to the bias of the sources, can be added the predominance of topographical names in some areas of known early settlement and their numerical superiority over habitative names for the administrative units

INTRODUCTION

which became parishes. I feel that a probability has been established in favour of the early prevalence of topographical settlement-names, but whatever their place in the chronological sequence of English name-giving the names discussed in this book deserved closer attention than was accorded to them in the earlier decades of organised place-name study.

It is fortunate that English place-names lend themselves to categorisation, since otherwise it would be more difficult to obtain an overview of such a great mass of material. The major classes of habitative and topographical can be broken up according to the broad themes of the generics. Words such as **well, ēg, pæth, denu, dūn, wudu** and **feld** place items comfortably into the seven chapters of this book. The classification of qualifiers is more complex. Most words used as generics can be used by themselves as settlement-names, as in Well LIN, Eye HRE and SFK, Pathe SOM, Dean and Down in various counties, but the commonest method of name-formation is to combine the generic with another word which provides the fine tuning to the general statement. The number of words used as generics is vast, but that of qualifiers is, for practical purposes, uncountable. Categorisation of qualifiers is attempted here in the reference sections which follow the discussions of the more prolific generics. This is a rewarding exercise, though bedevilled to some extent by uncertainties, obscurities and ambiguities. Uncertainty is most apparent in the matter of personal names.

Our knowledge of the stock of Old English personal names is far from complete, especially as regards those which are likely to have been in use in the 5th and 6th centuries. It is legitimate to postulate the existence of personal names for which there are parallels in Continental records, and to assume 'short' forms of recorded dithematic names, hypocoristic forms (like, in modern terms, Bill for William), and 'weak' forms of recorded 'strong' names and vice versa. There can be no doubt that personal names do play a very large part in place-name formation, and that many which are otherwise unrecorded manifest themselves in this material. In the early years of English place-name study, however, the shibboleth of the 'unrecorded personal name' was invoked much too frequently, and more convincing explanations have since been found for many place-names whose qualifiers were assigned to this source in Ekwall 1960 and in the pre-war volumes of the English Place-Name Society. The problems which attend the identification of personal names in place-names were the subject of Chapter 7 of *Signposts to the Past*, and they are frequently referred to in the reference sections of the present book. There is as yet no authoritative comprehensive study of this vexed aspect of place-name interpretation. It is a theme which could be recommended to a suitably qualified young scholar embarking on research. Rough justice is meted out in the present book by the giving of asterisks to personal names which are not on independent record but have some validation as parallels to recorded Continental names or as formations of regular types based on recorded Old

English items, while consigning to the 'obscure or ambiguous' category of qualifiers any personal names which are solely inferred from place-name evidence.

When allowance has been made for that and other problems, the identity of the majority of place-name qualifiers is sufficiently certain for analysis, and this often throws further light on the significance of the generics. Some landscape terms are particularly likely to be qualified by personal names: instances include **ēg, cumb, denu** and **halh**. For others the commonest qualifier is a descriptive term: this is so for **brōc, burna, well, ford, dūn** and **feld**, among others. With some generics the qualifiers are fairly evenly spread among a number of categories. The proportions of references to wild or to domestic creatures may be of interest, as may be those to wild vegetation or crops. Broad conclusions are drawn from the nature of the qualifiers in discussions of most of the generics included in the book, but this aspect is capable of much further development.

The main function of the reference sections is, of course, to tell readers what particular names mean. Where the qualifiers are descriptive, and sometimes when they refer to plants or wild creatures, readers will in some instances be able to observe the precise application, even in the modern landscape. The corrugated nature of the side of the **hōh** at Sharpenhoe BDF, the annual appearance of reed-canary grass in the streams of Husborne BDF and Hurstbourne HMP, the presence of a few ancient oaks at Oakhunger GLO, the stony surfaces at various Stanways and Stowells, the gravel in the river-crossing at Greatford LIN and the chalk at Chalford OXF are as visible to us as they were to the Anglo-Saxons. Perhaps more often than not, if a characteristic is referred to in a place-name, it will be discernible today.

Readers looking for a firm answer may be disappointed by the treatment of items classified as 'Names with obscure or ambiguous qualifiers', but it is hoped that these discussions will give an impression of the nature of the problems encountered in place-name interpretation. It remains to consider how, in the majority of cases, we know what a name means.

In England the study is based on written evidence. In the introduction to the third edition of *Signposts to the Past* I defined scientific place-name study as 'the deduction of etymologies from early spellings in accordance with established philological processes', and I pointed out that this method had been practised by philologists since the closing years of the 19th century. The amassing of the early spellings has, since 1923, been performed by scholars working for the Survey of English Place-Names, an ongoing project supported by the British Academy and the members of the English Place-Name Society. Important contributions have been made by Swedish scholars, to one of whom, Eilert Ekwall, we owe *The Concise Oxford Dictionary of English Place-Names*. There is now a vast quantity of material available in reference books, most of which are mentioned in the following chapters, with details set out in the Bibliography. Few place-names are transparent, and consultation of reliable reference books is essential for the reader who wishes to participate in this study. The subject has its own bibliography, *A Reader's Guide to the Place-Names of the United Kingdom*, by J. Spittal and J. Field, which covers all relevant publications from 1920 to 1989.

It is only possible to do synthetic work on place-names because so much material is available in print. The Survey of English Place-Names is far from complete, but its 73 volumes cover the whole or parts of 32 counties, and for some other counties there are reliable surveys which are not part of this series. Ekwall's Dictionary covers major names in all English counties.

The application of the term 'major' to the names in Ekwall's Dictionary requires explanation. It is a term which has been loosely interpreted, especially by the present writer, and both 'major' and 'minor' are employed without close definition or total consistency in the present book. Roughly speaking, 'major' names are those of ancient parishes or Domesday estates, but a settlement mentioned in an Anglo-Saxon charter might be included in this category even if it did not meet these two criteria. 'Minor' settlement-names are those of hamlets or single farms. This study is not concerned with the vast category of what are loosely designated 'field' names, or with names of landscape features which are not referred to in settlement-names, though such material is drawn upon in discussions of some generics. This book is primarily a study of a certain category of settlement-name for which, thanks to Ekwall's Dictionary, country-wide coverage is available. Where detailed county surveys exist it is possible to take other categories into account, and there is, of course, a fuller treatment of minor settlement-names in such counties.

While paying due tribute to *The Concise Oxford Dictionary of English Place-Names* it is necessary to note that this is an old book, which was first published in 1935. The fourth edition, published in 1960, supersedes earlier ones, and it is essential that this edition is the one to be used. The year 1960 is, however, a distant point in place-name studies, and much that is to be found in Ekwall's great work is now out of date. Developments in the subject have been rapid since the late 1960s, and they are moving very fast indeed at present. It is unfortunate that the dictionary nature of publication gives the impression that etymologies are definitive, and that this causes surprise and irritation at the shifts of opinion which characterise the subject now. Anyone comparing the etymologies offered in the present book with those in the glossarial index to its 1984 predecessor will note a good many differences. Constant revision and correction must be part of the nature of this study, and there is no way of avoiding the inconveniences caused by this process. The changes, though frequent, are, however, peripheral, and the core of the subject remains sound enough to carry the weight of historical interpretation placed upon it. The bibliography provided here should enable readers to track down the articles which have prompted revision of etymologies proposed in earlier works. The bibliography also constitutes a tribute to colleagues on whose work I have drawn. I hope all debts have been acknowledged in the text: if there are plagiarisms they were not consciously committed.

CHAPTER 1

Rivers and Springs: Ponds and Lakes

The words studied in this chapter comprise most of the terms used in Old English and Old Norse place-name formation for the various sources of water which enabled settlements to exist. Some refer to flowing and some to standing water. As frequently in topographical place-naming there is likely to be an inverse relationship between what is plentiful in the landscape and the objects selected as defining features; so that springs and ponds are particularly likely to be referred to in areas where water supply is a problem. In areas where few or none of the place-names refer to water it can be assumed that all settlements were adequately provided with it. The use of the name of a major river, like Cray, Darwen, Thame, or a great natural lake, like Ellesmere, Windermere, for a settlement has a contrasting significance; in these instances the settlement is probably the most important of a number which benefited from the resources and communication facilities of the river or lake.

It is clear from place-name evidence that Anglo-Saxon farmers attached great significance to springs. The most important 'spring' term is **well**. A **well** name may sometimes have been extended from the source to the river which issued from it, but it is probable that the original referent was always the source itself. It is very doubtful whether **well** could mean 'river' to Old English speakers, and it is unlikely that it ever had the modern sense, i.e. a shaft dug down to the water table. It is to be presumed that Anglo-Saxons did sometimes dig shafts to obtain water, but they were not so much given to this practice as were the inhabitants of Roman Britain. Old English **well** refers in the material studied here to places where water issues from the ground. If the spring was particularly prolific this could be noted by the use of the terms **ǣwylm** and **ǣwylle**. The Latin-derived term *funta* has been shown by the study in Cole 1985 to be mainly a 'spring' term, but she notes that there are several instances where the reference could be to shaft wells.

Springs are particularly persistent landscape features, and the water sources referred to in names containing **well, celde, funta, ǣwylm** and **ǣwylle** are worth looking for. No systematic attempt has been made to locate springs referred to by the Old Norse term **kelda**. In many instances springs at places with **well** names will be found to have the characteristics, such as 'broad', 'loud' or 'stone', referred to in names like Broadwell, Ludwell, Stowell.

THE LANDSCAPE OF PLACE-NAMES

Rivers and brooks, the other main sources of running water, are denoted in place-names by a hierarchy of terms based on size. Major rivers are **ēa** in Old English, **á** in Old Norse. Below this come **brōc** and **burna**, which are used for large streams, and below these a number of terms for rivulets. Bottom of the hierarchy, below **lacu**, **rīth** and **rīthig**, is probably OE *sīc*, well evidenced in field-names but not included in this chapter because the watercourses so called are too humble to give rise to settlement-names. Important new work on **brōc** and **burna** is summarised under those headings: they are synonymous as regards size, but they contrast in other respects.

The largest rivers mostly have their own individual names, many of them pre-English, and the use of these for settlements is an important aspect of place-name study about which something was said in the introduction to Chapter 1 of *Place-Names in the Landscape*.

The most important of the terms for bodies of standing water is **mere**. A great deal of work has been done on this term since 1984, leading to the recognition of several classes of **mere** names which is set out in Cole 1993 and represented cartographically on Fig. 8. Particular interest attaches to the small man-made ponds for which this term is used. Some are just stock ponds, but others performed the vital function of enabling settlements to exist in dry chalk land. These, like springs, are rewarding for field-workers.

á ON 'river'. Several rivers in CMB, LNC, WML and Yorkshire have names ending in *á*, e.g. Beela, Brathay, Greta, Liza, Rawthey, Rothay. The word is rare in settlement-names, but it does occur in Aby LIN, Acaster YOW, Ambleside WML (from *á* with ON *melr* 'sandbank' and *sǽtr* 'shieling'), and Ayresome in Middlesbrough YON ('at the houses of the river', from *ár*, gen. of *á*, and *húsum*, dative plural of *hús*). ON river-names in *á* were sometimes transferred to settlements in IOM, as Laxey, Cornaa ('salmon' and 'mill river'). Ayton YON (2) is probably OE *ēa-tūn* (see **ēa**) modified by Scandinavian speech.

ǣwylm and **ǣwylle** 'copious spring'. These terms have been regarded by place-name scholars as compounds of **ēa** 'river' with words meaning 'spring'. The Toronto Dictionary of Old English, on the other hand, considers them to be words for 'spring' with a prefix which was not **ēa**, but was confused with it. The manner in which both terms are used in place-names certainly indicates that they were considered to contain **ēa**.

ǣwylm and *ǣwylle* sometimes denote the source of a major river, as in Clyst William DEV (*Clistewelme* 1270) and Toller Whelme DOR. Ewen GLO (earlier *Ewelme*) is near the source of the Thames. Both words are also used for prolific springs which do not give rise to major rivers: Ewelme OXF, e.g., refers to a particularly fine spring used until recently for the production of water cress.

RIVERS AND SPRINGS: PONDS AND LAKES

ǣwylm is either used as a simplex name or combined with river-names: *ǣwylle*, in addition to simplex use, is compounded with *tūn* in Alton DOR, HMP, WLT, and in Carshalton GTL ('cress spring settlement', *ǣwylle* referring to a headspring of the river Wandle, as Ewell, nearby, refers to one of the sources of the Hogsmill River).

In charter boundaries and place-names the two words are not found north of GLO and ESX except for two outliers. Aldon in south SHR is 'hill with a copious spring', and Coundon near Coventry WAR (earlier *Cundelme*) is probably 'source of river *Cound*'.

bæc, bece 'stream, stream-valley'. The distribution of this term in place-names is patchy. It is probably not found at all in counties north of CHE, and there are some southern counties, e.g. BRK, OXF, where there is no trace of it in settlement-names, minor names, field-names or charter boundaries. Examples are concentrated in CHE, HRE and SHR, and in these counties (and to a lesser extent in GLO and WOR) the word occurs in minor names such as The Batch in a way which suggests that it had a continuous use in local dialect. It is also common in minor names in the Mendip area of SOM.

In SHR *bæc, bece* is the final element in the major settlement-names Colebatch, Pulverbatch and Welbatch, and it occurs in many minor names. In HRE it occurs in one major name, Evesbatch, and a number of minor names. In CHE there are the major names Betchton and Sandbach (from adjacent brooks), Bache and Comberbach. The term is rare in STF (Holbeche House in Himley being perhaps the only instance), and it has not been noted except for a few field-names in WAR. In GLO and WOR it occurs in minor names and field-names.

In the east midlands there are four major names which contain *bæc, bece*: these are Haselbech NTP, Hazlebadge DRB, Cottesbach LEI and Beighton DRB. The term does not appear to enter into minor names in the east midlands, and this contrasts with the way in which it is used in the west midland counties.

If it be accepted (as asserted in Chapter 5) that *bæc, bece* is not the final element in Landbeach, Waterbeach, Wisbech, Pinchbeck, Holbeach, then the term is rare in major names in East Anglia. A possible instance is Beccles SFK. Ekwall 1960 explains this as OE *bec-lǣs* 'stream pasture', and there is a tributary of the river Waveney to the west of Beccles which resembles the streams described by this word elsewhere. There is, however, evidence from Anglo-Saxon charters, and from some minor names and field-names, for the use of *bece* to denote a different type of stream in fenland areas.

The bounds of land near Whittlesey Mere CAM in a charter of 1022 refer to 'a narrow water three furlongs long called *Scælfremære bece*', and the charter mentions other stretches of water called *Trendmære bece* and *Deop bece*. *Scælfremære bece* ('stream by diver-bird lake') became Chalderbeach, the name of a farm in Holme HNT, and there is another minor name, Ashbeach, nearby in Ramsey. It will always be difficult in dealing with fenland names to decide between *bece* 'stream' and **bæc** 'ridge', and this problem is discussed under **bæc** 'ridge' in Chapter 5. It does, however, seem certain that *bece* was used for flat watercourses in deep fenland, and this usage is in marked contrast to that elsewhere, especially in the west midlands, for streams in narrow, very well-defined valleys.

The meaning of Batch in minor names, probably of recent origin, in the Mendips has not been ascertained. No correspondence has been noted with any clearly defined landscape feature.

As regards qualifiers, Colebatch SHR, Cottesbach LEI and Evesbatch HRE have personal names (*Cola*, *Cott, *Ēsa*), Holbeche STF and Sandbach CHE are 'hollow' and 'sandy', Haselbech NTP and Hazlebadge DRB refer to a tree species, Comberbatch CHE is 'stream-valley of the Britons' and Welbatch SHR had a wheel. The first element of Pulverbatch SHR is an unexplained *pulfre.

bæth, dat. pl. **bathum**, 'artificial pool'. Bath SOM derives from the dative plural, the earliest genuine charter reference being *æt Bathum* 796. It is called *æt thæm hatum bathum* in 864. Inscriptions on Anglo-Saxon coins minted here have *Bathan* or *Bathum*, and this name, which is obviously a reference to the elaborate constructions which made the waters of the hot springs available to Romano-Britons, was certainly the one in local use. Two grander names for the city occur in records dating from 965 to 973, a period within the reign of King Edgar and during the maximum influence of St Dunstan, who may have reformed the rule of Bath Abbey c.960. These names are discussed in Coates 1988. They are *Urbs Achumanensis* 965, *civitas Aquamania* 972 and *Acemannes ceastre, Acemannes beri* in the 973 annal of the Anglo-Saxon Chronicle which records King Edgar's coronation at Bath. One manuscript of the Chronicle has *Hatabathum*.

In Gelling 1984 *Acemannesceaster* was dismissed as an artificial construct due to reluctance to use a name meaning 'at the hot baths' in connection with the coronation. Professor Coates accepts that the name may have been a short-lived learned invention, but he points to a likely connection with *Aquae Sulis*, the Roman name of Bath, and suggests that there may have been a British name *Acumannā, meaning 'Aquae-place'. He agrees with Gelling 1984 that *Achuman* and *Acemannes ceaster* were not likely to have been in use in local speech in the 10th century. The source for them may, however, have been an ancient written record in which clerics found an early British name. The relation to the name of the Roman road, Akeman Street, has not been satisfactorily worked out.

There are a few major settlement-names containing *bæth* in addition to the SOM town. Baulking BRK supplies a good example of the use of the term for artificially constructed pools. Baulking is OE *Bathalacing*, which consists of a stream-name *Lācing* with the genitive plural of *bæth*. The stream, which is a small tributary of the river Ock, has a series of triangular pools along its course (Fig. 1). There is a large deposit of fuller's earth here, and this leaches into the pools. They may have been constructed for the purpose of soaking sheep fleeces, to which fuller's earth is beneficial: v. Cole 1994a.

Other settlement-names include Bale NFK, Bathley NTT (both with lēah); Bathampton WLT; Bampton and Morebath DEV; Bathe Barton DEV (where there is a deep pit in a field near the farm); Bathpool CNW; Moorbath DOR.

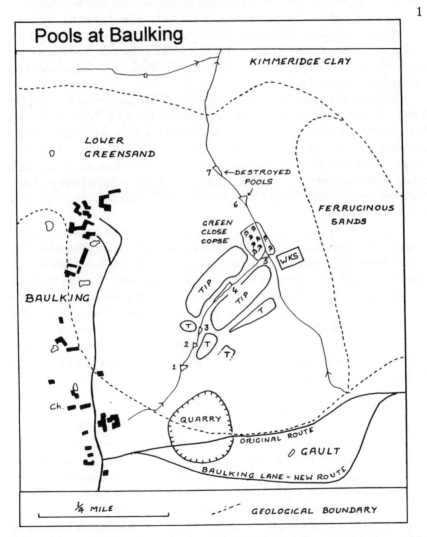

1

bekkr ON 'stream, beck', cognate with OE bæc, bece, is extensively used in place-names in some areas of Norwegian and Danish settlement, but the distribution is very uneven, both in northern England and in the east midlands.

In CMB and WML *bekkr* is much the commonest word for a stream in surviving place-names. It is common in Yorkshire, but less so in LNC. It is not used in early place-name formation in NTB and DRH. CHE has only one possible instance in a field-name. It is common in NTT but certainly rare in DRB and probably rare in LEI. No detailed place-name surveys are available for NFK and SFK. Recent volumes dealing with LIN reveal that *bekkr* is common in minor names and field-names in the wapentakes of Yarborough and Walshcroft. In NTP, CAM, BDF and HNT the term occurs occasionally in field-names.

Most occurrences of *bekkr* are in stream-names and field-names, but it is found in some ancient settlement-names. These include Caldbeck CMB, Ellerbeck YON, Fulbeck LIN, Hillbeck WML, Holbeck NTT, Maplebeck NTT, Skirbeck LIN, Skirpenbeck YOE.

Caldbeck and Fulbeck are 'cold' and 'foul', Holbeck runs in a hollow. Ellerbeck and Maplebeck refer to trees, the latter (like Fulbeck) being a hybrid OE/ON name. The qualifier in Ellerbeck is ON *elri* 'alder', and this compound recurs frequently in CMB, LNC, WML and Yorkshire. Skirbeck is 'bright', from ON *skirr* or a Scandinavianised OE *scīr*. Skirpenbeck probably means 'dried-up stream', from a derivative of the ON adjective *skarpr*. Hillbeck derives either from ON *hellir* 'cave' (referring to a deep valley) or ON *hella* 'flat stone'.

There are several instances of Holbeck. Ekwall 1960 translates this as 'hollow, i.e. deep, brook', but 'stream in a hollow' seems preferable. The stream which gives name to Holbeck parish NTT runs between rocks called Cresswell Crags. The stream at Holbeck Fm in Southwell NTT runs in a well-marked valley; this watercourse is *holan broc* in an OE document of 958, so in this instance the ON word has replaced an OE one. The stream called Holbeck near Ampleforth YON runs in a steep little valley, and so does the upper course of the stream which gives name to Holbeck on the southwest outskirts of Leeds. Beckermet CMB and Beckermonds YOW are ON names meaning 'junction of streams'.

brōc 'brook'. This word and **burna** are used in place-names for watercourses which are not quite large enough to qualify for the term *ēa*. There are only just over 100 major settlement names containing *brōc*, so in estimating its significance account must be taken of minor settlement-names and stream-names, especially the stream-names which appear in charter boundaries.

brōc is rare in NTB and DRH, where **burna** is the usual term, and in areas such as CMB, WML and parts of LNC and YON where **bekkr** is particularly frequent. Studies of the relative distributions of *brōc* and **burna** have led to differing conclusions. The distinction between them is certainly not related to size. The deciding factors have been considered to be ethnic (as between Saxon and Angle), chronological (with **burna** the earlier word), or geological. A discussion of the debate can be found in Cole 1991. In Gelling 1984 an ethnic basis was favoured, as against the preference of earlier authorities for a chronological one. Cole 1991 decided in favour of a geological basis, and there can be no doubt that the use of **burna** for clear streams and *brōc* for muddy ones is an important factor, even if it does not go back to the earliest stages of Old English speech (*v.* Fig. 2). The typical **burna** compound is with *scīr* 'bright' (as in Sherborne DOR, GLO, HMP, Sherburn DRH, YOE, YOW) while the typical *brōc* compounds are with *fūl* 'dirty' (Fulbrook BUC, OXF, WAR) and *hol* 'hollow' (Holbrook DRB, SFK, Holebrook DOR). There is no 'bright' brook (Shirebrook DRB is 'boundary brook' from *scīr* 'shire') and no 'foul' bourn (Fulbourn CAM is 'wild-bird stream'). Comprehensive study of the use of *brōc* in charter boundaries, however, has led Peter Kitson to the conclusion that this geologically-based distinction is the product of a dialectal distribution of the two words, rather than an aspect of their original meanings, and this leads back to an original 'ethnic' distinction. The debate will probably continue, but it does at least seem clear that the belief of earlier authorities that **burna** was an earlier word than *brōc* can be discounted.

Kitson (forthcoming) notes a special characteristic of *brōc* as used by charters in S.E. England. In flat marshy land in the general areas of the river Wantsum and Romney Marsh the word is used of very small streams, which indicates that in that dialect it was closer than it was elsewhere to meanings like 'marshland' which are found in West Germanic cognates.

brōc is common as a first element, particularly with *tūn*. When it is used as a generic the qualifiers are most frequently descriptive terms. There is a category of major settlement-names in which the qualifier is a personal name. Seven are listed in the reference section, and there are four possibles among names classified as having ambiguous first elements. This point needs stressing, as the omission of personal-name compounds from the discussion of *brōc* in Smith 1956 has led to later statements about this category being particularly rare.

Among compounds which refer to position, qualifiers meaning 'boundary', either literally or metaphorically, are relatively frequent, and Kitson notes that in charter names such references are very common with *brōc*. He also notes that more than half of the adjectivally qualified brooks in charter boundaries are either 'hollow' or 'foul'.

Reference section for brōc

As simplex name: Brook, IOW, KNT, Brooke NFK, RUT.

As first element with tūn: Bratton SHR, SOM, Brotton YON, Brocton STF(2), Brockton SHR(6), Broughton BUC(2), DRB(2), HNT, LNC(3), LEI(2), NTT, OXF(2), WAR, WLT, WOR, YON(2), YOW.

With hāmtūn: Brockhampton DOR, GLO(2), HRE(2), Brookhampton WAR, WOR(2). Brockhampton HMP and Brockington DOR are *brōchǣmetūn* 'settlement of the brook dwellers'.

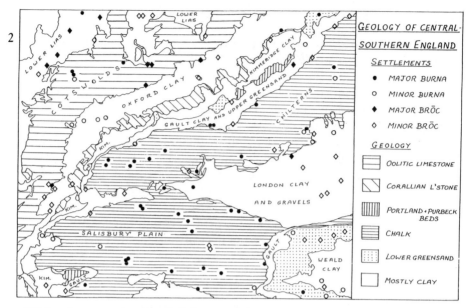

With other generics: Brockdish NFK (*edisc* 'pasture'); Brockley SFK (*lēah*); Brockweir GLO; Brockworth GLO; Brogden YOW (**denu**); Brookthorpe GLO; Brookwood SUR; Broxfield NTB.

As generic with descriptive terms: Braybrooke NTP ('broad'); Claybrook LEI ('clay'); Coalbrookdale SHR ('cold'); Dearnbrook YOW ('hidden'); Fulbrook BUC, OXF, WAR ('foul'); Holbrook, DRB, SFK, Holebrook DOR ('hollow'); Ludbrook DEV ('loud'); Sambrook SHR ('sand'); Sharnbrook BDF ('dung'); Skidbrook LIN ('dirt').

With personal names: Begbroke OXF (*Becca*); Binbrook LIN (*Bynna*); Budbrooke WAR (*Budda*); Cottesbrooke NTP (**Cott*); Didbrook GLO (**Dydda*); Forsbrook STF (*Fōt*); Ivonbrook DRB (*Ifa*); Ockbrook DRB (*Occa*).

With references to groups of people: Walbrook GTL ('Britons'); Whitsun Brook WOR (*Wixan*).

With references to vegetation: Bilbrook SOM, STF (*billere*, water plants); Ellenbrook LNC ('alder'); Rushbrooke SFK; Sedgebrook LIN; Tarbock LNC ('thorn'); Withybrook WAR.

With references to wild creatures: Beversbrook WLT; Birdbrook ESX; Bisbrooke RUT (earlier *Bittelesbroc* 'beetle'); Cornbrook LNC and Cranbrook KNT ('crane'); Kidbrooke KNT, SSX ('kite'); Sudbrooke (in Ancaster) LIN (*Suggebroch* 1168, 'sparrow').

With reference to a domestic creature: Swinbrook OXF ('swine').

With references to superstition: Ladbroke WAR (*Hlodbroc* 998, 'divination'); Purbrook HMP (*Pukebrok* 1248, 'puck'); Shobrooke DEV (*Sceocabroc* 938, 'goblin').

RIVERS AND SPRINGS: PONDS AND LAKES

With references to position: Lambrook SOM (*land*, probably in the sense 'boundary'); Marlbrook SHR, Meerbrook STF, Meersbrook DRB (*gemǣre* 'boundary'); Shirebrook DRB ('shire', it is on the DRB/NTT boundary); Sudbrooke (near Lincoln) LIN ('south'); Tachbrook WAR (believed to contain **tǣcels*, a derivative of *tǣcean* 'to teach', meaning 'boundary'); Westbrook IOW, KNT. There are two instances of Westbrook in BRK, both meaning '(place) west of the brook'.

With references to topography: Kelbrooke YOW (*ceole* 'gorge'); Polebrook NTP (*Pochebroc* 1086, 'pouch'); Wambrook SOM (*wamb* 'womb').

With references to buildings: Millbrook BDF, HMP; Skelbrooke YOW ('shieling').

With an earlier river-name: Bolingbroke LIN (*Bolin*); Carbrook NFK (*Kere*); Carisbrooke IOW (Lukely Brook may have been called *Cary*, a Celtic river-name found also in SOM).

Names with obscure or ambiguous qualifiers:

Bugbrooke NTP. The first element could be the personal name *Bucca* or a reference to bucks.

Colnbrook BUC, earlier *Colebroke*. The first element could be 'cool' or the personal name *Cola*.

Flashbrook STF. Early spellings indicate *Flotes-*. One of the meanings of OE *flot* may have been 'the act of floating', and a term for some sort of causeway may be very tentatively suggested.

Hinchingbrook HNT. Available spellings are late and inconsistent.

Shottesbrooke BRK. This could be either 'Scot's stream' or 'trout stream'.

Stitchbrook STF. Early spellings indicate *Sticeles-*. An OE **sticel(e)* has been postulated for Stittenham YON, with a meaning 'steep place'. Stittenham is in appropriately dramatic country, but it is doubtful whether the topography would justify such a term at Stitchbrook.

Tilbrook HNT. The first element could be the personal name *Tila* or the adjective *til* 'useful' from which the personal name derives.

Washbrook SFK. The first element could be 'flood' or 'place for washing'.

burna 'stream' occurs in about 155 names in Ekwall 1960. Various reasons have been adduced for supposing this to be a word which was favoured in the earlier part of the Anglo-Saxon period but which was replaced by **brōc**. Not all of these are sound. Cole 1991 shows that the appearance of five names containing *burna* and none with **brōc** in Cox's list of early-recorded names is likely to be an accidental result of the greater preservation of records for areas where *burna* is particularly common, and Gelling 1984 pointed to compounds with 'mill' in Milbo(u)rne DOR, NTB, SOM, WLT, Milburn WML, Melbourne DRB, which are relevant since the water-mill only became common in England in the 9th century. It is the obsolescence of *burna* which is to be questioned, not its earliness. The existence of four compounds with *hām* and the high incidence of major names in the corpus of *burna* material point to currency at the earliest date of English speech, but the word is likely to have remained current in areas where it was in common use, so there is no need to adduce influence from the ON cognate *brunnr* to account for its prevalence in northern counties (as Smith 1956).

The distribution of *burna* in settlement-names is uneven. It was in fairly common use in Saxon and Jutish areas (England south of the Thames and MDX, ESX), and remained so up to the date of the English takeover in DOR and DEV and the date of the diffusion of the water-mill to the south west. It seems clear that it was in common use in Northumbria before the Norse settlements. It is very common in Yorkshire and NTB, and there are four instances in major names in DRH. It is well represented in LNC and WML. In CMB, names in *burna* lie to the east of the county, adjacent to NTB and DRH, with the single exception of Stainburn.

Names in *burna* occur occasionally throughout East Anglia and LIN. Among major settlement-names there are four instances in CAM, four in NFK, three in SFK and seven in LIN. The term is also well represented in the area of the west midlands which formed the ancient sub-kingdom of Hwicce, but in most of the Mercian counties lying between and to the north west of these two areas *burna* is extremely rare. There is only one instance in the counties of HNT (Morborne), NTP (Lilbourne), NTT (Winkburn) and OXF (Shirburn). It is similarly rare, with only one or two instances, in CHE, HRE, LEI, SHR. There are three instances in STF (Bourne Vale in Aldridge, Harborne, Wombourn) but they are all in the south of the county, on the borders of Hwiccan territory.

The main exception to the scarcity of names in *burna* in those areas of Mercia which were not subject to Saxon influence is DRB, where 10 instances occur in DB, making a sharp contrast with NTT, where there is only one.

burna is sometimes used as a simplex name, the usual modern form being Bourne. A number of names which were simplex in DB later acquired prefixes: these include Eastbourne and Westbourne SSX. The little Stour river in KNT was alternatively *Bourne*, and Bekesbourne, Bishopsbourne, Littlebourne and Patrixbourne take their names from it. Three of them were simplex in DB, but Littlebourne ('small estate on the river Bourne') is so called in an Anglo-Saxon charter and in DB. In YOE the adjacent settlements of Battleburn, Eastburn, Kirkburn and Southburn probably represent divisions of a large unit called *Burn*.

burna is sometimes used as a first element, but it is commonest as the generic in compound names. As with **brōc** the largest category of compounds consists of those with words which describe the stream or its immediate surroundings. The qualifiers are quite frequently appropriate to the modern stream; at Whitbourne HRE, for instance, there is an efflorescence of lime in the stream bank. Apart from Winterbourne (discussed below) the commonest compound is with *scīr* 'bright'.

Again as with **brōc**, there are roughly equal categories which refer respectively to vegetation and to wild creatures. The vegetation qualifiers may still be appropriate, a striking instance of this being provided by Husborne BDF and Hurstbourne HMP. The qualifier in these is *hysse* 'tendril', and both streams are characterised by the trailing plant, reed canary-grass.

Personal names are not common with *burna*, and in this also there is a parallel with **brōc**. There are six names in which *burna* is the final element in an *-inga-* compound, and this, together with the number of instances with qualifiers of obscure meaning, could be held to support the status of *burna* as one of the earliest English place-name elements.

The distributional bias towards chalk country causes *burna* to be frequently used for intermittent streams. The recurrent settlement-name Winterbo(u)rne means 'stream dry in summer', and the two compounds with 'new' presumably have a similar significance. In some dialects *winterbourn* is used as a noun denoting such streams. In KNT they are called *nailbourns*, a generalised use of the name of the Nail Bourn, which is a tributary of the Little Stour. The Yorkshire term is *gipsey*, a generalised use of the name of the Gipsey Race, a notable intermittent stream on the Wolds.

The cognate ON *brunnr* 'spring, stream' is probably found in Burnham LIN(2) and Nunburnholme YOE, which are from the dative plural *brunnum*, and in Bourne LIN, Burnby YOE. The great majority of the eastern- and northern-England names listed in the reference section are, however, likely to be OE coinages, with spellings such as *-brunne* reflecting the influence of ON speech.

Reference section for **burna**

As simplex name: Bourn CAM, Bourne HMP, LIN. Other instances are noted in the article.

As first element with hām: Burnham BUC, ESX, NFK.

With tūn; Brunton NTB(2).

With other generics: Burnham SOM (**hamm**); Brundish NFK (*edisc* 'pasture').

As generic with descriptive terms: Blackburn LNC; Brabourne KNT and Bradbourne DRB ('broad'); Colburn YON ('cool'); Glusburn YOW ('bright'); Harborne STF ('dirt'); Holborn GTL ('hollow'); Honeybourne GLO/WOR; Meaburn WML and Medbourne LEI, WLT ('meadow'); Melchbourne BDF ('milk'); Morborne HNT ('marsh'); Newbourn SFK, Newburn NTB; Oborne DOR (earlier *Woburn*, 'crooked'); Saltburn YON; Sambourn WAR ('sand'); Shalbourne WLT ('shallow'); Sherborne DOR, GLO, HMP, WAR, Sherburn DRH, YOE, YOW, Shirburn OXF (all 'bright'); Shernborne NFK ('dung'); Sleekburn NTB ('glossy'); Stainburn

THE LANDSCAPE OF PLACE-NAMES

CMB, YOW and Stambourne ESX ('stone'); Whitbourne HRE ('white'); Wimborne DOR ('meadow'); Winkburn NTT (?'with corners'); Woburn BDF, SUR, Wombourn STF, Wooburn BUC (all 'crooked'). For Winterbourne *v. supra.* Harberton DEV is 'settlement by a pleasant stream'.

With references to vegetation: Albourn SSX and Auborn LIN ('alder'); Ashbourne DRB (Ashburnham SSX and Ashburton DEV contain *æsc-burna*); Ellerburn YON ('alder'); Fairbourne KNT and Fairburn YON ('fern'); Golborne CHE, LNC ('marsh marigold'); Hurstbourne HMP and Husborne BDF (*hysse* 'trailing water plants'); Iburndale YON ('yew'); Melbourn CAM (*melde*, a plant-name); Nutbourne (in Westbourne) SSX; Radbourn WAR, Radbourne DRB, Redbourn HRT, Redbourne LIN, Rodbourne WLT(2) (all 'reed'); Selborne HMP ('sallow willow').

With references to wild creatures: Auburn YOE ('eel'); Barbourne WOR ('beaver'); Broxbourne HRT ('badger'); Cabourn LIN ('jackdaw'); Cranborne DOR and Cranbourne HMP ('crane'); Enborne BRK (*ened* 'duck'); Exbourne DEV (*gēac* 'cuckoo'); Faulkbourne ESX ('falcon'), Fishbourne SSX and Fishburn DRH; Fulbourn CAM ('bird'); Hartburn DRH, NTB; Otterbourne HMP and Otterburn NTB, YOW; Pickburn YOW ('pike'); Rockbourne HMP ('rooks'); Roeburn LNC; Swanbourne BUC.

With domestic creatures: Lambourne ESX and Lambourn BRK; Shipbourne KNT ('sheep'); Swinburn NTB and Somborne HMP ('swine'); Tichborne HMP ('kid').

With personal names: Aldbourne WLT and Aldingbourne SSX (*Ealda*); Bedburn DRH (*Bēda*); Easebourne SSX (*Ēsa*); Ebbesborne WLT (**Ebbel*); Hagbourne BRK (*Hacca*); Nailsbourne SOM (**Nægl*); Ogbourne WLT (*Occa*); Simonburn NTB (*Sigemund*); Tedburn DEV (**Tetta*); Walleybourne SHR (*Walh*).

With -ingas group names: Bassingbourn CAM ('**Bassa's people'); Bathingbourne IOW ('**Beadda's people'); Collingbourne WLT ('Cola's people'); Hollingbourne KNT (the basis of the *-ingas* name is uncertain); Pangbourne BRK ('Pǣga's people'); Sittingbourne KNT ('people of the slope').

With other group names: Englebourne DEV (Angles); Walburn YON (Britons).

With references to position: Melbourne YOE ('middle'); Northbourne KNT and Nutbourne (in Pulborough) SSX ('north'); Sudbourne SFK ('south'); Tyburn GTL (**tēo* 'boundary'). Eastbourne and Westbourne SSX were simplex in DB. Eastburn YOW and Westbourne GTL are '(place) east/west of the stream'.

With topographical terms: Cliburn WML (**clif**); Ditchburn NTB; Morborne HNT (**mōr**); Slaidburn YOW ('sheep pasture'); Washbourne GLO (**wæsse**); Welbourn LIN, Welborne NFK, Welburn YON(2) (**welle**); Woodburn NTB.

With references to structures: Ledburn BUC ('leet'); Melbourne DRB, Milborne DOR, SOM, Milbourne NTB, WLT, Milburn WML ('mill'); Osborne IOW ('sheep-fold').

With a reference to sanctity: Holybourne HMP.

RIVERS AND SPRINGS: PONDS AND LAKES

Names with obscure or ambiguous qualifiers:

Barbon WML. This has been considered to contain *bera* 'bear', but it is doubtful whether bears were extant in WML when English speech became current there. Another possibility is ON *bjórr* 'beaver', in which case the generic is probably ON *brunnr*. The early forms are divided between *-brunne* and *-burne*.

Brinkburn NTB. The first element could be the personal name *Brynca* or 'brink'.

Calbourne IOW. The first element could be the plant-name *cawel*, 'kale', perhaps referring to sea-cabbage, or it could be another instance of the pre-English river-name *Cawel*, which has become Cale in SOM.

Duntisbourne GLO. A personal name **Dunt* has been conjectured from this place-name and two others, Dunsfold SUR and Downstow DEV.

Gisburn YOW, earlier *Gisleburn*. First element perhaps an adjective **gysel*, 'gushing', but there may have been a personal name **Gysla*, which would suit this name and Gisleham and Gislingham in SFK.

Kilburn DRB, YON. These are likely to have the same first element. A personal name **Cylla* has been conjectured to suit some place-names. Ekwall 1960 prefers *cylen* 'kiln', but there are no spellings with *-ln-* for either name.

Kilburn GTL. Possibly identical with the preceding names. But there are three spellings with *-n-*, *Cuneburna*, *Keneburna*, *Kyneburna*, which predate the series with *Kele-*, *Kelle-*, *Kylle-*, *Kulle-*, and these suggest 'stream of the cows', with *cȳna*, gen. pl. of *cū*. The letters *-n-* and *-l-* sometimes interchange in ME place-name spellings.

Legbourne LIN. Available spellings are *Lecheburne* DB, *Lecheburna* c.1115, *Lecceburne* 1158. Formally this could be from *læcc*, *læce*, with abnormal development to Leg- due to ON influence. But a word meaning 'boggy stream' seems inappropriate with *burna*.

Leybourne KNT (earlier *Lillanburna*, *Lilleburna*), Lilbourne NTP, Lilburn NTB. The easy answer is the personal name *Lilla*, but the triple occurrence raises doubts.

Leyburn YON. The ME spellings of the first element, *Le-*, *Lai-*, *Lay*, resemble those for Layham SFK, for which there is an OE spelling *Hligham*. A side-form **hlīeg* of *hlēo* 'shelter' has been suggested, but this is very tentative.

Naburn YOE. No satisfactory suggestion has been offered for the first element.

Warnborough HMP, *Weargeburnan* in a 10th-century charter. This has been considered to mean 'felons' stream', but in Hough 1995c it is argued convincingly that *wearg* may have had an earlier meaning 'wolf'.

Washbourne DEV. 'Flood' or 'place for washing'.

Wellesbourne WAR, *Welesburnan* and *Walesburnam* in OE sources. Perhaps the likeliest first element is *wēl* 'deep place in a river', but the word is rare in place-names and other examples are in more northerly counties. The variation of *-e-* and *-a-* is puzzling.

Weybourne NFK. ME spellings are *Wabrun(n)*, with a single *Walbruna* in 1158. No convincing etymology has been offered.

celde 'spring'. This is a rare term. The only major settlements which contain it are Bapchild and Honeychild KNT. Smith 1956 lists also Absol in Great Waltham ESX and Chillmill in Brenchley KNT. The former is a safe instance, but Chillmill is as likely to contain *cild* 'young nobleman'. Learchild NTB, which Smith classifies here but also lists under **helde** 'slope', is certainly a **helde** name, the topography offering a perfect example of the use of that term.

Bapchild is OE *Baccancelde*, from the personal name *Bacca*, and Absol (*Apechild* 13th cent.) also contains a personal name, probably *Eapa*. Honeychild (*Hunechild* c.1150) may have the plant-name *hūne*, 'hoarhound', as first element.

ēa 'river'. This word and ON **á** are cognate with Latin *aqua*. In OE *ēa* was the standard word for a major river. In the fenland of eastern England it became *Eau*, which appears to represent a combination of the OE and ON forms with influence from French; hence such river-names as Eau and Great Eau LIN. In DEV and SOM *ēa* became Yeo (but there are rivers called Yeo in both counties which have different etymologies). Elsewhere, as a river-name, it appears as Ray and Rea, with R- from ME *atter e* 'at the river'. Romney Marsh KNT is named from a river called *Rumenea* in OE.

ēa appears in settlement-names as a first element, most commonly with **tūn**. This compound becomes Eton, Eaton, Yeaton (SHR) and Ayton (YON). There are about 20 places with this name, mostly situated on major rivers, though some of the SHR examples are by large brooks. When all the *ēa-tūn* names are located and considered in the context of their immediate settlement-patterns it becomes clear that the significance of the name is functional, not locational. These places probably performed a special function in relation to the river, such as providing a ferry service or keeping a limited stretch of water open for the carriage of goods. Several are within the orbit of notable composite estates: Eaton Hennor HRE is by Leominster, Eaton Hastings BRK by Faringdon, and Water Eaton STF by Penkridge. The most telling consideration, however, is that the banks of the rivers by which most of them stand have numerous other settlements for which 'place by the river' would be just as appropriate if all that was intended was a geographical statement. Along the Thames are Eton BUC, Eaton and Eaton Hastings BRK, Castle and Water Eaton WLT, whose proximity to the river is no closer than that of villages to either side. Eaton NTT is one of a line of settlements – East

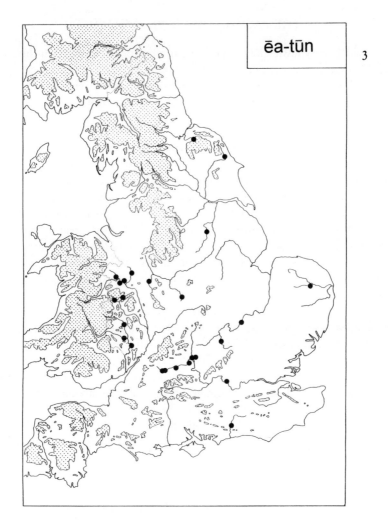

Retford, Ordsall, Eaton, Garmston – beside the Idle. Nuneaton WAR is one of numerous settlements beside the Anker.

In two instances the settlement is a double one: Wood and Water Eaton OXF and East and West Ayton YON lie one on each side of their rivers – Cherwell and Derwent.

It will be seen from Fig. 3 that *ēa-tūn* names are concentrated in the west midlands.

OE *ēa-gemōt* 'river-junction' is the source of Eamont WML and Emmott LNC, and other compounds with *ēa* as first element include two instances of Ewood in LNC, Ewart in NTB and Eythrope BUC. Ealand and Elland are discussed under **land**.

ēa is rare as a generic in settlement-names, but it does occur in Graveney KNT, Pevensey SSX and Welney NFK, and probably in Friskney LIN. Rivers of the status required for this designation had individual river-names, and it is

THE LANDSCAPE OF PLACE-NAMES

these which are transferred to the more important settlements on their banks (e.g. Bladon, Thame OXF, Frome DOR, HRE, SOM, Blyth NTB, Roden SHR).

A phrase meaning 'between the rivers' has become Tinhay, Tinney and Twinyeo in DEV, and Twineham in SSX. Another instance was *Twynham*, later Christchurch, HMP.

flēot 'creek, tributary of an inland river flowing through flat land'. Settlement-names containing *flēot* are mostly by streams flowing to the sea or into wide estuaries, which are tidal in their lower course. They occur on the east coast from Fleetham NTB to Ebbsfleet KNT, and along the south coast from several names on IOW to Flete on the river Erme DEV. Inland settlement-names with *flēot* occur mostly in eastern counties. Fleet HMP (near Farnborough) being probably the furthest west. Ekwall 1960 gives *flēot* as the first element of Flashbrook STF, but the early spellings do not suit, and Flashbrook would be a solitary outlier in the west midlands. As an occasional element in minor names and field-names *flēot* is more widespread, and it occurs in charter boundaries in BRK, NTT and SOM, as well as along the east and south coasts. The distribution is shown on Fig. 4, and the relationship of some names to ancient coast-lines on Fig. 5. Cole 1997 is a detailed discussion.

There is a concentration of *flēot* names in Yorkshire. Six examples – Adlingfleet, Broomfleet, Faxfleet, Ousefleet, Swinefleet, Yokefleet – are on either side of the Ouse and Humber, in the six-mile stretch east of Goole. Marfleet is to the east and Stillingfleet to the west of this cluster. Hunslet is by the river Aire on the outskirts of Leeds. Surfleet and Fleet near Holbeach LIN are now in the marshes inland from the Wash, and Fletton near Peterborough is on the edge of the Fens. Fletton is not certain to be a compound of *flēot* with *tūn*: it may contain the obscure stream-name *Flyte*, which occurs in the OE bounds of nearby Yaxley. Fleet Marston BUC is by a tributary of the Thames and Fleet HMP is between the headstreams of the Whitewater.

Cox 1990, 1992 presents evidence for an OE appellative *biflēot* meaning 'small piece of land cut off by the changing course of a stream', and this may be the origin of Byfleet SUR.

The river Fleet GTL was originally the lower end of river Holborn, but the name was later extended to the whole stream. The town of Fleetwood LNC was created in 1836 by Sir Peter Hesketh Fleetwood, whose surname presumably derives from an unidentified place in southern or western England.

Reference section for flēot

As simplex name: Fleet BUC, DOR, HMP(2), LIN, Flete DEV.

As first element: Fleetham NTB, YON (*hām*); ?Fletton HNT (*tūn*).

As generic: Adlingfleet YOW (*ætheling* 'prince'); Benfleet ESX ('tree'); Broomfleet YOE (pers.n. *Brungar*); Ebbsfleet KNT (uncertain); Faxfleet YOE (ON pers.n. *Faxi*, or OE *feax* 'hair' used of course grass); Herringfleet SFK ('Herela's

16

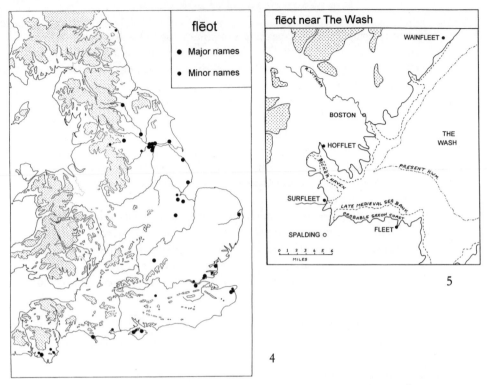

people'); Hofflet LIN ('hollow'); Hunslet YOW (pers.n. *Hun*); Marfleet YOE (*mere*); Ousefleet YOW; Purfleet ESX (obscure); Saltfleet LIN; Shalfleet IOW ('shallow'); Southfleet KNT; Sticelett IOW (uncertain); Stillingfleet YOE ('Styfel's people'); Surfleet LIN ('sour'); Swinefleet YOW (probably 'swine'); Trumfleet YOW ('circle'); Wainfleet LIN ('waggon'); Yokefleet YOW (ON pers.n. *Jókell*). The minor name Elphaborough, in Morley YOW, is *Elfletburgh* 1326; probably 'eel fleet' with *burh* 'fortified place' added.

funta 'spring', an OE adaptation of Latin *fons*, gen. *fontis*, or of late Latin *fontana*. Names containing *funta* are shown on Fig. 6, with other names containing Latin-derived elements. Some have modern -hunt owing to an interchange of *f* and *h* which is a regular phenomenon in place-names; Bedford and Bedmond have corrupt modern forms; Havant and Fovant show lack of stress.

The Latin word was borrowed also by British speakers (it is Welsh *ffynawn*), and Fonthill WLT and Fontmell DOR, in which it is combined with British terms, are best regarded as British rather than OE names. The combination with OE elements, however, together with the distribution pattern, points to the OE origin of most names in the corpus. In Bedfont, Bedmond and the lost Bedford in Eastbourne *funta* is qualified by *byden* 'vessel, tub', a word which has been noted more than 30 times with **well**.

17

It is surprising that OE speakers sometimes used this foreign word in preference to their native **well**. The neighbouring parish to Bedfont MDX is Stanwell, which suggests that there was a distinctive meaning. Fig. 6 shows a close spatial relationship to other names containing Latin loan-words. A convenient explanation for the use of *funta* might have been that these springs had Roman building work, but since this suggestion was mooted in Gelling 1977 no evidence has come to light which would support it. Like **ǣwylm** and **ǣwylle**, *funta* is often used of particularly copious springs, though not all instances fall into this category. This aspect is fully explored in Cole 1985. A number of minor names collected in that paper have been included in the reference section and on the map.

Funtington SSX presents a problem. It is difficult to resist derivation from *funta*, as there are copious springs at the place, but the *-ingtūn* formula, usually employed with personal names, seems inappropriate. Perhaps *tūn* has been added to a *funta* derivative, **funting*, meaning 'place of springs'.

The Cornish derivative of *fontana* is *fenten*, often Fenter- or Venton- in modern forms.

Reference section for **funta**

As simplex name: Founthill SSX (in Newick, *Funte* 1296), Funthams CAM (in Whittlesey, *Funtune* 13th cent., *Funtum(e)welle* 1423, 1446, *Fountains Well* 1602, *Funtams* 1636, dative plural).

As first element: Frontridge SSX (in Burwash, **hrycg**), Funtley HMP (**lēah**).

As generic with personal names: Chadshunt WAR (**Ceadel*); Fovant WLT (**Fobba*); Havant HMP (*Hama*); Pitch Font SUR (in Limpsfield, *Pīc*); Tolleshunt ESX (**Toll*); Urchfont WLT (**Eohrīc*); Wansunt GTL (*Wont*). Mottisfont HMP contains *mōtere* 'speaker', possibly used as a personal name, but perhaps indicating the holding of assemblies at this outstandingly fine spring.

With other qualifiers: Bedfont MDX, Bedford SSX (in Eastbourne, *Bedefonte* 1486), Bedmond HRT (in Abbots Langley, *Bedfunte* 1433) (all *byden* 'tub'); Boarhunt HMP (*byrh*, gen. of *burh* 'fort, manor'); Bonhunt ESX (probably *bān* 'bone'); Chalfont BUC ('calf'); Cheshunt HRT (*ceaster* 'Roman site'); Teffont WLT ('boundary').

In British names: Fonthill WLT ('fertile upland'), Fontmell DOR ('bare hill').

kelda ON 'spring' is only found in northern counties and Scotland. It is more frequent in minor names than in those of ancient settlements, but Kellet LNC, Kelleth WML (with **hlīth** as generic), Kendale YOE (DB *Cheldale*), Creskeld YOW ('cress spring') and Threlkeld CMB ('thralls' spring') are in the latter category. Minor names include several instances of compounds with **berg** (Kelber, Kelbarrow) and of 'cold spring' (e.g. Cawkeld in Watton YOE, Calkeld Lane in Lancaster). There are also several instances of Hal(l)ikeld 'holy spring', an

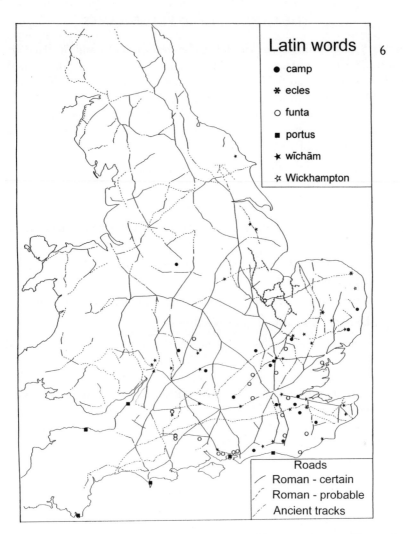

OE/ON hybrid. Keldholme in Kirkby Moorside YON is a well-recorded minor name.

kelda is cognate with OE **celde**, and Smith 1956 suggests that Keld- and -keld may sometimes be an ON adaptation of **celde**. This is very unlikely in view of the rarity and restricted provenance of the OE word. In the EPNS survey of WML (2, p. 186) Smith gives **celde** as the source of a ME field-name *le Chelde*, but this is probably an alternative spelling for *kelda*.

lacu OE 'small, slow-moving stream, side-channel'. This is not the source of Modern English *lake*, which was taken into ME from French. The OE word does, however, survive in modern dialects, or did until recently.

lacu is commoner in minor names, field-names and OE charter boundaries than it is in major settlement-names. In charter boundaries it only occurs S.E. of a line from the lower Severn via Avon and Welland to the Wash (Kitson, forthcoming), but it is more widely distributed in settlement-names and field-names, occurring, e.g. in Lacon SHR (from dat.pl., probably referring to drainage channels) and Potlock DRB. It may sometimes have been used for tributaries of major rivers, such as river Medlock LNC and this usage may also be seen in Mortlake GTL. The specialised use was, however, for a side-channel of a river, and this is clearly seen in the following names:

Bablock Hythe OXF, where Bablock was probably the name of one of the side-channels of the Thames;

Fishlake YOW, on a side-channel of the river Don;

Lake WLT, south of Amesbury, on a stretch of the Avon which has many side-channels;

Potlock DRB, S.W. of Derby, where the present course of river Trent may have been a side-channel;

Shiplake OXF, on a stretch of the Thames where there are side-channels;

Standlake OXF, on an arm of river Windrush.

As regards qualifiers, Bablock contains the personal name *Babba*, Fishlake refers to fish, Medlock contains **mǣd**, Shiplake refers to sheep and Standlake to stones. The first element of Potlock is 'pot', perhaps referring to depressions in the stream-bed. In Mortlake *lacu* is compounded with an unexplained element *mort* found also in Morthoe DEV (**hōh**) and a number of minor names. The compound with *lacu* is repeated in Mortlock's Fm in Radwinter ESX (*Mortelake* 1201). In Morcombelake DOR *lacu* has been added to an earlier name *Mortecumbe* (**cumb**).

lād, 'dyked water-course, canal', Modern English *lode*. In Gelling 1984 an attempt was made to draw a distinction of meaning between this word and the related term **gelād**, which is discussed in Ch. 3. The two terms were discussed together in Forsberg 1950, Smith 1956 and Ekwall 1960, though Ekwall made, rather tentatively, the distinction between senses, 'water-course' and 'passage over a river', which is asserted more firmly here.

lād is found mainly in minor names in the eastern fens and in the Somerset Levels. Whaplode LIN (with first element referring to a kind of fish) is probably the only example in a major name. Lode CAM, though now a parish, was formerly a district of Bottisham. The term is very common in the CAM fens, applied to innumerable drainage channels.

Dr Bryony Cole has suggested to me that *lode* in SOM minor names (such as Cogload and Curload near Lyng, Long and Little Load near Martock, Northload Bridge near Glastonbury) may refer to artificial channels used for transport. In literary OE *lād* means 'way, course, journey, conveyance', and from the last of these senses has developed modern *load*. The notion of load-bearing canals links the senses nicely. There is a notable instance of *lād* in an OXF charter dated 1005, which states that two weirs belong to Shifford, 'one above the *lād*, one below'. The reference is likely to be to the large channel called Great Brook which runs into the Thames here, and Blair 1994 (p. 121) commenting on this says 'Anglo-Saxon canals are a neglected topic'.

A meaning 'ferry' has been ascribed to *lād*, but this is not clearly evidenced in names of pre-Conquest origin. In Early Modern English *lode* was used for ferries on the river Severn, in the counties of GLO, WOR and SHR. This development may be a specialised use of *load* 'burden carried', influenced by the presence of a number of **gelād** names by the Severn. The complex problems of the 'ferry' names are discussed under **gelād** in Ch. 3.

Also discussed in Ch. 3 are a few minor names and field-names in which *lād* has the sense 'path, lane'.

There are several instances recorded in OE on the north coast of Kent of a term *gegnlād*, surviving in one example as Yantlet Creek near the Isle of Grain. P. Kitson (forthcoming) translates this as 'counter-lode', suggesting that it describes creeks where the ebb and flow of the tide would sometimes produce contrary movements of water.

læcc, læce OE 'stream, bog'. This relative of **lacu** is considered to be the ancestor of dialect *latch, letch*, used in the N.W. midlands of 'a stream flowing through boggy land, a muddy hole or ditch, a bog'. It is used as a river-name in river Leach GLO, this being transferred to the settlements of Eastleach and Northleach and combined with **(ge)lād** in Lechlade. It occurs as a simplex name in Lach Dennis (southeast of Northwich) and Lach (south of Chester) CHE. It is the first element of Latchford CHE, OXF (in Great Haseley), Lashbrook OXF, DEV (in Bradford), and the second element of Blacklache LNC (in Leyland), Brindle Heath LNC (in Eccles), and Cranage, Shocklach and Shurlach CHE. Brindle Heath is named from *Le Brendlache* 1324, with ME *brend* 'burnt' in some sense. Cranage, Shocklach and Shurlach refer to crows, goblins and scurf.

lœkr ON 'brook'. This ON word is the probable origin of Leake NTT, YON, LIN, Leck LNC, Leek STF. An OE **lece* has been postulated as the source of these, with ON pronunciation influencing the development. In Gelling 1984, however, it was noted that the distribution of the names is appropriate to an ON origin, and that there is no substantial evidence for the hypothetical OE **lece*.

mere OE 'pond, lake, pool', also 'wetland'. This word occurs in at least 120 names in Ekwall 1960. The size of the feature referred to varies from that of a duck pond in southern counties to that of extensive lakes in north SHR, CHE and the Lake District. The Anglo-Saxons do not appear to have felt the need of a different word for the larger areas of inland water. In so far as *mere* had a specialised meaning this was probably that the pond or lake was not part of a larger feature. For a land-locked bay of the sea, a wide estuary, or a pool in a major river, the Anglo–Saxons were more likely to use **pōl**.

Besides being used for standing-water features, *mere* was applied to areas of wetland, a sense first noted in Cole 1993 (p. 40) for what she designated 'a less cohesive group of *mere* names'. Bradmore and Gibsmere NTT, Fowlmere CAM, Fulmer BUC, for instance, probably refer to such areas. This may be the sense also in Sturmer ESX. The 19th-century OS map shows a wide pool on the river Stour, which is a feature for which **pōl** rather than *mere* would be expected. But the map also shows a wide belt of marsh on either side of this stretch of the river, so Sturmer is probably '(place overlooking) the Stour wetland'. This may be the sense in Mersea ESX.

Some settlements like Windermere WML, Combermere, Hanmer and Pickmere CHE, Ellesmere SHR, derive their names from large natural lakes. Contrasting with these there is a particularly interesting class of names in which the feature is a small man-made pond fed by surface water. In some areas (notably in chalkland) it is the creation and maintenance of these ponds which enabled the existence of settlements there. This was noted by the archaeologist O. G. S. Crawford (1953, 123–5) in a perceptive discussion which suggested that although *-mere* names are of Anglo-Saxon origin many of the ponds are likely to be prehistoric; and excavations of ponds in DOR and SSX have shown that in some cases they go back at least to Roman times. The names reflect the appreciation by English speakers of the crucial importance of the ponds to the settlements. Good examples, chosen from many, are Fimber on the Yorkshire Wolds, Catmore BRK and Stanmer SSX. Fimber and Catmore are shown on Fig. 7. Fig. 8 plots examples of *mere* names showing the various types of features to which they refer.

mere occurs frequently as a first element, most often in the compound *meretūn*, which gives modern Marten, Martin, Marton, Merton. There are about 30 examples, and these have been subjected to detailed study in Cole 1992. Some are by natural lakes (e.g. Marton near Worthen SHR, see Fig. 7). Some are by wetlands or seasonally flooded river-valleys. A few (Marten WLT, Martin Hussingtree WOR, Martin in East Langdon KNT) are by man-made ponds of the sort described above, and with these may probably be counted a few compound *-mere* names to which *tūn* has been added. Three names on the S.E. boundary of GLO, Rodmarton, Didmarton and Tormarton, have sometimes been considered to contain *(ge)mǣre* 'boundary', but the ponds at Rodmarton ('reed-pond settlement') and Didmarton are still there, and one is known to have survived until recently at Tormarton. In Farmington GLO (which despite the modern form is a 'thorn-pond settlement' like Tormarton) the pond may have been at Bittam Copse, ½ mile below the village. Tadmarton OXF is beside a stream with a very wet valley floor, where a 'toad

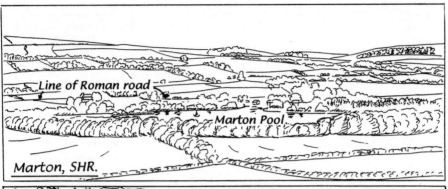

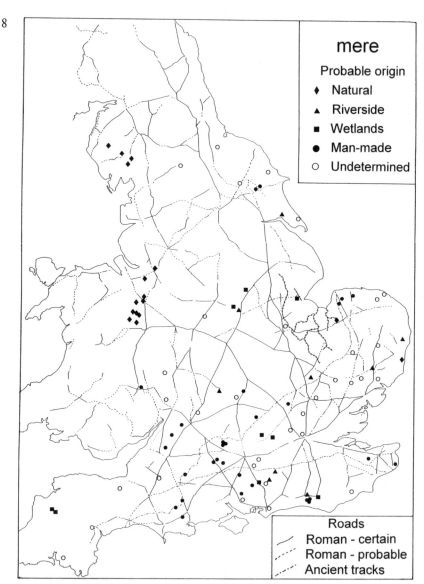

pool' would be a likely feature. It is probable that in most instances *mere-tūn* is a 'functional' name, given by English speakers to settlements where abundant water supplies could serve travellers as well as inhabitants, or where the resources of the *mere* were of outstanding importance in the economy of a composite estate.

mere-tūn names are mapped on Fig. 9. Cole 1992, 1993 should be consulted for more detailed discussion of all *mere* names.

The recurrent name Frogmore, in which -more is usually for earlier -*mere*, is discussed by Margery Guest in EPNS *Journal* 28, with the suggestion

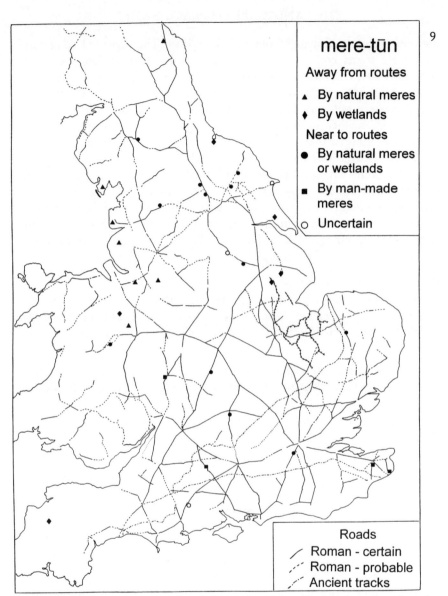

mere-tūn

Away from routes
▲ By natural meres
♦ By wetlands

Near to routes
● By natural meres or wetlands
■ By man-made meres
○ Uncertain

Roads
╱ Roman - certain
╱ Roman - probable
╱ Ancient tracks

that it refers to seasonally flooded areas which afforded breeding-sites for frogs.

A special use of *mere* is found in the ssx river-name Cuckmere, probably originally the name of the former estuary, from Alfriston south to where the river goes into the shallow inlet of Cuckmere Haven via a dramatic series of bends. Cuckmere is not recorded until 1275, so a certain etymology is not possible, but the name may consists of *mere* qualified by OE *cucu*, which is a form of *cwicu* 'living', modern *quick*. This word is recorded in OE as a qualifier for **wæter**, and NED (*quick*, a., sb.[1], and adv.9) gives watery instances meaning 'running, flowing' from c.1000 to 1889. The suggested compound with *mere* seems contradictory, since a *mere* is by definition a body of standing water, and the term is well-evidenced in SSX with that meaning. The regular place-name term for a body of flowing water is pōl. On the south coast, however, pōl is used for large natural harbours, as in Poole DOR, or for tidal creeks, neither of which senses is appropriate at Cuckmere Haven. Before the construction of Arlington reservoir, five miles up river, the Cuckmere estuary was liable to extensive flooding, possibly similar to the type of flooding which led to the use of the term **wæsse** in Anglian dialects. Lacking this word the early Saxons of Sussex may have constructed a term meaning 'live pond' to describe the phenomenon of a shallow body of water which formed temporarily in the estuary at flood time. It is possible that the river had the pre-English name Exe, and that this is preserved in Exceat in Westdean, which may be a group-name, 'dwellers by the Exe'.

Reference section for mere

As simplex name: Meare SOM, Mere CHE, LIN, WLT, Mareham-le-Fen LIN (dat.pl.). Meertown STF was earlier *Mere*. Delamere CHE is *mere* with French *de la*, which became prefixed because the name usually occurred in references to the 'forest of Mere'.

As first element with tūn: Marten WLT, Martin HMP, KNT, LNC(2), LIN(2), NTT, WOR, Marton CHE(2), LIN, LNC, SHR, WAR, WML, YOE(2), YON(4), YOW(2), Merton DEV, NFK, OXF.

With other generics: Marbury CHE(2) (*burh*); Mardale WML (**dalr**); Marfleet YOE (**flēot**); Marham NFK and Marholm NTP (**hām**); Marland DEV, LNC (**land**). Maresfield SSX and Mersea ESX contain the genitive *meres*, with **feld** and **ēg**. Marley, which occurs several times as a minor name, may sometimes be 'pond clearing' rather than 'boundary clearing', as may Mearley near Clitheroe LNC.

As generic with words for wild creatures: Almer DOR ('eel'); Anmer NFK and Enmore SOM ('duck'); Bridgemere CHE ('bird'); Catmore BRK; Cranmere SHR and Cranmore SOM ('crane'); Cromer NFK ('crow'); Eldmire YON (*elfetu* 'swan'); Finmere OXF ('woodpecker'); Fowlmere CAM and Fulmer BUC ('fowl'); Frogmore BRK, DEV, DOR, HMP, HRT; Hurtmore SUR ('stag'); Ilmer BUC ('hedgehog'); Peamore DEV ('peacock'); Pickmere CHE ('pike'); Swanmore HMP; Woolmer Forest HMP ('wolf'). Tadmarton OXF is probably 'toad-pool settlement'.

With domestic creatures: Boulmer NTB and Bulmer ESX, YON ('bull', this compound occurs also in the minor names Bomere SHR, Bowmoor GLO, Bulman Strands WML); Keymer SSX ('cow').

With references to vegetation: Ashmore DOR; Buttermere CMB, WLT; Dogmersfield HMP ('dock'); Grasmere WML; Haslemere SUR and Hazelmere BUC; Oakmere CHE; Peasemore BRK, Radmanthwaite (earlier *Rodemerthwayt*)

NTT, Redmarley WOR, Redmarshall DRH, Redmire YON and Rodmarton GLO (all 'reed'); Rushmere SFK(2); Tormarton and Farmington GLO ('thorn').

With descriptive terms: Blackmoor HMP and Blackmore HRT; Bradmore NTT ('broad'); Colmore HMP ('cool'); Falmer SSX ('fallow-coloured'); Fenemere SHR ('mouldy'); Holmer BUC, HRE ('hollow'); Livermere SFK ('liver-shaped' or 'liver-coloured'); Ringmer SSX ('circle'); Stanmer SSX and Stanmore GTL ('stone'); Tangmere SSX ('tongue-shaped'); Widmerpool NTT ('wide').

With personal names: Badlesmere KNT (*Bæddel*); Bosmere SFK (*Bōsa*); Cadmore BUC (*Cada*); Colemore SHR (*Cūla*); Didmarton GLO (*Dyddi*); Egmere NFK (*Ecga*); Ellesmere SHR (*Elli*); Gibsmere NTT (*Gyppi*); Imber WLT (*Imma*); Linchmere SSX (*Wlenca*); Monmore STF (*Manna*); Patmore HRT (*Patta*); Tedsmore SHR (*Tedd*); Windermere WML (Old Swedish *Vinandr*).

With river-names: Ismere WOR (*Ouse*); Kentmere WML (*Kent*); Sturmer ESX (*Stour*).

With topographical terms: Dummer HMP (*dūn*); Minsmere SFK (ON *mynni* 'river-mouth'); Seamer YON(2), Semer NFK, SFK, YON (*sǣ*); Sledmere YOE ('valleys'); Wetmoor STF ('river-bend').

With other qualifiers: Fimber YOE ('wood-pile'); Fishmere (earlier *Fiskermere*) LIN ('fishermen'); Southmere NFK; Walmer KNT, WOR ('Britons'); Wickmere NFK ('settlement'). In Throckmorton WOR *tūn* has been added to a compound of *mere* with *throcc*, a word which was probably used for a structure such as a platform.

Names with obscure or ambiguous qualifiers:

Bridmore WLT. On the evidence of the spellings, which start in 1185, this is one of several names which could contain *brȳd* 'bride'. There is, however, a feature called *Brȳdinga dic* in the bounds of a charter of 955, which was in this vicinity, and this appears to demand either a personal name or a place-name *Brȳd*.

Tusmore OXF. Most of the spellings indicate *Tures-* or *Tores-*, but *T-* could be for original *Th-*. The Anglo-Scandinavian personal name *Thōr*, *Thūr* may be the likeliest candidate. In Gelling 1984 an etymology 'pond of the thorn-ford' was tentatively suggested, but the reasoning behind this has proved impossible to recover.

Udimore SSX. This may be 'Udi's pond', with connective *-ing-*, but a comparison with Odiham HMP suggests another possibility. Odiham is *Wudiham* in the Anglo-Saxon Chronicle annal for 1116, indicating a first element *wudig* 'wooded'. Otherwise it is *Odiham* from DB onwards. There may have been a comparable disappearance of *W-* in Udiham, so 'wooded pond' is a possible etymology.

pōl 'pool', **pull, pyll** 'pool, tidal creek, small stream'. A degree of confusion is apparent in the treatment of these terms in dictionaries and place-name reference books, and their relationship to Celtic words (W *pwll*, Corn *pol*, Breton *poul*, Irish *poll, pull*, Gaelic *poll*, Manx *poyll*) has not yet been worked out. Smith 1956 and Ekwall 1960 treat *pōl* and *pull* as variants of the same word, which strictly speaking they cannot be, though spellings appropriate to both words interchange with each other in early forms for numerous English place-names.

pōl/pull is occasionally used of a harbour, as in Poole DOR, Hartlepool DRH. Skippool LNC (with ON *skip* 'ship') was formerly a harbour formed by the junction of two brooks near Poulton le Fylde. The *Pool* referred to in Netherpool and Overpool near Ellesmere Port in Wirral CHE was described as a port in 1366. This may be partly the significance in Welshpool MTG.

In some instances *pōl/pull* refers to creeks or coastal pools. North and South Pool DEV refer to a tidal arm of Kingsbridge Harbour, Poulton cum Spital and Poulton cum Seacombe in Wirral CHE refer respectively to the creek at Port Sunlight and to Wallasey Pool, which was formerly a great tidal inlet from the Mersey. Poulton (in Awre) GLO is near an inlet of the Severn, and Poulton le Sands LNC is near the coast. Blackpool LNC refers to a peaty pool about half a mile from the sea, and Liverpool LNC to a tidal creek with turbid water. Walpole NFK is in drained marshland by the Wash. Radipole DOR is by Radipole Lake, which opens into the sea at Weymouth.

Many names containing *pōl* are inland. Some instances refer to surviving pools. At Polstead SFK there is a large pool in the river Box, and at Claypole LIN a wide place in the river Witham. At Walpole SFK a large pool appears at times of flood in an ox-bow of the river Blyth. At Harpole NTP and Wimpole CAM the pools may have been converted into ornamental lakes in the grounds of the Hall. In many of the instances listed in the reference section, however, the reference must be to pools long since drained.

Cornish *pol* (which means 'pit' as well as 'pool, stream, cove, creek') is used frequently in Cornish names (Padel 1985, 187–89), but in some names (e.g. Polperro and Polruan) *pol* has replaced earlier *porth* 'harbour'.

pyll occurs in a few names in the south west in the sense 'tidal creek'; here belong Pilland DEV, Pilton, DEV, SOM, Pylle (near Pilton), Huntspill and Uphill SOM.

In OXF and BRK a ME derivative of *pyll* is used in field-names for a small stream. In Pilton RUT it is apparently used of the upper reaches of the river Chater.

Reference section for **pōl, pull**

As simplex name: Nether- and Overpool CHE, Pool DEV, YOW (near Byram), Poole CHE, DOR, GLO, Poolham LIN (dat.pl.).

As first element: Polehanger BDF (**hangra**); Polsloe DEV (**slōh**); Polstead SFK (*stede* 'place'); Poolhampton HMP ('settlement of the dwellers at the pool'); Poulton CHE(3), GLO, KNT, LNC(3); Pulborough SSX (**beorg**); Pulford CHE; Pulham DEV, DOR, NFK (*hām* or **hamm**).

As generic with descriptive terms: Blackpool DEV, LNC; Bradpole DOR ('broad'); Claypole LIN; Harpole NTP ('dirty'); Liverpool LNC; Radipole DOR ('reedy').

With references to animals and birds: Cople BDF (*cocc* 'cock'); Hampole YOW (*hana* 'cock'); Otterpool KNT.

With an ethnic term: Walpole NFK, SFK ('Britons': the NFK name has been considered to contain *wall* 'wall' with reference to a Roman embankment, but the available spellings, beginning with *Walepol* c.1050, indicate the gen.pl. of *Walh*).

With a personal name: Wimpole CAM (*Wina*).

Added to an earlier place-name: Widmerpool NTT.

pyll
As simplex name: Pylle SOM.

As first element: Pilton DEV, RUT, SOM.

As generic: Huntspill SOM (personal name *Hūn*). Uphill SOM is '(place) above the creek'.

rīth, rīthig 'small stream'. The short form (which has become *rife* in SSX dialect) is used as a simplex name in Ryde IOW and in a number of minor names in southern England. It is the generic of a number of major names which are listed in the reference section.

The longer form *rīthig* occurs in Cropredy OXF and Fulready WAR. In charter boundaries *rīthig* occurs north of the area where *rīth* is found but not in fully Anglian areas (Kitson, forthcoming).

rīth has been considered to be the source of Reeth YON and the first element of Ritton NTB, but it is very unlikely that either *rīth* or *rīthig* was used in northern England, so other etymologies should be sought for these names.

Reference section for rīth, rīthig
As simplex name: Ryde IOW.

As generic with descriptive terms: Childrey BRK (*cille 'gully'); Fingrith ESX (?'finger'); Sawtry HNT ('salt'); Fulready WAR ('foul').

With reference to birds: Hendred BRK ('wild birds'); Rawreth ESX ('heron').

With a personal name: Chaureth ESX (*Ceawa).

With other qualifiers: Coldrey HMP ('charcoal'); Meldreth CAM ('mill'); Shepreth CAM ('sheep'); Shottery WAR ('Scots'); Tingrith BDF ('assembly').

With obscure or ambiguous qualifiers:
Cottered HRT. The spellings indicate **cōd*, apparently unique to this name. A word meaning 'fish-spawn' has been suggested, on the basis of an ON word which would be a cognate, but it seems unlikely that this tiny stream would be characterised by an unusual quantity of fish-spawn.

Cropredy OXF. OE *cropp*, modern *crop*, is evidenced in a number of place-names. In Cropwell NTT (**hyll**) *cropp* clearly means 'hump', referring to the distinctive shape of the hill. As no other sense has been clearly identified in place-names this is perhaps the likeliest meaning in Cropredy, in which case the reference may be to some feature on the ridge west of the village, where the little stream rises.

Seagry WLT. This has been explained as 'sedge stream', but **Sedgery* might have been expected from this. Also, the little stream flows in a gully, not an obvious 'sedge' situation. No other etymology is available.

sǣ, sā 'lake, sea'. The meaning 'sea' occurs in Seacombe CHE, Seaford SSX, Seaham DRH, Seascale and Silloth CMB, and most instances of Seaton. The meaning is 'lake' in Saham NFK, Soham SFK, Seacroft YOW, Seathwaite LNC and Seaton on Hornsea Mere YOE. In Seamer YON(2) and Semer NFK, SFK, YON *sǣ* was glossed by **mere** when the sense 'lake' was no longer obvious.

sǣ meaning 'lake' or the corresponding ON *sǣr* is the generic in Haddlesey YOW (?'heath'), Hornsea YOE ('horn-shaped promontory'), Kilnsea YOE ('kiln'), Rotsea YOE (personal name **Rōt(a)*) and Withernsea YOE (Withern- from an earlier place-name, *Wythorn*, of uncertain etymology). Meaux YOE (earlier *Melse, Mealsa, Mealse*) is from ON *mel-sǣr* 'lake with sandy shores'.

Soham CAM (*Sǣgham* c.1000) is thought to contain a related word **sǣge* 'swamp, lake'. Kilnsey YOW, which has slightly different spellings from Kilnsea YOE, may be a compound of **sǣge* with *cylen* 'kiln'.

wæter 'water'. This word is an alternative to **mere** in the names of some of the lakes of the Lake District. It is occasionally used in ancient river-names, as Blackwater SUR, Freshwater IOW (believed to refer to a headwater of river Yar).

As first element in a compound name, such as Watercombe DOR, Waterden NFK, Waterdine SHR, *wæter* probably means 'wet', as it does when used as an affix to originally simplex names, like Waterperry, Waterstock OXF, Waterbeach CAM.

As a generic in names which do not refer to lakes *wæter* may be used for areas subject to flash flooding. Broadwater occurs as a settlement-name in HRT (near Stevenage) and SSX (near Worthing), as well as giving rise to a number of minor names and field-names. At Broadwater HRT Mrs M. Guest informs us

RIVERS AND SPRINGS: PONDS AND LAKES

that 'there are still people in the district who can remember buses trundling along through inches of water on the main roads after heavy rain'. Local papers (*Hertfordshire Mercury* and *Stevenage Gazette*) reported flooding in March 1947 and September 1968.

Several names derive from *wætergefeall* 'waterfall'. Watership HMP is from *wæterscipe* 'water-supply'; and there is another instance in HRT, where Brocket Park near Welwyn was earleir *Watershepe*.

well (-a, -e), wiell, will, wyll (-a, -e), wælle 'spring'. The gender and declension of this word as evidenced in charter boundaries vary from region to region, as does the nature of the vowel. In west-midland counties the form *wælle* (which gives modern -wall) co-existed with the West Saxon form *welle*. Kitson (forthcoming) gives details of the forms in which the word appears in charters.

As regards meaning, in charter boundaries and in settlement-names this term usually denotes a natural spring, which may or may not be the source of a river. It is sometimes an element in a stream- or river-name, and it then has a secondary meaning 'stream'. Welney NFK is OE *Wellanea* 'the river called Well'. The modern sense 'shaft sunk down to a water-bearing layer' is ME, except for biblical references to Jacob's well.

well is the commonest of the terms discussed in this chapter, occurring in at least 280 names in Ekwall 1960. It is poorly represented in names recorded before *c*.730 (Cox 1976), and may not have become the usual word for a spring until after that date.

well is evidenced as a simplex name and as a first element, but it is much more frequent as a generic in compounds. The wide range of qualifiers can be seen from the analysis presented in the reference section. Descriptive terms are the most frequent, with some adjectives (black, bright, broad, cold, hollow, loud, white) recurring fairly frequently. There are also a number of 'stone' wells.

A notable feature of the corpus is the unusually large number of settlement-names in which *well* is combined with a first element which cannot be identified. These are listed in the reference section with some comments on possible etymologies.

well names are a rewarding target in field work. In many instances the springs are extant, and descriptive qualifiers, such as 'stone' or 'broad', are still clearly applicable.

Reference section for well (-a, -e), wiell, will, wyll (-a, -e), wælle

As simplex name: Wall under Heywood and Walltown SHR, Well KNT, LIN, YON, Wool DOR, Outwell and Upwell CAM, Coffinswell and Edginswell DEV, Wells NFK, SOM, Welham (dat.pl.) NTT, YOE.

As first element with habitative terms: Welby LIN; Welton CAM, LIN(3), NTP, YOE, Wilton SOM, WLT; Welwick YOE; Wilcot WLT.

THE LANDSCAPE OF PLACE-NAMES

With other generics: Welborne NFK, Welbourn LIN, Welburn YON(2) (**burna**); Welbury YON (**beorg**); Welcombe DEV, WAR, Woolacombe DEV, Woolcombe DOR (**cumb**); Weldon NTP (**dūn**); Welford GLO, NTP, Walford HRE, SHR (**ford**); Wellow NTT (*haga* 'enclosure'); Wellow LIN (**hōh**); Wallhope GLO, Wallop HMP, SHR (**hop**); Woolley SOM (**lēah**).

As generic with descriptive terms: Blackwell DRB(2), DRH, WOR; Bradwall CHE, Bradwell BUC, DRB, ESX(2), SFK, STF, Broadwell GLO, OXF (all 'broad'); Brightwell BRK, OXF, SFK; Caldwell DRB, YON, Caudle GLO, Cauldwell BDF, Chadwell ESX, LEI, Belchalwell DOR, Cholwell SOM, Coldwell NTB(3) (all 'cold'); Cawkwell LIN ('chalk'); Colwall HRE and Colwell DEV, NTB ('cool'); Farewell STF ('fair'); Framwellgate DRH ('strong'); Fulwell DRH, OXF ('foul'); Greetwell LIN ('gravel'); Harwell NTT ('pleasant'); Hollowell NTP, Holwell DOR, HRT, LEI ('hollow'); Ludwell DRB, OXF, WLT ('loud'); Prittlewell ESX ('babbling'); Radwell BDF, HRT ('red'); Scaldwell NTP, Shadwell GTL ('shallow'); Shirwell DEV ('bright'); Stanwell MDX, Stawell SOM, Stowell GLO, SOM, WLT ('stone'); Twywell NTP ('double'); Wallingwells NTT ('welling'); Warmwell DOR; Wherwell HMP ('cauldron'); Whitwell DOR, DRB, HRT, IOW, NFK, RUT, WML, YON(2) ('white'); Wombwell YOW ('womb-shaped'); Wordwell SFK ('twisted').

With personal names: Adwell OXF (*Eadda*); Amwell HRT (**Æmma*); Askerswell DOR (*Ōsgār*); Badwell SFK (*Badda*); Bakewell DRB (*Badeca*); Bawdeswell NFK (*Baldhere*); Berkswell WAR (*Beorcol*); Bracewell YOW (ON *Breithr*); Brauncewell LIN (ON *Brandr*); Cadwell OXF (*Cada*); Canwell STF (*Cana*); Chatwall SHR and Chatwell STF (*Ceatta*); Dodwell WAR (*Dodda*); Dunkeswell DEV (*Duduc*); Epwell OXF (*Eoppa*); Etwall DRB (*Ēata*); Fradswell STF (*Frōd*); Gabwell DEV (**Gabba*); Harswell YOE (**Hersa*); Hudswell YON (**Hūdel*); Loddiswell DEV (**Lod*); Mongewell OXF (*Mund*); Netteswell ESX (**Nēthel*); Offwell DEV (*Offa* or *Uffa*); Pedwell SOM (*Pēoda*); Perdiswell WOR (**Perd*); Shotteswell WAR (*Scot*); Sizewell SFK (*Sigehere*); Sopwell HRT (**Soppa*); Tideswell DRB (*Tidi*); Wirswall CHE (*Wighere*).

With references to vegetation: Ashwell ESX, HRT, RUT; Boxwell GLO; Bunwell NFK ('reed'); Carswalls GLO, Carswell BRK, Caswell DOR, NTP, OXF, SOM, Craswall HRE, Cresswell DRB, STF, Kerswell DEV(3) (all 'cress'); Elmswell SFK; Feltwell NFK (OE *felte*, a plant-name); Haswell DRH, SOM, Heswall CHE ('hazel'); Mapledurwell HMP ('maple-tree'); Miswell HRT and Muswell BUC, GTL ('moss'); Nutwell DEV; Pirzwell DEV ('pease'); Sparkwell DEV ('brushwood').

With references to wild creatures: Andwell HMP ('duck'), Barwell LEI ('boar'); Cornwell OXF and Cranwell LIN ('crane'); Crowell OXF ('crow'); Greywell HMP (probably 'wolf'); Hartwell BUC, NTP, STF; Hauxwell YON and Hawkwell NTB, SOM ('hawk'); Roel GLO ('roe'); Roxwell ESX ('rook'); Snailwell CAM ('snail'); Swalwell DRH ('swallow'); Tathwell LIN ('toad'); Tranwell NTB (ON *trani* 'crane'); Wigwell DRB ('beetle'); Winswell (earlier *Wyveleswell*) DEV ('beetle').

RIVERS AND SPRINGS: PONDS AND LAKES

With references to domestic creatures: Bulwell NTT ('bull'); Goosewell DEV; Hanwell GTL ('cock'); Titchwell NFK ('kid').

With references to religion and superstition: Elwell DOR and Holywell LIN (*hǣl* 'omen'); Halliwell LNC, Halwell, Halwill DEV, Holwell DOR, OXF, Holywell HNT, KNT, NTB (all 'holy'); Runwell ESX and Rumwell SOM (*rūn* 'secret'). Seawell NTP, Sewell BDF, Showell OXF and Sywell NTP, all meaning 'seven springs', may belong here. Dewsall HRE refers to St David.

With references to topography: Backwell SOM (**bæc**); Bywell NTB (*byge* 'river-bend'); Cromwell NTT (*crumbe* 'river-bend'); Harwell BRK (*Hāra* 'Grey One', a hill-name); Hopwell DRB (**hop**); Kettlewell YOW ('kettle', perhaps in the sense 'narrow valley'); Orwell CAM (**ord** 'point', i.e. hill-spur); Pickwell LEI (**pīc**); Rothwell LIN, NTP, YOW (***roth**); Scarthingwell YOW ('gap in the hills'); Shorwell IOW ('steep slope').

With categories of people: Childwall LNC and Chilwell NTT ('children'); Churwell YOW ('peasants'); Clerkenwell GTL ('clerics'); Gagingwell OXF ('kinsmen'); Maidenwell LIN and Maidwell NTP; Nunwell IOW; Sunningwell BRK (*Sunningas*). Sugarswell ('robber's spring') occurs as a minor name, once in WAR and twice in OXF.

With adjectives of direction: Astwell NTP and Eastwell KNT, LEI; Norwell NTT; Southwell NTT; Westwell KNT, OXF.

With references to structures: Burwell CAM, LIN ('fort'); Crockernwell DEV ('pottery factory'); Hemswell LIN and Elmswell YOE (*helm*, probably 'shelter'); Lewell DOR (*hlēo* 'shelter'); Wyville LIN ('heathen temple'); Yarwell NTP ('fish enclosure'). Shingleswell (in Ifield) KNT perhaps refers to a spring over which there was a shingled roof.

With terms for boundaries: Marwell HMP (*gemǣre*); Shadwell NFK, YOW (*scēad*); Shawell LEI (*scēath*).

With references to conduits and water containers: Bedwell HRT and Bidwell NTP (*byden*, 'tub', this compound has been noted many times in minor names and field-names; it was probably a compound appellative); Pipewell NTP; Thatto LNC (*Thotewell* 1246, *thēote* 'waterpipe').

With other qualifiers: Botwell GTL ('healing'); Chopwell DRH ('commerce'); Hardwell BRK ('hoard'); Letwell YOW ('obstruction'); Trowell NTT ('tree'); Watchingwell IOW ('wheat place').

Names with obscure or ambiguous qualifiers:
Banwell SOM. In addition to this well-recorded major name there are several instances of what appears to be the same compound in minor names and field-names. The first element may be *bana* 'killer' used for a contaminated spring, prefiguring the sense of ME and modern *bane*.

Bardwell SFK. Possibly OE *brerd*, suggested in Ekwall 1960, meaning 'bank'. This could refer to the position on the 100ft contour overlooking a stream.

Barnwell CAM. Personal name *Beorna* or gen.pl. of *beorn* 'man, warrior'.

Barnwell NTP, *Byrnewillan c.*980. Possibly *byrgen* 'burial place'.

Bitteswell LEI. The spellings suggest **bythme**, but the topography tells against this.

Britwell BUC (in Burnham), OXF. Early spellings indicate *bryt-*. Brightwells Fm in Watford HRT is another instance, though with late assimilation to the transparent name Brightwell. Formally *bryt-* in these three names could be *Bryt* 'Briton', but British people are usually called *Walh* in place-names, and a triple occurence of 'Briton' with *well* seems unlikely. The same problem is presented by Britford WLT.

Camberwell GTL. The same first element occurs in Camerton CMB, YOE, but no plausible suggestion has been made.

Cobhall HRE, earlier *Cobbewell*. There was probably an OE personal name *Cobba. Alternatively an OE antecedent of modern *cob* 'round lump' may have been used as a hill-term.

Coxwell BRK. Early spellings indicate *coc*. The single -*c*- tells against the bird-name *cocc* and the hill-term *cocc*, and the short vowel rules out *cōc*. Coxford NFK presents the same problem.

Crabwall CHE. Usually rendered 'crayfish spring', but information is needed about the likelihood of any creature designated by OE *crabba* being found here.

Crudwell WLT. An OE ancestor of ME *crudde*, modern *curd*, would suit, but the word is not recorded until the 14th century.

Digswell HRT. This is one of several names for which the spellings indicate an element *dicel*, of unknown meaning.

Dowdeswell GLO, *Dogodeswellan* in an 8th-cent. charter. The word or personal name *dogod* appears to be unique to this place-name.

Ecchinswell HMP, *Eccleswelle* DB. This presents the same problems as Ashford MDX, discussed under **ford**.

Eriswell SFK. The available spellings are inconsistent. DB *Hereswella* suggests personal name *Here, but *Evereswell* 1249 may indicate *eofor* 'boar'.

Fritwell OXF. Derivation from *freht*, *firht* 'augury' is uncertain as early spellings show no sign of -*h*-.

Glapwell DRB. Either *Glappa* personal name or *glæppa* 'buck-bean'.

RIVERS AND SPRINGS: PONDS AND LAKES

Harpswell LIN. Possibly an abbreviated form of *herepæth* 'main road', but it is outside the area where that term is common.

Hawkwell ESX. The first element is *haca* 'hook', which could refer to a bend in a small stream or to some constructional feature at a spring.

Herringswell SFK. OE *Herningwelle, Hyrnincwylle*, first element perhaps a derivative of *horn*, referring to the curving hill which overlooks the settlement.

Hinderwell YON, earlier *Hilderwell*. Probably **hyldre* 'elder tree', but there is a St Hilda's well here, and it has been suggested that the saint's name, OE *Hild*, was replaced by ON *Hildr*.

Hipswell YON, earlier *Hippleswell*. Ekwall 1960 suggests **hyppels* 'stepping-stones', a derivative of *hoppian* 'to hop'. This would suit the marshy terrain, but there is no other evidence for the word.

Howell LIN. Early forms require *hu-* which is unexplained.

Ickwell BDF, earlier *Gikewelle*. Possibly *gēoc* 'help', in which case the name is analogous to Botwell.

Knapwell CAM. Either *Cnapa* personal name or the noun meaning 'boy, servant'.

Ogwell DEV, OE *Wogganwylle*. A personal name **Wocga* has been conjectured to suit this place-name.

Ridgewell ESX. The early spellings exhibit an unusually high degree of incompatibility.

Shelswell OXF. The first element could be personal name *Scyld* or a noun meaning 'shallow place'.

Sotwell BRK. The early spellings are inconsistent, but the balance of evidence perhaps favours an OE **Suttanwell*. If that is the genuine form, the same first element is found in Sutcombe, discussed under **cumb**.

Tinwell RUT. EPNS Rutland survey suggests *Tyningawelle* 'spring of Tyni's people', with an unrecorded but plausible personal name. The *-inga-* etymology is, however, only supported by one spelling, mid 13th century. *Tiningewelle*. The regular form from mid 12th to 14th cent. is *Tinewell(e)* or *Tinwell(e)*, and the etymology is best left open.

Tuxwell SOM. See the note on Tuxford NTT in Chapter 3.

Wiswell LNC. Early spellings have *Wise-* or *Wyse-*, and the absence of *-ss-* probably precludes **wisse*. Ekwall 1922 suggested *wise* 'wise': this could be another 'divination' spring.

CHAPTER 2

Marsh, Moor and Floodplain

The place-name elements studied in this chapter all have predominantly or exclusively watery connotations. Some are words for marshes, others refer to patches of drier ground which provided settlement-sites in wetland areas. In the second category the most important words are the Old English terms ēg and **hamm**, and Old Norse **holmr**.

ēg tops the list of elements used in names recorded between *c*.670 and *c*.730, but this does not, of course, mean that it was the commonest place-name-forming term in use at that early date. It was noted above that the 224 names occurring in that list include 50 which refer to monastic foundations. Nine of these are ēg compounds, so allowance has to be made for the predilection of Dark Age ecclesiastics for 'island' sites. The number noted, nineteen, is nevertheless a high score when considered in relationship to the comparatively modest frequency of ēg in the whole corpus of English place-names, and it is unlikely that the word was as commonly used after 730 as it was before. Many ēg names may well go back to the earliest phase of English speech, and the word probably lost its 'dry ground in marsh' sense before the end of the Anglo-Saxon period.

Settlements with names containing ēg probably provide the most perfect examples we have of exact correspondence between name and site, and the informed observer can derive satisfaction from this in the wetlands of the eastern fens, the Somerset Levels, and north-west Berkshire. There is seldom any difficulty in discerning the slightly raised sub-circular features on which ēg settlements stand. Some are still surrounded by water at times of flood. In 1992 *The Times* published a photograph of a coracle being paddled over the flooded fields around Muchelney SOM.

When the piece of dry ground is linear rather than sub-circular it is more likely to be called **hamm**, so the senses of these two elements overlap. 'Dry ground in marsh' is only one of the meanings of the protean element **hamm**, but the word belongs in this chapter because most of its other senses are also watery. The even more protean word **halh** can mean 'dry ground in marsh', but it is relegated to Chapter 4 because most of its other senses are not watery.

Many of the marshes referred to in names containing **mersc** and **mōr** have been subject to centuries of drainage and they now give the appearance not of marsh but of fertile farmland, though the expert eye can often detect traces of their former state. Limited patches of marsh, such as those referred to by the terms **slæp**, **slōh** and **sol**, are liable to be obstinately wet even in modern conditions, and these are worth looking for if the settlement has remained small, as in the two instances of **sol** in the parish of Wolverley WOR. Linear marshes denoted by **fenn** are also likely to be still clearly apparent.

The common use of **mōr** and **mersc** as qualifiers in **tūn** names reflects the high value of wetland resources to early farmers. Moreton BRK was, in the 10th century, a component of a large estate centred on Blewbury. The other settlements in this unit are on the slopes of the Downs, and the vital hay crop would come from the wet ground around Moreton. Even areas too wet for ploughing were of great value, and their eventual reclamation was sometimes much resented, as witness the Otmoor riots when this piece of primeval landscape in Oxfordshire was tamed in the 19th century.

It is gratifying to note that the suggestion made in 1984 about the meaning of the highly specialised term **wæsse** has gained general acceptance, leading to the substitution of 'alluvial land' in recent reference books for the earlier translation 'swamp'.

ēg (West Saxon **īeg**) 'island'. This word is used for islands in the modern sense (e.g. Brownsea DOR, Canvey and Mersea ESX, Portsea HMP), but the characteristic usage in settlement-names is for an area of raised ground in wet country. This sense predominates heavily over all others, and names in which it occurs are likely to belong to the earliest stratum of OE toponymy. In names of late OE origin *ēg* may sometimes mean 'well-watered land'. In the northern tip of DRB and regions north of that it seems necessary to postulate other meanings such as 'hill jutting into flat land' and 'patch of good land in moor'. In southern England, also, *ēg* is occasionally used of high promontory sites, Kersey SFK being probably the most unequivocal example. The part of Stowmarket SFK which was anciently called *Thorney* occupies a low promontory.

The use of *ēg* in settlement-names overlaps with that of **dūn**, **hamm**, **halh** and **ness**. Some low-lying settlements with names in **dūn**, such as Hedon YOE and Mundon ESX, look to the modern observer more deserving of names in *ēg*. If it be accepted that Holbeach and some other fenland names contain the locative of **bæc**, this term also encroaches on the meanings of *ēg*. On the east bank of the Severn estuary **hamm** appears to be used rather than *ēg* for farms in very low ground, and there are other areas where **halh** usurps this function.

ēg was probably not used for land in a river-bend. Bungay SFK stands at the neck of a great loop of the river Waveney, but the reference is probably more precisely applied to the slightly raised strip of land within the loop which is clearly shown by the hachuring on the 19th-century Ordnance Survey map. The meaning 'land between rivers', sometimes adduced for *ēg*, is not clearly established.

There are two instances in which ēg appears to have been used in a district-name which was applied to several adjacent settlements. In NFK there are three fenland villages – Tilney All Saints, Tilney cum Islington and Tilney St Lawrence – in a 2½ mile line near King's Lynn. In GLO there are four villages – Down Ampney, Ampney St Mary, Ampney St Peter and Ampney Crucis – spread over an area east of Cirencester. Previous reference books hold that Ampney is 'Amma's river', the settlements being named from Ampney Brook, but it is unlikely that this brook would have been called ēa, and the spellings are more consistent with ēg. Tilney and Ampney may have been applied in the first instance to one 'island' in the two areas. The process whereby several adjacent settlements came to use the same name requires study in these and many other cases.

There are about 180 names containing ēg in Ekwall 1960, and the element is not very common in minor names or field-names. In view of this moderate frequency it is remarkable that ēg heads the list of OE terms in names recorded by c.730 (Cox 1976, p. 58). There are 19 instances, and even when some allowance has been made for the liking of early saints for island bases it must be concluded that ēg was more commonly used in place-name formation in the years before AD 730 than it was after. As with settlement-names in dūn, the sites of villages with ēg names are often the best which a region has to offer, and it might be worth the archaeologist's while to look for continuity of settlement from pre-English times.

Smith 1956 says 'the element is widespread'; but the distribution is not even, nor is it entirely dictated by geography. Names in ēg become less common north of a line from the Mersey to the Humber, and the true frequency in northern counties is difficult to determine. Early spellings of possible instances show confusion between ēg and other elements, and different reference books give different origins for a number of names. In YOW, for instance, the possible candidates among major names are Arksey, Bardsey, Hessay, Embsay, Methley, Pudsey and Wibsey. EPNS survey considers Hessay and Methley to be probably ēg names, but rejects this for Embsay, Pudsey and Wibsey, for which Smith prefers 'enclosure' (gehæg) or 'high place' (hēah). Ekwall 1960 accepts four of the possible examples as names in ēg but ascribes Hessay to sǣ 'lake' and Methley to lēah. My own feeling is that Arksey and Bardsey are safe examples, Hessay is probably an ēg since the 50ft contour makes an appropriate feature, while Embsay, Pudsey and Wibsey may be ēg names if the senses 'hill jutting into flat land' and 'good land in moor' are allowed. Methley, however, seems more likely to contain lēah. One of the possible CHE examples, Aldersey, is ascribed in EPNS survey (4, p. 82) to gehæg 'enclosure', but the site is appropriate to ēg in the sense 'dry ground in marsh', and the spellings with -hey(e), -hay are not so numerous as to compel acceptance of the alternative etymology. It must be noted that there is a high degree of uncertainty about the incidence of ēg in northern counties.

The use of ēg for places in high situations, which is partly the cause of this uncertainty, is found in the northern tip of DRB, where Edale, Abney and Eyam are fairly close together near the YOW border. It is possible that in these names ēg refers to isolated patches of good land in moorland. A similar sense development is postulated *infra* for **hamm** in some south-western names.

Some other sites in DRB and counties further north suggest a meaning 'hill-spur'. Cold Eaton DRB is on a spur overlooking the river Dove and Little Eaton, north of Derby, has a clear 'hill-spur' site. At Corney CMB the hall and village occupy a long promontory with wet ground on either side. Bardsea LNC is on the tip of a hill which juts out into flat coastland. This sense is occasionally found further south. As noted above, Kersey SFK is a clear example, and it has recently been suggested that the ēg of Lindsey was the hill on which the Roman *colonia* of Lincoln was built (Yorke 1993, 143). Tebay WML might be suspected of having one of these 'hill-spur' sites, but in fact Old Tebay (located with some difficulty in this area dominated by the M6 service station) probably occupies a 'raised area in wet ground' site.

The counties with the greatest number of major names in ēg are NFK(15), CAM(13), SOM(13), LIN(11), SSX(8) and BRK(8). There are 11 in ESX, but four of them (Brightlingsea, Canvey, Mersea and Osea) are islands in the modern sense, so the term is rather poorly represented in this large county in the sense 'dry ground in marsh'. SHR and WLT have seven examples, CHE, DRB, GLO, OXF and SFK have six. There are five in BDF, GTL and WOR, four in BUC, HMP, KNT and NTT. YOW has three/six. HRT, LEI, NTP, STF, SUR (excluding part of GTL) and YON have three. WAR has only two instances (Binley and Maney). There are two major names in HRE, one of which, Eye near Leominster, has a classic ēg site. The southern counties of HNT and MDX have only one example each. More typically there is only one example in the northern counties of CMB, LNC, WML, and the word does not occur at all in major names in NTB or YOE. DRH has Hartlepool (*Heruteu* 'stag island' in Bede), and Hexham was earlier *Hagustaldesēg. Watts 1994 discusses the use of ēg in Hexham and in Lastingham YON (earlier *Læstingaēg), concluding that in both these names ēg was used for a patch of good land in moorland. The element dies out in the south west, after being frequent in SOM. In DOR it is used in the name of Brownsea Island, but not in settlement-names. There is a single instance in DEV (Babeny in Lydford, perhaps a southern example of 'good land in moor'), but probably none in CNW.

As regards distribution within counties there are some noteworthy clusters. The densest concentration (Fig. 10) is in north-west BRK and the adjacent part of OXF. Charney, Goosey, Hanney, Pusey and Tubney are parishes on the river Ock or its tributaries. Cholsey and Hinksey are parishes on the river Thames, and Mackney is a hamlet near Cholsey. Two names on the north bank of the Thames, Witney and Chimney OXF, belong with this group, Witney being one of the few places with ēg names which have developed into towns. Binsey, Hinksey, Medley and Osney cluster round Oxford. This area has some of the earliest Anglo-Saxon archaeology in the country, and it is likely that these *ēg* names arose in the mid-5th century. There are other clusters of *ēg* names in CAM, NFK and SOM. CAM is the only county in which *ēg* is frequent in minor names.

ēg is used as a simplex name and is well-evidenced as a first element, but much the commonest use is as generic in a compound.

Analysis of the types of qualifiers used with *ēg* is attempted in the reference section. The exercise is bedevilled to some extent by the perennial problems of distinguishing some personal names (e.g. *Hengest, Mūl, Nǣgl*) from the object or creature from which they derive, and of deciding which otherwise unrecorded personal names may reasonably be inferred from place-name evidence. A re-examination of the corpus has resulted in the reclassification of some names which were listed in 1984 (p. 38) as containing personal names, but no classification is ever likely to be definitive. It is still clear, however, that personal names form by far the largest category of qualifiers used with *ēg*, and that these names are mainly those of male Anglo-Saxons.

There are only two feminine names in the corpus examined: *Ēadwynn* (Adeney) and *Hildeburg* (Hilbre). There is one British name, *Cerot* (Chertsey) and there are two Old Norse, *Hákr* (Haxey) and *Maccus* (Maxey). The main division is between monothematic and dithematic OE masculine names.

Dithematic names are *Ælfrīc, Aldhere, Baldhere, Beadurīc, Beorhtrīc, Beornmund, Beornrǣd, Cyneheard, Dōmgēat, Mæthelhere, Waldhere, Wigbēd, Wilhere*. Two of these *Dōmgēat* and *Wigbēd*, are not evidenced except in the place-names Dauntsey and Wibsey.

Some of the much greater number of monothematic names may be shortened forms of dithematic ones (like *Ceomma* in Chimney, thought to be a short form of *Cēolmǣr* or *Cēolmund*), but most of them are likely to be the sort of single-theme names which are considered to have been commonest in the early centuries of the Anglo-Saxon period. Some items listed here could contain significant words, e.g. there may have been something pointed (OE *pīc*) at Pitsea or something resembling a nail (OE *nægl*) at Nailsea. Throughout the work of the English Place-Name Society runs the assumption that a first element in the genitive singular is most likely to be a personal name. This has been shown to be mistaken as regards words for living creatures, where the genitive singular is likely to be collective (Kitson forthcoming), and in response to this Blankney, Hinksey and Molesey have here been classified as referring to horses and mules. Most of the monothematic personal names here adduced as first elements of *ēg* compounds are, however, unlikely to be challenged.

No other category of qualifiers used in *ēg* compounds approaches the size of that of personal names.

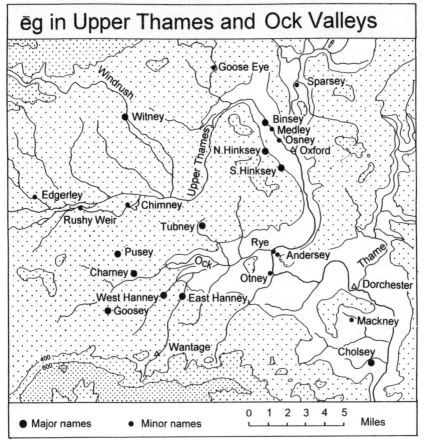

ēg in Upper Thames and Ock Valleys 10

Reference section for ēg, īeg

As simplex name: Eye HRE, NTP, OXF, SFK, Eyam DRB (dat.pl.); Rye HRT, SSX (with R- from definite article); Kingsey and Towersey BUC.

As first element with tūn: Eaton BDF, CHE(2), DRB(4), LEI, STF, Eyton SHR(5).

With other generics: Ebury GTL (*bury* 'manor' is a later addition to a simplex name); Edale DRB (**dæl**); Emstrey SHR (*ēg-mynster* 'minster church on an island'); Eyecote GLO; Eyeworth BDF.

As generic with personal names: Abney DRB (*Abba*); Adeney SHR (fem. *Ēadwynn*); Aldersey CHE (*Aldhere*); Alney GLO (*Ælle*); Ampney GLO (**Amma*); Arlesey BDF (*Ælfrīc*); Babeny DEV (*Babba*); Badsey WOR (*Bæddi*); Bardney LIN (**Bearda*); Bardsea LNC and Bardsey YOW (*Beornrǣd*); Battersea GTL (*Beadurīc*); Bawdsey SFK (*Baldhere*); Beckney ESX (*Becca*); Bermondsey GTL (*Beornmund*); Binsey OXF (*Byni*); Bodney NFK (**Beoda*); Bolney SSX (*Bola*); Brightlingsea ESX (*Beorhtrīc*); Brownsea DOR (**Brūnoc*); Cadney LIN (*Cada*); Chebsey STF (**Cebbi*); Chedzoy SOM (*Cedd*); Chertsey SUR (**Cerot*); Chimney OXF (**Ceomma*);

THE LANDSCAPE OF PLACE-NAMES

Cholsey BRK (*Cēol*); Colney NFK (*Cola*); Costessey NFK (**Cost*); Cuckney NTT (**Cuca*); Dauntsey WLT (**Dōmgeat*); Dengie ESX (*Dene*, -ing- alternating with -es-); Doxey STF (**Docc*); Embsay YOW (*Embe*); Epney GLO (*Eoppa*); Gedney LIN (**Gǣda* or **Gydda*); Godney SOM (*Gōda*); Harlsey YON (**Herel*); Haxey LIN (ON *Hákr*); (Gate) Helmsley YON (*Hemele*); Hilbre CHE (fem. *Hildeburg*); Impney WOR (*Imma*); Kempsey WOR (*Cymi*); Kinnersley SHR (*Cyneheard*); Lindsey SFK (**Lelli*); Lulsley WOR (*Lull*); Mackney BRK and Makeney DRB (**Maca*); Mattersea NTT (*Mæthelhere*); Maxey NTP (ON *Maccus*); Nailsea SOM (**Nægl*); Oaksey WLT (**Wocc*); Olney BUC (**Olla*); Osea ESX (*Ufi*); Osney OXF (*Ōsa*); Partney LIN (**Pearta*); Patney WLT (**Peatta*); Pewsey WLT (**Pefe*); Pitney SOM (*Peota*); Pitsea ESX (*Pīc*); Romsey HMP (**Rūm*); Sibsey LIN (*Sigebald*); Tebay WML (*Tibba*); Tetney LIN (*Tǣta*); Titsey SUR (**Tydic*); Tubney BRK (**Tubba*); Waldersea CAM (*Waldhere*); Whittlesey CAM (*Witel*); Wibsey YOW (**Wīgbēd*); Willersey GLO (*Wilhere*); Witney OXF (*Witta*); Yeoveney MDX (**Geofa*).

With descriptive terms: Blakeney GLO, NFK ('black'); Bradney SOM ('broad'); Caldy CHE ('cold'); Fotheringhay NTP ('grazing'); Henney ESX ('high'); Langney SSX and Longney GLO ('long'); Muchelney SOM ('large'); Sandy BDF; Stanney CHE and Stonea CAM ('stone'); Stuntney CAM (EPNS survey suggests that OE *stunt* could mean 'steep'); Turvey BDF ('turf').

With references to topography: Barway CAM (**beorg**); Billinghay LIN ('ridge'); Cooksey WOR (cōc); Coveney CAM (*cofa* 'bay'); Horningsea CAM ('horn-shaped hill'); Wendy CAM ('bend'); Winchelsea SSX ('bend').

With references to wild creatures: Bawsey NFK ('gadfly'); Corney CMB, HRT ('crane'); Dorney BUC ('bumblebee'); Goosey BRK; Hartlepool DRH and Harty KNT; Iltney ESX ('swan'); Selsey SSX ('seal').

With references to domestic creatures: Blankney LIN (probably *blanca* 'white horse'); Chickney ESX ('chicken'); Hannah LIN and Hanney BRK ('cock'); Hinksey BRK ('stallion'); Horse Eye SSX and Horsey NFK, SOM; Molesey SUR ('mule'); Oxney KNT, NTP; Quy CAM ('cow'); Sheepy LEI and Sheppey KNT.

With references to vegetation: Bunny NTT ('rush'); Hessay YOW ('hazel'); Kersey SFK ('cress'); Minety WLT ('mint'); Pusey BRK ('pease'); Ramsey ESX, HNT ('wild garlic'); Stiffkey NFK ('tree-stump'); Thorney CAM, SOM, SSX, and as a lost name in GTL and SFK.

With references to groups of people: Athelney SOM ('princes'); Bungay SFK (*Buningas* 'Buna's people'); Denny CAM ('Danes'); Hilgay NFK (*Hillingas*, with uncertain base); Lydney GLO ('sailors'); Nunney SOM ('nuns'); Shingay CAM (*Scēningas* 'Scēne's people'); Wallasay CHE ('Welshmen'); Wormegay NFK (*Wyrmingas* '*Wyrm's people').

MARSH, MOOR AND FLOODPLAIN

With references to direction or position: Eastrea CAM; Medley OXF ('middle'); Middleney SOM; Modney NFK (*meoduma* 'middle'); Southery NFK and Southrey LIN. Othery SOM, 'the other island', is named in relation to Middlezoy.

With earlier place- or river-names: Campsey SFK (*Camp* is probably a district-name, 'open land'); Charney BRK (river *Cern*); Colney HRT (river Cole); Middlezoy and Westonzoyland SOM (-zoy is earlier *Soweie* 'island on river Sow'); Portsea HMP.

Names with obscure, uncertain or ambiguous qualifying elements:
Anglesey CAM. 'Island of the Angle' is formally possible but unconvincing in Anglian territory. Perhaps OE *angel* 'fish-hook' in some transferred topographical sense.

Arksey YOW. Ekwall 1960 suggests *Arnketill*, but early forms have *Arkes-* and it is unlikely that this Scandinavian personal name would be reduced so consistently.

Barney NFK. Pers.n. **Bera*, *berern* ('barn') or *beren* 'growing with barley'.

Bilney NFK(2), Binley WAR. The occurrence of three instances makes it difficult to accept the etymology 'Billa's island', and derivation from *bile* 'beak' is unlikely as that word did not have an *-n* genitive.

Canvey ESX. This has been explained as a much-reduced reflex of an OE *Caninga ēg* 'island of Cana's people'. But the forms (of which the earliest is *Caneveye* 1254) suggest a first element *canef*, or perhaps *cnef* with *-a-* developing as in Canute from Cnut.

Eisey WLT. The first element (?*ēs*) is unexplained.

Fleckney LEI. This name shares an obscure first element (?*flecca*) with Flecknoe WAR.

Hackney GTL. Either *haca* 'hook' or *Haca* personal name.

Hilsea HMP. Ekwall 1960 suggests **hyles* 'holly', which would correspond to an Old High German word, but there is no other evidence for such a word in English, and holly is regularly referred to in place-names as *holegn*. Coates 1989 suggests a reference to sea-holly.

Kelsey LIN. Perhaps a doublet of Cholsey BRK with Scandinavian *K-* for *Ch-*.

Manea CAM. There is a doublet, Maney, in Sutton Coldfield WAR. Both names have been ascribed to *gemǣne* 'common', but the spellings suggest *Man-*.

Palstre KNT. OE *palstr* 'spike' is formally suitable. This word has not otherwise been noted in place-names, however, and the precise meaning is unclear.

Pentney NFK. The first element could be a river-name *Pante*.

Poultney LEI. A similar first element is found in Poltimore DEV. There is a ME verb *pilte, pulte* 'to thrust out' and there may have been OE words from the same stem which would suit the place-names.

Stickney LIN. Stickford, two miles north, has the same first element. Ekwall 1960 suggests *sticca* 'stick', used in some topographical sense.

Tiln NTT, Tilney NFK, (Thorpe) Tilney LIN. These are three instances of a compound of *ēg* with a first element *tilan-*, which is probably an oblique case of the adjective *til* 'useful'. There was a personal name *Tila* from the same source, but it is less likely that it would occur three times with *ēg*.

Torksey LIN. A personal name **Turec* has been conjectured to suit this name and Torkington CHE. As will be seen from the number of asterisks in the 'personal name' category of generics listed above, the criteria for unrecorded personal names have been accorded generous treatment in the analysis of *ēg* compounds; but as evidence for **Turec* rests entirely on these two place-names the etymology is best left open.

Whitney HRE. Either *hwīt* 'white' or *Hwīta*, personal name. Whitton LIN (*Witenai* DB) is identical.

fenn 'fen' has an uneven distribution in settlement-names, as can be seen from Fig. 11. It rarely occurs in ancient names in the districts of eastern England now known as The Fens. There appears to have been an early use in ESX (see below), then a preference for other more or less synonymous terms until a considerably later date. Perhaps *fenn* was brought back into use to replace *mōr* when the meaning of that term was felt to be 'barren upland'.

The precise relevance of *fenn* in any name has to be ascertained by field-work. Investigations in NTB, for instance, suggest that the word was sometimes used there for narrow strips of wet ground in districts where there could never have been large areas of marsh.

Fenham NTB(2) is from the dative plural. Fenham near Holy Island may refer to the soggy salt marshes on the coast. Fenham in Newcastle upon Tyne adjoins the great common called Town Moor. Apart from these two instances of the dative plural, *fenn* only occurs uncompounded in minor names and field-names. In major settlement-names it is more frequent as a qualifier than as a generic. Both *fenn* and the adjective *fennig* are occasionally used as affixes, as in Fen Ditton CAM, Fenny Drayton LEI.

The East Saxon form *fænn* is the first element of Vange and Fambridge and the second element of Bulphan ESX. North and South Fambridge are on either side of the Crouch estuary, and *fænn* was probably the name of the district. The north bank of the river Thames in London appears to have been called *fenn*. The OE bounds of land granted to Westminster Abbey refer to *bulunga fen* near the abbey and to 'London fen' near Blackfriars bridge, and Fenchurch is named from this stretch of marshy ground.

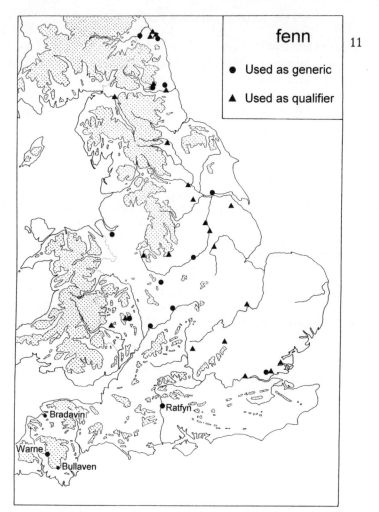

There are detailed place-name surveys of a sufficient number of counties for it to be confidently asserted that *fenn* is rare or absent in place-names likely to be of pre-Conquest origin in large areas of England. The EPNS survey of CAM, for instance, gives on p. 324 a number of field-names recorded in the 13th and 14th centuries, but the list on p. 301 of elements used in place-names of higher status shows that *fenn* was not used at all in settlement-names. In counties south of the Thames the loneliness of Ratfyn WLT (shown on Fig. 11 by the river Avon near Amesbury) is supported by the statement on p. 431 of the EPNS survey of WLT that '*fenn* is very rare', and by the observation in the BRK survey (pp. 867–8) that it is only found occasionally in field-names and modern minor names. The only southern county for which this is not true is DEV, where, although only three examples are shown on Fig. 11, *fenn* is well represented in minor names, some with early ME spellings. The general distinction between 'major' and 'minor' names is less relevant in

DEV than elsewhere, and a number of other *fenn* names, besides Bradavin in Westleigh and Bullaven in Harford, are likely to be ancient settlement-names. This higher frequency in DEV is consistent with the suggestion that *fenn* came back into fashion at a relatively late date.

Other areas in which *fenn* is rare or absent are the central part of the Welsh Marches and the north west. It is not common in Yorkshire. The EPNS survey of YOE (p. 322) comments that in field-names '*fen* is very rare considering the character of the county'.

In the relatively small corpus of *fenn* names the compounds with *tūn, cot* and *wīc* are disproportionately common, perhaps because the area of wet ground played an important part in the economy of settlements which lay beside it.

Reference section for **fenn, fænn**

As simplex name: Fenham NTB(2) (dat.pl.).

As first element with habitative terms: Fancott BDF, Fencott OXF, Fencote HRE, YON; Fenhampton HRE (*hāmtūn*); Fenton CMB, DRB (lost), HNT, LIN(2), NTB, NTT, STF, YOW; Fenwick NTB(2), YOW. In Fenby LIN the generic is ON *bý*.

With other generics: Fambridge ESX; Fenchurch GTL; Vange ESX ('fen district').

As generic with personal names: Edvin HRE (*Gedda*); Matfen NTB (*Matta*); Pinvin WOR (*Penda*); Ratfyn WLT (*Hrōtha*).

With references to domestic creatures: Mousen NTB ('mule'); Swinfen STF ('swine'); Wervin CHE ('cattle').

With reference to a wild creature: Mason NTB ('marten').

With other qualifiers: Bulphan ESX ('fort'); Pressen NTB ('priest'); Redfern WAR ('bushes'); Sinfin DRB ('wide'); Warne DEV ('quaking').

hamm 'land hemmed in by water or marsh; wet land hemmed in by higher ground; river-meadow; cultivated plot on the edge of woodland or moor'.

This place-name element has been the subject of much discussion in recognition of which the latest contributor (Kolb 1989) called his paper '*Hamm*: a long-suffering place-name word'. The course of the debate up to 1984 was charted in *Place-Names in the Landscape*. Since that date no-one has disputed that while the basic meaning is 'something enclosed' the enclosing factor is predominantly water or marsh, and the referent is seldom, if ever, a man-made enclosure which has no special topographical characteristic. The assertion made in Gelling 1960 (the paper which started the argument) that the original meaning of *hamm* in England was 'land in a river-bend' has recently received strong support from Kolb (1989) and Kitson (forthcoming).

There is still some disagreement about the geographical limit of *hamm*. It is certainly predominantly south midland and southern and probably commonest in SSX; but Kolb's assertion that it is not likely to occur beyond the northern boundaries of HRE, WOR, WAR, NTP, HNT and CAM is contradicted by clear evidence for its use in ancient names in SHR. Two districts of Shrewsbury have names in -ham, these being Coleham, outside the English Bridge, and a lost *Romaldesham*, inside the Welsh Bridge. The status and situation of these places strongly support derivation from *hamm*, and *Romaldesham* has spellings in *-hom*. The significance of this last point is discussed below. An unequivocal SHR example is a minor settlement called The Home in Wentnor, 5 miles south-west of Church Stretton. This farm, which occupies a narrow promontory with streams on three sides, is recorded in 1255 as *Homme*. Ham is also well-evidenced as a SHR field-name, often for a field in a river-bend.

Since the distribution of the element has been carried as far north as Shrewsbury it is possible that it should be allowed for in Cheshire. A strong candidate in CHE is Frodsham, which has 13th-century spellings *Frodshum* and *Frotheshamme*, and which occupies a site which is a perfect example of 'promontory into marsh', one of the senses of *hamm* frequently evidenced in southern counties.

Kolb's definition of the area in which *hamm* was current in OE is based on the distribution of the word in ME sources and as a modern dialect term. Kitson's discussion is based on occurrences in charter boundaries; he says 'mainly in Wessex' and 'so rare in predominantly Anglian areas that it is probably to be regarded as a non-Anglian word', but he is content to allow its occurrence in a few names in the north west, which he takes to be evidence of settlers from the mixed Anglian/Saxon stock of the Hwicce. It should probably be accepted that apart from the Shropshire and Cheshire examples *hamm* is not to be found north of the line suggested by Kolb.

Uncertainty about the geographical limits of *hamm* is one aspect of the problem of distinguishing it from the habitative word *hām* 'village'. This, as Kolb says, 'is an old crux'. There are several possible criteria. it seems clear that *hām* 'village' was not used uncompounded in place-names, and for this reason simplex Ham names and variants like the SHR Home discussed above are always ascribed to *hamm*. In ME spellings a double consonant (*mm*) is much more likely from *hamm*, and early spellings with *-e* should derive from *hamm* since *hām* is uninflected in the dative. The long vowel of *hām* was shortened in compound names before the ME development of $\bar{a} > \bar{o}$, so that (although the word *home* is the modern reflex of *hām*) place-name spellings like *-hom(m)*, *-hom(m)e* are generally taken to be an indication of *hamm* and to exhibit the rounding of short *a* before nasal consonants. Hence the significance attached to spellings like *Rumbaldeshom* 1298 and *Romaldesham* in Shrewsbury. Kolb 1989 does not accept the criteria of *-mm-* and *-o-*, but these are so often found in names which on other grounds are much more likely to contain *hamm* than *hām* that it seems perverse to reject them.

Attempts to separate *hām* and *hamm* must depend mainly on spellings and topography; status and likely chronology are also relevant, though not conclusive. It seems clear that *hām* was a habitative term characteristic of the earliest stages of English name-giving (Cox 1976) and that it referred to settlements of some importance. Many places with *hamm* names also have a high administrative status, but *hamm* continued in use throughout the Anglo-Saxon period and was therefore much more likely than *hām* to be used in minor names, some of which may have arisen in the later Anglo-Saxon and post-Conquest centuries. *hamm* survived as a dialect word in the south-western counties and as a field-name term in a wider area. It is not at all likely that *hām* will be found in field-names.

It will never be possible to compile a definitive list of compound names containing *hamm*, but there are many certain instances of simplex names, and of compounds with OE spellings inappropriate to *hām* which refer to places in unmistakeable *hamm* situations. Examples include Ham (near Petersham), Fulham and Twickenham in bends of the river Thames, Birlingham, Pensham, Evesham in bends of the Worcestershire Avon, and Passenham and Buckingham in bends of the river Ouse. Spellings for these names are: *Hamme* 958, *Fullanhamme* c.895, *Tuican hom* 704, *Byrlingahamme* 972, *Pedneshamme* 972, *Homme* 709, *Passanhamme* c.925, *Buccingahamme* c.925. The 'river-bend' sense is clearly established by these and other examples, but it will never be possible to say whether later-recorded names of places in comparable situations contain *hām* or *hamm*. Lower down the Bedfordshire Ouse from Passenham and Buckingham there are Blunham, Biddenham, Bromham, Pavenham, Felmersham, Haversham and Tyringham. All these are enclosed by loops of the river, but there are no OE forms for any of them and no ME forms with spellings other than -*ham*. It seems more likely than not that they all contain *hamm*, but there is obviously no reason why a -*hām* name should not occur in a *hamm* situation, so the matter has to be left open in these and many other cases.

Examples in which both early spellings and situations indicate derivation from *hamm* can be supplied for all the postulated senses of the element, and this aspect of the subject is set out in detail in Gelling 1960. There can never, however, be a firm choice between *hamm* and *hām* for places which occupy *hamm* situations but lack diagnostic spellings. Another cause of uncertainty is the uneven distribution of the habitative term *hām*. We do not know how far west it is likely to occur. Birmingham is an almost certain *hām*, and so is Pattingham, west of Wolverhampton, but *hām* is very rare in the west midland counties, and the few possible examples in Shropshire, Herefordshire and Worcestershire have situations highly suggestive of *hamm*. As regards the south west, it was argued in Gelling 1960 that all -ham names in Devon (not just those like Abbotsham and Brixham for which there are diagnostic spellings) are more likely to contain *hamm* than *hām* since the habitative element is likely to have become obsolete before the English takeover of the region.

Readers wishing to pursue these problems can only be advised to study the extensive literature on *hamm*, details of which can be found in Gelling 1984.

The procedure adopted here is to list in this article some unequivocal examples of *hamm* names under the various senses in which the element is used, and to classify a larger corpus of *hamm* names according to their first elements in the reference section.

The common name Hampton requires comment. This modern form has three different origins. It can be a habitative compound of *hām* and *tūn* (as in Northampton, *Hamtun* 917), it can derive from OE *hēan tūn* 'at the high settlement' (Wolverhampton is *æt Heantune* 985), or it can be a compound of *tūn* with our topographical word *hamm*. Spellings usually afford decisive evidence for one of these derivations: Southampton, e.g. has an OE spelling *Homtun*, *hamm* being used of the promontory situation, and Hampton Bishop HRE, enclosed by the rivers Wye and Lugg, is *Homptone* 1240.

The four main senses ascribed to *hamm* in this article could be subdivided into a more elaborate classification, such as was presented in Dodgson 1973, one of the major contributions to the debate. For present purposes, however, it seems best to simplify, and examples of unequivocal *hamm* names are presented under four main headings.

1. *Land hemmed in by water or marsh*
The following are in river-bends:

Thames: Fulham, Twickenham, Ham near Petersham, Hampton, Ham in Chertsey, Mixnams, Ham Fields, Wittenham, Culham, Inglesham. There are other -ham names along the Thames (e.g. Remenham and Cookham) for which the spellings give no indication of *hamm* rather than *hām*.

BDF *Ouse*: Passenham, Buckingham. As noted above there are many other names in bends of this river for which there are no spellings indicative of *hamm*.

WAR and WOR *Avon*: Birlingham, Pensham, Evesham, Hampton Lucy.

WLT and SOM *Avon*: Chippenham, Keynsham.

DOR *Stour*: Hampreston, Hammoon.

Wye: Holme Lacey, Hampton Bishop. Here again there are other -ham names in loops of the river, such as Ballingham and Lulham, for which there are no early spellings which support a *hamm* derivation.

Severn and Teme: These rivers have simplex minor names – Corn Ham, Port Ham, Mean Ham, Maisemore Hams, Hasfield Ham, Severn Ham GLO, Upper Ham, Lower Ham, Hamcastle, Ham Fm WOR – in river-bends. There are two names of higher status, Walham GLO (*Walehamme* 1248) and Eastham WOR (*Esthamme* 1258), among these. The minor names illustrate the findings of the Survey of English Dialects (quoted by Kolb 1989) that Ham was the answer to the question 'What do you call that low-lying flat land in the bend of a river, generally very fertile?', though this modern usage was only found in south-western counties.

Situated on promontories on the south coast or in wide estuaries are Topsham DEV, Hamworthy DOR, Southampton HMP, Bosham SSX, Iham (in Winchelsea) SSX, Ham Green near Upchurch KNT. In High and Low Ham SOM *hamm* is used for a dramatic inland promontory surrounded by the marshes of Sedgemoor, and many minor names in the Somerset Levels contain *hamm* used of low ridges. There are examples of this usage applied to inland promontories in SSX, in the marshes along the rivers Brede, Rother and Ouse. Beddingham, for instance (OE *Beadingahamme*) occupies a low, narrow promontory by a tributary of the Ouse, and Icklesham (OE *Icoleshamme*) is on a promontory between coastal marshes and those by the river Brede. The promontory occupied by Wittersham KNT overlooks the Rother Levels, and Methersham and Kitchenham SSX are on smaller projections on the other side of the river and its marshes. Dodgson 1973 should be consulted for a detailed study of the numerous -ham names of this area, where 'promontory into marsh' is the usual sense of *hamm*. See Fig. 12. Witcham CAM occupies a promontory jutting into the Bedford Levels, and Ham near Berkeley GLO is another good example of this meaning.

Under the heading of 'land hemmed in by marsh' may be included a number of names in the Severn estuary and the adjacent marshes of Sedgemoor and Wedmore. Burnham SOM is the largest of the places, but some small settlements are early-recorded: Bilsham Fm near Aust GLO is *Billeshamme* 955-9, and Hamp in Bridgwater SOM is *Hame* in DB. Distinguishing a *hamm* from an *ēg* would be a tricky exercise in these parts.

2. *Wet land hemmed in by higher ground*

This meaning was suggested in Dodgson 1973 as an addition to the senses advocated in Gelling 1960. Dodgson only found it in a few of the names in KNT, SUR, SSX examined in his paper, the most convincing being Withyham SSX. It seems to me suitable for a number of major names outside these counties, particularly Cheltenham GLO, Chesham, BUC, Dyrham GLO, Ham WLT, Portisham DOR and Todenham GLO.

3. *River meadow*

There are some ancient settlement-names containing *hamm* which denote places surrounded by small streams, often draining into nearby larger rivers, but where the topography is not sufficiently clearly defined to produce a promontory, and in these names a more general sense of 'river meadow' is appropriate. Eynsham OXF, credited in Domesday Book with 255 acres of

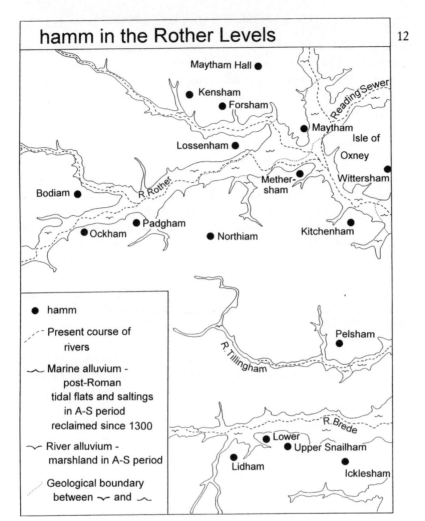

meadow, is a particularly convincing example. Farnham SUR had 35 acres of meadow in the survey, and Damerham HMP had 26. In BRK this sense is appropriate for Benham, Shrivenham and Marcham, all with large acreages of meadow in 1086. It is possible in some GLO names, including Highnam, Churcham and Kingsholm near Gloucester. Other likely instances are East and West Ham GTL, Woodham BUC, Colham in Hillingdon MDX and Cuxham OXF.

4. *Cultivated plot on the edge of woodland or moor*
In the literary language of the late OE period *hamm* occurs with the sense 'cultivated enclosure'. In Ælfric's *Lives of Saints*, dated c. AD 1000, there is a reference to a *wyrtigan hamme*, which is a garden plot. This late sense is manifested in some minor place-names and in charter boundaries in certain areas. The relevant place-names and charter boundary occurrences have a clear

THE LANDSCAPE OF PLACE-NAMES

correlation with marginal land, which may be on the edge of high moorland or of forest. The charter names which belong in this category are heavily concentrated in the hilly, wooded country on either side of the BRK/HMP border, an area in which there may well have been individual plots of pasture or cultivated ground, isolated from each other by surrounding woodland. The BRK names include *rige hamm*, *flex hammas*, *mint hammas* and *sceaphammas*, suggesting plots used for specific purposes. If it be accepted that *hām* 'village' is unlikely to be found in names of minor settlements on the edge of moorland in DEV, then a number of -ham names (Didham, Callisham, Gnatham, Dittisham) near the south-west edge of Dartmoor, and others (Winsham, Viveham, Fernham, Frizenham, Smytham, Craneham) in high ground in the north-west of the county could be ascribed to *hamm* in the sense 'cultivated plot in marginal ground'. It was suggested in Gelling 1960 that the bright green appearance of such plots against drab moorland might have triggered an association with riverside meadows. There are no diagnostic spellings for these DEV *-ham* names, but there is no doubt that the district-name South Hams borne by a large area of south DEV derives from *hamm*. In Gelling 1960 it was suggested that this district-name contained *hamm* in the late sense 'cultivated ground', and that South Hams meant the cultivated lands south of Dartmoor. This definition is accepted in Mills 1991.

Probable examples of this non-watery sense of *hamm* can be found in minor names elsewhere, e.g. Ham Green in Mathon HRE, at a height of 450ft in hilly country west of the Malverns, and some KNT names such as Bentham Hill in Southborough, Ham Fm in Barham and Tickham in Lynsted.

The meaning 'enclosed plot' did not oust the older watery meanings of *hamm*: these survived in dialect.

The great majority of *hamm* names can be classified in the above categories. There are, however, a few major names which cannot confidently be accorded this treatment. Tidenham GLO is a particularly difficult case. It occupies a great promontory by the river Severn, but this is on a different scale to the promontories of such names as Southampton. A survey of the manor dated 956 speaks of land 'outside the *hamm*', which suggests that we are here dealing with an artificially rather than a naturally enclosed site, and which makes the 'river meadow' sense unlikely. Perhaps the estate was known by its Welsh name *Istrat Hafren* ('Severn valley', recorded in charters of *c.*703 and *c.*878 in the Book of Llandaff) until the mid 10th century, in which case it would be possible for *hamm* to be used in its latest OE sense.

Feckenham WOR (*Feccanhom* 804) is surrounded by streams and part of the parish is called Ham Green, so it is probably a 'river meadow' name, but no meadow is recorded in its DB entry. Harrietsham KNT is a possible but not unequivocal 'river meadow' name, as are Barkham BRK, Littleham DEV.

hamm is a place-name element which offers scope for local studies, both of specimens with diagnostic *hamm* spellings, and of those which lack such spellings but which occupy sites suggestive of *hamm*. It will probably continue to be a long-suffering word.

Reference section for hamm

As simplex name: Ham GLO, GTL, KNT, SOM, SUR, WLT (and *passim* in minor names and field-names within the geographical limits indicated above), Hamp SOM, Holme Lacey HRE.

Simplex, but with late OE or ME affix: Hammoon, Hampreston DOR, Hamsey SSX, Hamworthy DOR, Abbotsham and Georgeham DEV, Churcham and Highnam GLO, Woodham BUC. Evesham WOR, which occurs in OE sources once as *Homme* otherwise with a variety of first elements, is perhaps best classified as originally simplex. The qualifying element which prevailed is a personal name *Ēof*.

As first element with tūn: Hampton GTL, Hampton Bishop and Hampton Court HRE, Hampton Lucy WAR, Southampton HMP.

As generic with personal names: Benham BRK (**Benna*); Bilsham GLO (*Bill*); Bodiam SSX (*Boda*); Bosham SSX (*Bosa*); Brixham DEV (*Beorhtsige*); Colham MDX (*Cola*); Culham OXF (**Cūla*); Cuxham OXF (**Cuc*); Dittisham DEV (**Dyddi*); Feckenham WOR (**Fecca*); Frodsham CHE (*Frōd*); Fulham GTL (**Fulla*); Hankham SSX (**Haneca*); Harrietsham KNT (*Heregeard*); Icklesham SSX (*Icel*); Inglesham WLT (**Ingīn*); Ockham (in Ewhurst) SSX (*Occa*); Padgham (in Ewhurst) SSX (*Pada*); Passenham NTP (*Passa*); Pensham WOR (**Peden*); Polsham SOM (*Paul*); *Romaldesham* SHR (*Rumbeald*); Tidenham GLO (**Dydda*); Todenham GLO (*Teoda*); Wittenham BRK (*Witta*); Wittersham KNT (*Wihtrīc*).

A rather high proportion of these names is conjectural. *Benna, Cūla, Cuc, Dydda* and *Dyddi, Fecca, Fulla, Haneca, Ingīn, Peden* are not recorded. However, *Dydda, Dyddi, Haneca, Ingīn* and *Peden* are fair inferences from the recorded names *Dudd, Hana, Inga, Peada*, and there is an Old German name which would correspond to *Fulla*. *Cuc* may be a short form of compound names in *Cwic-*. *Cūla* (or a formally identical word) occurs in a number of place-names, for which see Ekwall 1960 s.n. Cowlinge. *Benna* and *Fecca* are probably the most difficult to justify.

With references to special persons or groups: Damerham HMP ('judges'); Kingsholme GLO; Methersham SSX ('mower'); Walham GLO ('Welshmen'). There are several *-inga-* compounds: Beddingham SSX, Birlingham WOR, Buckingham BUC, Etchingham SSX ('people of **Bēada/*Byrla/Bucca/ Ecci*'). Wellingham SSX is '*hamm* of the spring-dwellers'.

With descriptive terms: Northiam SSX ('high *hamm*' with 'north' prefixed); Littleham (by Exmouth) DEV. OE *sīdan hamme* 'spacious *hamm*' is a frequently occurring name. It has become a parish name in DEV and OXF and occurs in at least seven minor names and a number of field-names. There is

THE LANDSCAPE OF PLACE-NAMES

one OE example, *sidanhamme* in the bounds of Wylye WLT. The usual modern form is Sydenham, but Sideham and Sidnums also occur. Sydenham GTL is a corrupt form of Chippenham, discussed below. Felpham SSX contains *fealh*, perhaps 'newly cultivated land'.

With topographical terms: Burnham SOM (**burna**); Cheltenham GLO (?hill-name *Celte*); Chesham BUC (*ceastel* 'stone-heap' referring to the puddingstone boulders on which the church is built, overlooking the *hamm*); Petham KNT ('pit'); Portisham DOR ('town'); Pulham DEV (**pōl**); Topsham DEV (the highest point of the promontory was probably called *topp* 'summit'); Twickenham GTL (perhaps an unrecorded *twicce 'river-fork').

With references to vegetation: Barkham BRK ('birch'); Farnham SUR and Fernham BRK, DEV ('fern'); Marcham BRK ('smallage'); Witcham CAM ('wych-elm'); Withyham SSX ('willow').

With reference to fauna: Dyrham GLO ('deer'); Elham KNT ('eel'); Snailham SSX.

Names with obscure or ambiguous qualifiers:

Chippenham WLT. This is *Cippenhamme* in the Anglo-Saxon Chronicle's annals for 878 and 879. Chippenham CAM, Cippenham in Burnham BUC and Sydenham GTL (*Cyppenham* 1319) are probably identical, though without diagnostic *-hamm* spellings. Another *Cippanhamm* occurs in OE charter boundaries near Bishop's Cleeve GLO. A noun *cippa seems to be indicated, but no such word is on record.

Eynsham OXF. OE spellings include *Egonesham* and *Egeneshomm*. No convincing explanation has been offered for these.

Iham SSX. This was the name of part of the cliff on which New Winchelsea was built c.1280. The old name survives in Higham Street. Early spellings (1200–1428) have *I-* or *Y-* for the first element. This is not the manner in which *īeg* 'island' manifests itself as a first element, and absence of *H-* makes *hēah* 'high' unsuitable. The absence of *Iw-* spellings probably rules out *īw* 'yew'. A similar problem is posed by Iden, four miles north of Winchelsea, which is discussed under **denn**. The nature of the qualifier in *Iham* must be left open. Northiam, six miles north-west of *Iham*, was *Hiham* in DB, *Hyham* c.1210, and part of the village is still called Higham. This is probably 'high village', a different name from *Iham*, but the two were liable to confusion in the 13th century, and 'north' was prefixed to Higham to obviate this. There is one spelling for *Iham* which has 'south' prefixed – *Suthyhomme* 1339.

Keynsham SOM. The first element has been explained as a personal name *Cægin, a diminutive of *Cæg. It is doubtful, however, whether *Cæg ever existed as a personal name. The supposed manifestations in place-names could be explained as a topographical use of *cæg* 'key', used as a hill-name, for which see the discussion under **hōh**. The etymology of Keynsham must be left open.

Kitchingham, Kitchenham SSX. There are three identical names in SSX, Kitchingham in the parish of Etchingham, and two examples of Kitchenham, in Ashburnham and Peasemarsh. They are all ME *Kechen-*, which can only be ascribed to OE *cycene* 'kitchen'. The two Kitchenhams, both of which have *-mm-* spellings, both occupy perfect 'promontory into marsh' sites. The topography of Kitchingham, which has only *-ham* spellings, is less clearly defined. The name may refer to the growing of vegetables and herbs. The names are listed in this section not because the nature of the first element is uncertain, but because its significance is not perfectly clear.

Mixnams SUR. A word **mixten* meaning 'dung' has been conjectured to explain this name and Muxworthy DEV, early spellings being *Mixtenhammes* and *Mixtewurthe*. This element has not been noted in any other names, however, whereas *mixen*, the recorded word for 'dung' is fairly common.

Shrivenham BRK. The first element is the past participle of the verb *scrifan* 'decree, allot, pass sentence on, impose penance', modern *shrive*. In EPNS survey, p. 376, it is suggested that the reference might be to land given to the church in satisfaction of an ecclesiastical claim.

holmr ON 'island, inland promontory, raised ground in marsh, river-meadow'. This is probably the commonest ON topographical term in England. It was adopted into late OE, and there is some interchange between *holm* and **hamm** in the west and north midlands. As with *ēg*, the characteristic use in settlement-names is for an island of firm ground in a marsh.

Hulm(e), -hulm(e), found in CHE, LNC, YOW (most conspicuously in the Manchester area) was once considered to be a specifically Danish, as opposed to Norwegian, form of the word. This view, which is obviously at odds with the distribution, was challenged in Hald 1978, and the matter has since been considered in EPNS CHE survey 5 (pp. 237–9) and Fellows-Jenson 1985 (pp. 313–16). The tentative conclusion is that the *u-* forms represent a dialectal development with no bearing on ethnic origin.

The fact that *holm* became an English word means that minor names and field-names which contain it are not direct evidence of Scandinavian settlement. Many such names recorded in ME sources are likely to have been coined by English-speaking people using this borrowed term. Major settlement-names, however, especially those recorded in DB, are likely to date from the 10th century, and may in some cases result from an expansion of settlement soon after the Viking invasions.

There are about 50 names containing *holmr* in Ekwall 1960. Most of the places have remained small, and they are often rather isolated; but Oldham (near Manchester) and Durham are notable exceptions.

holmr is used in the naming of coastal islands, smaller than those designated by ON *ey*, round Wales and Scotland. In settlement-names the characteristic use is for a patch of slightly raised ground in a wet area. Instances which can be easily identified on OS maps as having this sense include Holme HNT (west of Ramsey), Holme NTT (north of Newark on Trent), Holme Pierrepoint NTT, Holme upon Spalding Moor YOE, Holmes Chapel (alias Church Hulme) CHE. 'River-meadow' is often given as a meaning of *holmr*, and many of the sites are in low ground by rivers, but a large-scale map or local knowledge would probably reveal that some of the settlements are on slightly raised ground. Balkholme YOE, which has OE *balca* 'ridge' as first element, probably occupies a low ridge like that noted in the discussion of Holbeach, under **bæc**.

Sometimes the raised ground is a promontory. Holme YON, west of Thirsk, is a good example of this, and the sense may also be found in Cheadle Hulme CHE. There are two dramatic promontory sites: Oldham LNC (*Aldhulm* 1227) and Durham (*Dunholm* c.1000). Neither name is easily explained; both are discussed in Gelling 1970.

Oldham and Durham have high sites which are promontories. For some high sites with *holmr* names which are not promontories it would be reasonable to postulate a meaning 'cultivated land on the edge of moors', with the same sense-development as is suggested for **hamm**. Two instances of Sleightholme in YON may have this sense (though another Sleightholme in CMB, east of Silloth, is a typical 'island in marsh' site). Another high site for which a meaning 'enclosure in marginal land' would be suitable is Hulme, east of Stoke on Trent STF, and there are some minor names on the Yorkshire Wolds for which it seems the only possible sense. Such a meaning was envisaged for an Orkney name, Holmes in Paplay, in Marwick 1952 (p. 91): 'There is no island in Paplay to give rise to such a name, and it would thus appear to be ON *holmar*, plur. of *holmr*, a small island, but used in the secondary sense of a patch of ground situated like an island in the midst of ground of a different nature'.

Names with modern Holme, -holme forms do not always derive from *holmr*. As a simplex name Holme can have several other origins, OE *holegn* 'holly' being perhaps the commonest. Nunburnholme YOE contains *brunnum*, dat.pl. of ON *brunnr* 'spring'. Lealholm YON is the dat.pl. of OE *lǣl* 'twig', and Moorsholm YON contains *hūsum* 'at the houses'.

Reference section for ON **holmr**

As a simplex name: Holme BDF, DRB(2), HNT, LNC, LIN(2), NFK(4), NTT(2), WML, YOE(2), YON(3) and many minor names, Hulme CHE(3), LNC(3), STF(2).

As an affix: Holme Cultram CMB.

As generic with personal names: Broadholme NTT (ON *Broddi*); Gauxholme LNC (ON *Gaukr*, or the noun which means 'cuckoo'); Kettleshulme CHE (ON *Ketil*, or 'kettle' used in a topographical sense); Levenshulme LNC (OE *Lēofwine*); Torrisholme LNC (ON *Thóraldr*); Tupholme LIN (ON *Túpi*).

MARSH, MOOR AND FLOODPLAIN

With terms for vegetation or crops: Almholm YOW ('elm', ON *almr*); Bromholm NFK ('broom'); Haverholme LIN ('oats', ON *hafri*); Hempholm YOE ('hemp').

With topographical terms: Balkholme YOE ('balk'); Broxholme LIN (probably 'bog', v. Fellows-Jensen 1978, p. 153); Durham DRH (dūn); Keldholme YON (**kelda**); Sookholme NTT ('gully', OE *sulh*).

With descriptive terms: Langham LIN ('long'); Oldham LNC ('old').

With earlier place-names: Axholme LIN (Haxey); Burtholme CMB (river *Burth*).

With a bird-name: Trenholme YON ('crane', ON *trani*).

A minor name Sleightholme occurs four times. One instance is in CMB, east of Silloth. The others – two surviving, one lost – are in YON: Sleightholme south-west of Barnard's Castle, Sleightholme Dale in Kirkby Moorside, and a lost place in Barnby, west of Whitby. The topography of the sites varies considerably. The first element has been explained as ON *slétta* 'level field' or a related adjective *sléttr* 'smooth'. OE *sleget* 'sheep pasture' might also be considered.

kjarr ON 'brushwood'. The ME derivative *kerr* was used of 'a marsh, especially one overgrown by brushwood', and it is likely that this is the sense of *kjarr* in English place-names. The word occurs mainly in minor names and field-names in the Danelaw and in north-west England, but it is the second element of a few settlement-names. In Altcar LNC it is added to a British river-name. In Broadcar NFK and Holker LNC it is qualified by OE *brād* 'broad' and *holh* 'hollow'. Redcar YON is another hybrid, with OE *hrēod* 'reed' as first element. Ellerker YOE also refers to vegetation but contains an ON word, *elri* 'alder'. Walker NTB is 'marsh by the Roman Wall'.

Bicker LIN and Byker NTB are either 'village marsh', with ON *bý*, or 'place by the marsh', with OE *bī*.

mersc, merisc 'marsh' occurs in about 60 names in Ekwall 1960, leaving aside those, like Burgh le Marsh LIN and Marsh Baldon OXF, in which it is used as an affix. In over half the instances it is the first element, the compound with *tūn* accounting for 33. It is possible that these *mersc-tūn* settlements had a specialised function in relation to the marsh, or that their economy depended heavily on its products. They do not, as a class, however, bear a significant relationship to Roman roads and ancient routeways (as is sometimes the case with recurrent *tūn* compounds).

Major names containing *mersc* are listed in the reference section.

Reference section for **mersc**
As a simplex name: Marche SHR, Marsh Gibbon BUC, Marishes YON, Marske (dat.pl.) YON(2).

As first element with tūn: Marston BDF, BUC(2), CHE, DRB(2), GLO, HRE(2), HRT, LEI, LIN, NTP(2), OXF, SOM(2), STF(2), WAR(6), WLT(3), YOW, Merston IOW, KNT, SSX, Morston NFK.

With other generics: Marsham NFK (*hām*); Marshwood DOR.

As generic with descriptive terms: Lamarsh ESX and Lamas NFK ('loam'); Michelmarsh HMP ('large'); Rawmarsh YOW ('red'); Saltmarsh(e) HRE, YOE.

With references to living creatures: Crakemarsh STF (?'raven'); Crowmarsh OXF; Henmarsh GLO; Stodmarsh KNT ('herd of horses').

With personal names: ?Bickmarsh WAR (*Bica*, or a noun meaning 'projection'); Killamarsh DRB (*Cynewald*); Titchmarsh NTP (*Ticcea*).

With references to groups of people: Burmarsh ('city-dwellers', OE *burgware*, referring to Canterbury); Tidmarsh BRK ('people', OE *thēod*, referring to communal use).

With reference to vegetation: Peasmarsh SSX.

With OE *cegel 'pole': Chelmarsh SHR, Kelmarsh NTP.

mōr 'marsh' and 'barren upland' (i.e. modern *moor*). The word is well-evidenced in both senses in literary OE, but neither its ultimate etymology nor the relationship between the meanings has been ascertained with certainty. NED (though without total conviction) cites derivation from a root meaning 'to die', which suggests an original meaning such as 'barren land'. Other authorities (e.g. Holthausen 1934, Onions 1966) suggest a connection with **mere** ('sea, lake'), and this could indicate that the 'marsh' sense is the basic one. Kitson (forthcoming) takes wetness to be the primary meaning, and suggests that **mersc** is a formation from *mōr* + *-isc*.

In place-names there are numerous instances of *mōr* used for both high and low features. It is particularly well-evidenced in the names of large areas of low-lying wetland, like Sedgemoor SOM, The Weald Moors SHR and Otmoor OXF. Some of the high ground to which *mōr* is applied is very wet, this being exemplified by the great moors of the WML/YOW border, such as Stainmore. The moors of the south west – Dartmoor and Exmoor – are also boggy in parts. It is not certain, however, that all stretches of barren upland for which the word is used were characterised by soddenness. In Gelling 1984 (p. 54) it was stated erroneously that Snelsmore BRK was an area of dry heathland, whereas in fact this 'moor' also has notable bogs. Other instances of dry land for which *mōr* was used which were cited in that discussion still appear valid, however. It seems unlikely that the small upland called Dunsmore Heath in WAR was ever marshy. In BRK the neighbouring settlements of Draycott Moor and Southmoor are on top of the Corallian ridge which separates the Thames valley from that of the river Ock, and there are minor names in *-mōr* along this ridge. While 'low-lying wetlands' and 'boggy uplands' are certainly the commonest senses, there is evidence to suggest that *mōr* was sometimes applied to uplands which were barren because they were dry.

This dual, or perhaps triple, significance prevents the *mōr* names from being a homogeneous corpus.

mōr is used most frequently as the first element of compound place-names. In the corpus of about 100 names included in Ekwall 1960, 64 have *mōr* as first element, 44 of these being compounds with *tūn*. This leaves a relatively small category in which *mōr* is the generic, and some of these (Dartmoor DEV and Exmoor DEV/SOM, Dunsmore WAR, Otmoor OXF, Sedgemoor SOM, Snelsmore BRK, Stainmore YOW/WML, Weald Moors SHR) are district-names which were not used for any individual settlement. For place-name-forming purposes a *mōr* was probably seen primarily as a factor in the economy of a farming community. Some lowland areas designated *mōr* were a major resource for settlements on their fringes or on island sites within them. Maids Moreton BUC adjoins Chackmore and Moorton OXF is by Northmoor. Murcot OXF is similarly sited on the edge of Otmoor. Some *mōrtūn* settlements may be component parts of multiple estates. North and South Moreton BRK were in the 10th century part of a 100-hide estate centred on Blewbury, and it is easy to see them as the suppliers of hay for the whole unit (Gelling 1988, pp. 197–200). A contrasting sense is seen in Moretonhampstead DEV, where the reference is to Dartmoor, and the name was perhaps primarily a comment on the bleakness of the site. Murton WML has a similar situation.

There is a recurring compound with OE *wicga*, found in Wigmore HRE, SHR, and a number of minor names and field-names (Gelling 1984, p. 56). At Wigmore HRE there is a large area of unstable marsh in which blister bogs appear and disappear, and OE *wicga*, which is recorded as an insect name but may have a more general sense 'something which wiggles', probably refers to this characteristic.

The corresponding ON word *mór* cannot be distinguished in place-names from the OE one.

Reference section for **mōr**

As simplex name: Moor WOR, Moore CHE, More SHR. Allensmore HRE was originally simplex, as were Northmoor OXF and Southmoor BRK, which are respectively north and south of the Thames.

As first element with tūn: Morton DRB, DRH(3), IOW, LIN(3), NFK, NTT(2), SHR, WAR, WOR(2), YON(2), YOW, Moreton BRK, BUC, CHE(2), DEV, DOR, ESX, GLO(3), HRE(2), NTP, OXF, SHR(4), STF(3), WAR, Murton DRH(2), NTB, WML, YON(2), Moorton OXF (Moreton on recent O.S. maps).

THE LANDSCAPE OF PLACE-NAMES

With other habitative terms: Moorby LIN, Moreby YOE (ON *bý*); Moorsholm YON (*Morehusum* DB, 'at the marsh houses'); Morcott RUT, Murcot OXF, Murcott WLT; Morestead HMP; Morwick NTB; Mosborough DRB.

With topographical generics: Morborne HNT (**burna**); Morden CAM, DOR, SUR, Moredon WLT, Mordon DRH (all **dūn**); Morebath DEV (**bæth**); Moreleigh DEV, Morley DRB, DRH, NFK, YOW (**lēah**); Morland WML (**lundr**).

As generic with personal names: Carlesmoor YOW (ON *Karl*); Cottesmore RUT (**Cott*); Dunsmore WAR (*Dunn*); Gransmoor YOE (*Grante*); Kexmoor YOW (ON *Ketil*); Mentmore BUC (**Menta*); Otmoor OXF (*Otta*); Pedmore WOR (*Pybba*); Snelsmore BRK (*Snell*).

With descriptive terms: Blackmoor DOR, Blackmore ESX, WLT; Littlemore OXF; Radmore STF ('red'); Ringmore DEV(2) ('cleared'); Stainmore YOW/WML ('stone'); Swarthmoor LNC ('black'); Weald Moors SHR and Wildmore LIN ('uncultivated'); Whitmore STF ('white'); Wigmore HRE, SHR ('unstable').

With references to vegetation: Barmoor NTB ('berry'); Breamore HMP and Brymore SOM ('broom'); Sedgemoor SOM; Twigmore LIN.

With references to a living creature: Podmore STF ('toad').

With river-names: Dartmoor and Exmoor DEV.

With reference to a product: Quernmore LNC (probably high moorland where there was rock for quernstones).

Names with obscure or ambiguous qualifiers:

Chackmore BUC. The group of names containing an element *ceacca* is discussed under Checkendon (**denu**).

Poltimore DEV. The first element is discussed under Poultney (**ēg**).

Silkmore STF. The same first element occurs in Silkstead HMP and Selkley Hundred WLT. Ekwall 1960 identifies this with that of *sioluchamm* in a HMP charter boundary and suggests a diminutive of a term meaning 'drain' or 'rill'. This is convincing, but it is not clear whether the feature would be natural or man-made, and this makes classification difficult.

Wedmore SOM. OE spellings are *Wethmor* and *Wedmor*, so the evidence is conflicting.

mos OE, **mosi** ON 'moss, bog'. The OE word is only evidenced once outside place-names, in a STF charter boundary of 975, where it occurs in the boundary-marks *hedenan mos* and *micle mos*. The meaning in the Madeley boundary is obviously 'bog', but it is reasonable to suppose that OE *mos* was also used for moss and lichen, since it is the ancestor of the modern word.

In north-west England *moss* became and remains a term for a bog. This widespread use may be due to the presence in that area of the ON word, but in place-names the use spreads as far south as SHR and STF. There is an isolated example in Moze ESX. Derivation from the OE word is more likely in ESX and also in Mose near Quatford SHR. Further north *mos* occurs as the first element of Moston in CHE, LNC and north SHR, where Norse words are possible, but where settlements with names in *-tūn* are likely to have been established before the Norse invasions. Mostyn in Whitford FLI is another *mos-tūn*. The ON word is likely in names with ON generics, such as Mosedale and Mosser CMB, Mozergh WML (the last two with *erg* 'shieling'). In the names of LNC bogs, like Rathmoss, Chatmoss and Wirples Moss, however, the word could have been common late OE. Rathmoss has ON *rauthr* 'red' as first element, the other two have English qualifiers, 'cat' and 'path'.

When *mos* is the first element in a compound with other generics than *tūn* it is obviously possible that it is the plant-name which is involved. This is not so for Moseley YOW, however, as that place is in a township called Moss.

The related OE word **mēos** occurs occasionally in place-names. This survived as an alternative dialect word for the plant until the 19th century in the form *mese*. Place-name evidence suggests that it was also used in the sense 'bog'. It gives rise to two river-names, Mease LEI/DRB/STF and Meese Brook STF/SHR, and the settlements of Millmeece and Coldmeece STF take their names from the latter. Meesden HRT (a **dūn**) and Muswell Hill near Brill BUC are both boggy without being notably characterised by moss. The compound with **welle** occurs again in Muswell Hill GTL and Muzwell Fm in Fingest BUC. In the compound with **brōc** found in Misbrooke Fm in Capel SUR and Mixbrooks Fm in Cuckfield SSX the meaning may be 'moss'; there are certainly patches of moss on the banks of the former stream. *mēos* occurs as a simplex name in Muze in Tawstock DEV.

Names with modern forms like Mose-, Mows-, Mus-, Mouse- most frequently contain *mūs* 'mouse'.

mýrr ON 'swamp' is the source of ME and modern *mire*. It occurs mainly in minor names, most of them probably of ME origin. It is the first element of Myerscough LNC, and there was another instance of this compound in IOM, where it was the ON name for the marsh now called The Curraghs. ON *mýrr-skógr* is the equivalent of the English name Marshwood DOR. *mýrr* is used as an affix in Ainderby Mires YON.

slæp 'slippery place' (only evidenced in place-names and charter boundaries) occurs uncompounded in two SHR names – Sleap in Crudgington and Myddle – and in *Slepe* HNT, the former name of St Ives. Compounded with *-tūn* it occurs in Slapton, BUC, DEV, NTP, and is the generic in Hanslope BUC, Islip NTP, OXF. (In spite of their identical modern forms the last two have different river-names, *Ise* and *Ight*, as first elements). Smith 1956 and Ekwall 1960 give *slæp* as the generic in Postlip GLO and Ruislip GTL, but these are equally likely to contain *hlēp* 'leaping place'.

THE LANDSCAPE OF PLACE-NAMES

The nature of the *slǣp* at each of the settlements could probably be discovered by local investigation. At Slapton NTP the church and manor house occupy a knoll, with the village lying between the knoll and the river Tove. A spring rises near the church, and the water from this must have made the tracks up the hill from the village very treacherous. At Slapton DEV the road from the coast to the village has a steep rise. The road down to the crossing of the river Ray at Islip OXF is steep on both sides.

slǣp was probably used of limited patches of slippery ground, rather than being a term for a marsh.

slōh 'muddy place', modern *slough*, is fairly well evidenced in minor names and field-names in counties as far north as BDF and CAM. It is the source of Slough BUC, and occurs uncompounded also in minor names in BDF, ESX, SSX and SUR. Smith 1956 lists a number of names in which it is the generic, but only one of these, Polsloe DEV, achieved sufficient status to be included in Ekwall 1960. The growth of a major settlement in BUC with the name Slough was presumably due to economic factors which overrode the unpromising nature of the site.

Slaughter GLO and some minor names in SSX and SUR (listed in Gelling 1984, p. 58) derive from a word *slōhtre*, only evidenced in place-names, which is related to *slōh*. The meaning of *slōhtre* is not, however, certain to be 'muddy place'. Löfvenberg 1942 (pp. 190–1) points out that words related to OE *slōh* in other Germanic languages had meanings such as 'ditch, ravine, deep channel', and this is topographically appropriate to the stream valley near the GLO villages of Upper and Lower Slaughter. It should be noted that Slaughterford WLT contains *slāh-thorn* 'sloe-tree', not *slōhtre*.

slōh is related to **sol**, *infra*.

sol(h) 'muddy place, wallowing place for animals'. This word is a variant of **slōh** (Kitson 1990, 196–9). Both words are limited to England south of the Wash, with *solh* occurring further north than **slōh**. The distributions are to some extent complementary.

In KNT, as has been demonstrated by Professor Everitt (1986, p. 169), a *sol* was sometimes developed into a pond for watering stock. Charter-boundary names show association with harts, boars and bullocks. The BRK settlement-name Grazeley is first recorded c.950 as *grǣgsole*. This and a few other place-names indicate a substantive use of *grǣg*, 'grey', probably as an animal-name. The creature in question has been considered to be the badger, but in a recent paper (Hough 1995a) Dr Carole Hough argues persuasively that the wolf is a likelier candidate.

sol is most common in minor names, especially in KNT, but there are a few names of higher status in addition to Grazeley BRK, Bradsole and Soles KNT being such. It is the first element of Solton in West Cliffe KNT.

MARSH, MOOR AND FLOODPLAIN

There are two minor names in the parish of Wolverley WOR which offer excellent surviving instances of the sort of feature termed *sol*. Solcum Fm overlooks a steep valley which has a spring in its bottom, resulting in a patch of thick black mud which covers the track, and at Blakeshall (*Blakesole c.*1190) an old farm overlooks a valley-end in which there is a mud-filled hollow.

Solway Moss CMB, earlier *Solome Mosse* from a lost hamlet of *Solum*, has generally been explained as deriving from the dat.pl. of *sol*, but in view of the otherwise southern distribution this seems improbable. *Solum* may be the dat.pl. of a corresponding ON word. The French-derived word *soil* may be found in some other minor names in northern England for which *sol* has been adduced.

*****sele**, a variant of *sol(h)*, occurs in some KNT names, including Selgrove in Faversham and Selstead in Swingfield.

strōd, generally rendered 'marshy land overgrown with brushwood', is only recorded in charter boundaries and OE place-name spellings, and the sense is inferred from Germanic cognates. It occurs fairly frequently in the south of England as a simplex name. Strood KNT and Stroud GLO have become major settlements, but most instances are farms or hamlets. Stroud Green occurs in several southern counties. Another modern form is Strode.

In Bulstrode (near Gerrards Cross) BUC *strōd* is compounded with *burh* 'fort', referring to the large hill-fort by Bulstrode Park. It is doubtful whether 'marsh' is applicable here, and it also seems inappropriate to the hamlet of Bulstrode (first element 'bull') in King's Langley HRT. Possibly *strōd* denoted a type of vegetation not always associated with marsh.

In counties to the north and west of CAM and HNT *strōd* is very rare, possibly absent, but it occurs in two names of identical meaning in the north country: Langstrothdale YOW and Long Strath (near Keswick) CMB. In the early spellings for Langstrothdale there is interchange between *strōd* and *strōther*, which is a derivative, and *strōther* is found occasionally in minor names in the north (e.g. Haughstrother east of Haltwhistle NTB).

The cognate ON word, **storth**, which occurs in some names in the east midlands and the north (e.g. Dalestorth in Mansfield NTT, Storth, northeast of Arnside CMB, Storrs, east of Carnforth LNC) is rendered 'brushwood' or 'young growth of trees', and is not considered to have any connotation of marsh. Storiths, northeast of Skipton YOW is translated 'the plantations' in EPNS survey 5, p. 74.

wæsse 'land by a meandering river which floods and drains quickly'. The word is only known from place-names and charter boundaries. Prior to Gelling 1984 the accepted definition was 'swamp'. Since 1984 (e.g. in Mills 1991) it has been translated 'alluvial land'.

wæsse is a rare place-name element, examples being heavily concentrated in the west midland counties, with outliers in DRB, HNT, NTB. It is the likely source of the river-name Gwash LEI/RUT/LIN (earlier *Wass(e)*). It is not always possible to distinguish with certainty between *wæsse* and the other place-name elements *gewæsc* 'washing, flood' and *wæsce* 'place for washing (of sheep or clothes)'. *wæsse* is almost certainly absent from southern counties, however, and from East Anglia.

The west-midland *wæsse* names are Alrewas and Hopwas STF, Bolas, Buildwas and Westley SHR, Rotherwas and Sugwas HRE, Broadwas WOR, Wasperton WAR and Washbourne GLO (OE *Wassanburna*). The outliers are Alderwasley DRB, Allerwash NTB and a lost *Arnewas* HNT (in Sibson cum Stibbington parish). Most of these are on major rivers, either by meanders or by right-angle bends. Bolas SHR and Alrewas STF are also by confluences, respectively where the Meese joins the Tern and where the Tame goes into the Trent. In the case of Alderwasley DRB the exact reference is uncertain. The parish extends to the Derwent, but the *wæsse* may have been a miniature one on the stream which flows past Alderwasley Hall. At Westley SHR the *wæsse* is by a small stream, and Washbourne GLO is by a meandering stream which joins the Carrant Brook near Beckford. A charter of 951 relating to Marchington SHR mentions *pirewasse* on Pur Brook, but at a point higher up the brook than seems appropriate for a *wæsse*.

The most spectacular site is that of Buildwas SHR, where the village overlooks the eastern part of a long stretch of the river Severn which meanders in great loops over a broad flood-plain immediately before being confined in the Ironbridge Gorge. The flood-plain can be a lake one day and a stretch of firm grazing ground on the next, and similar phenomena have been observed near other settlements with *wæsse* names. At Allerwash NTB a long bank has been constructed to control the flooding of the South Tyne. This behaviour on the part of a river does not create swamp, so, although the semantic reference is certainly to wetness, 'alluvial land liable to sudden floods' or the like should replace the earlier definition.

Three of the *wæsse* names have *alor* 'alder' as first element. The flood-plain at Broadwas is broad, that at Sugwas was frequented by sparrows, and that at Rotherwas grazed by cattle. Bolas and Buildwas have less-easily interpreted first elements, for which suggestions made in Gelling 1990 are respectively an unrecorded OE word *bogel* 'little bend' and a word related to modern *bold*. Hopwas contains **hop**, perhaps in its 'enclosure in marsh' sense. The lost *Arnewas* may have been frequented by eagles: if so, probably the fish eagle (Gelling 1987).

CHAPTER 3

Roads and Tracks: River-Crossings and Landing-Places

In the Introduction to this book attention is drawn to the possibility that travellers' perceptions of landscape features might have played a significant part in the naming of settlements. The roads and tracks from which defining landforms could be observed are themselves sometimes referred to in settlement-names, and countrywide study of the vocabulary leads to an understanding of the specialised meanings of some of the words in this category.

The least specialised road-term is **weg**, and, as with other not very specific generics, this leads to the qualifiers in **weg** compounds being predominantly descriptive terms. If the roads were not metalled in later centuries it is still possible to see why they were called 'red', 'white' or 'stony'. Other terms, like **pæth** and **stīg**, would convey more precise information about the nature of the route.

The commonest road-term in settlement-names is **strǣt**, which is used either as a simplex name or as the first element in a compound. The oft-repeated *strǣt-tūn* is probably more than a simple observation that the settlement referred to is beside a Roman road. Like many recurrent *tūn* compounds Stratton/Stretton may be suspected of having functional connotations. These places may have offered some facilities to people using the roads.

A large category of settlement-names refers to places where roads cross rivers, and most of these have as generic the word **ford**. If a settlement was associated with a crossing-place on a river this must often have been perceived as the most significant factor in its environment. **ford** is another not very specific generic, and here, as with **weg**, the qualifiers are predominantly descriptive. In many instances the fords can still be seen beside the bridges which have superseded them, and very often the characteristics of stoniness, chalkiness, sandiness and so on which are referred to in the names are still clearly visible.

It is suggested in this chapter that Old English speakers had more specialised terms than **ford** for crossings which presented exceptional difficulties, the ones discussed here being **fær**, **gelād** and **gewæd**. The evidence for a specialised meaning is set out in the articles on these words. Anyone who has seen Ann Cole's slides of flooding by the Severn and Thames near places with **gelād** names in the winters of 1989, 1990 and 1992 will understand the use of a more specialised term than **ford** to describe these crossing-places.

The inclusion of the 'landing-place' terms **hȳth** and **stæth** could be queried on the grounds that they refer to man-made installations rather than to natural features, or to facilities like the road system which were mostly inherited by Anglo-Saxon settlers. The siting of landing-places is, however, very much a matter of topography, and the **hȳth** names interconnect with wetland terms in an illuminating fashion. It is clear from the distribution shown on Fig. 13 and the siting of some wetland **hȳth** names shown on Figs 14 and 15 that this element relates to organised transport systems in a manner which justifies its inclusion in this chapter.

ānstīg or **anstig**. The settlement-names Anstey DEV, HMP, HRT, LEI, Anstie SUR, Ansty DOR, SSX, WAR, WLT and Ayngstree WOR derive from a road-term the origin and meaning of which are currently under discussion. In addition to the names listed there are six occurrences in charter boundaries in DOR, HMP and WLT, and there was an Ainsty Wapentake in YOW.

In literary Old English there is a single occurrence of a related word *anstiga*. The Greek name Thermopylae is glossed by *fæstin vel anstigan*. The writer of the gloss may have understood *anstiga* to mean 'fort', since that is the meaning of *fæsten*, or he may have felt that *anstiga* complemented *fæsten* in his conception of the site of Thermopylae. The term *anpath* in *Beowulf* may be analogous. With such slight literary evidence, however, identification and interpretation of this word depends on the evidence of place-names.

Place-name scholars have traditionally explained *ānstīg* as 'single-file uphill path', from *ān* 'one' and *stīg*. This was challenged in Gelling 1984, where a suggestion was put forward of a meaning 'single path', i.e. 'short stretch of road used by at least four routes which converge on it at either end'. This was based mainly on the situations of Anstey HRT, LEI and Ansty WAR where, in addition to the presence of an appropriate road-pattern, the gradients are slight, and 'steep, single-file track' is inappropriate. This suggestion has not met with general acceptance, however, and it should probably be abandoned.

There are two subsequent discussions, one of which (Atkin 1998) is in print. Peter Kitson observes that while my remarks about the absence of steep gradients at the three midland examples are correct, the topography of Anst(e)y names in southern counties is markedly different. Ansty DOR, HMP, WLT and Anstie SUR are in hilly country, as are all six charter instances. He envisages the possibility of two distinct terms, the traditional feminine *ānstīg* 'one path', i.e. 'link road', which fits the HRT, LEI and WAR places (though not in the 'single-file' interpretation), and another term in which a neuter noun *stig* 'climbing path' is combined with a prefixed preposition *an* 'on' to give a term appropriate to steep tracks.

The second discussion, by Mary Atkin, presents a thorough survey of the topography of all the relevant sites, combined with an investigation of their situation in relation to natural and administrative boundaries (i.e. watersheds and county divisions), and their status and ownership as revealed by later records, particularly the Domesday survey. Her assemblage of material would support a meaning such as 'strategic point for the defence of a region', and the coherence is such as to suggest that all the names have the same origin. The words *stīg* or *stig* must imply climbing, though, as stated above, the ascent would be gradual in a few cases.

On the whole it seems likely that Kitson's neuter **anstig* 'ascent' is the best bet. The link between this and military strongpoints is to some extent obscure, but the general bearing of Mary Atkin's material accords well with the Thermopylae gloss.

Notes on the location of the places will be useful: some of them are single farms. East and West Anstey DEV are adjoining parishes near the SOM border. Anstey HMP adjoins Alton in the east of the county. Anstey HRT is south-east of Royston, and Anstey LEI is north-west of Leicester. Higher and Lower Ansty DOR lie on either side of a road junction called Ansty Cross at GR ST 770036. Ansty WAR is north-east of Coventry. Ansty WLT is north of Cranborne Chase, and Ansty SSX is south-west of Cuckfield. Anstie Grange SUR is south of Dorking. Ayngstree WOR is in Clifton upon Teme. Ainsty Wapentake, YOW, met at Ainsty Cliff in Bilbrough, where Mrs Atkin informs me that there is a sharp rise. Anser Gallows Fm in Hempstead ESX (GR TL 629366, *Anstye Gallows* 1825, and perhaps referred to in the surname of Nicholas *de Anestie* 1221) is a probable example.

brycg 'bridge' occurs in over 60 major place-names. No instances occur among names recorded by c.730 (Cox 1976), and probably most of the bridges were built to supplement the fords which were of major importance in the early years of the Anglo-Saxon kingdoms. Bridgham NFK and Brigham CMB, YOE must refer to early bridges, however, as *hām* was probably not used to form place-names later than AD 800, and these bridges may be Roman. Bridgham is on the river Thet, between two Roman roads, and not too far from the crossing of the Peddar's Way to have been named from a bridge there. Brigham near Cockermouth CMB is not far from the supposed Roman road running from Papcastle to Egremont. Brigham YOE is on the river Hull, in a region where only short stretches of Roman road have been identified, and the pattern cannot be reconstructed.

In one instance a change of name from -ford to -bridge is documented. Redbridge HMP is referred to as *Hreutford* c.730, *Hreodford* c.890, *Hreodbrycge* 956, and Coates (1989, p. 137) comments 'it is hard to escape the conclusion that a bridge was built to carry the Dark Age ancestor of the A35 over the river Test sometime close to AD 900'.

brycg was sometimes felt to be adequate as a simplex name, as in Bridge KNT, and in the earliest forms of Bridgnorth SHR, Bridgerule DEV, Bridge Sollers HRE, Bridgwater SOM. The last three of these had a personal name or surname added in the medieval period, and Bridgnorth probably acquired its affix because it denoted the most northerly name of this group. Handbridge CHE and Swimbridge DEV were simplex names in DB. In Swimbridge the prefixed personal name (*Sǣwine*) is that of the pre-Conquest tenant of the manor. In Handbridge ME *hone* 'rock' refers to the outcrop of sandstone on which the hamlet stands.

In compound names *brycg* occurs as first element in the *-hām* names discussed above, in four compounds with **-ford**, and in a few other names listed in the reference section.

As a generic, *brycg* is combined with terms which show a fairly even spread over a number of categories, under which they are listed in the reference section. The largest is that which refers to the material or method of construction. There are very few instances in which the identity of the first element is doubtful, and this accords with the likelihood of relatively late origin.

There are a few names in which *brycg* means 'causeway': Slimbridge GLO may be one, and it is certainly the sense in Bridgend LIN which is at the end of a causeway across the fens. There was probably a compound appellative *hrīsbrycg*, 'brushwood causeway', which occurs in minor names in ESX (Risebridge in Romford, Rice Bridge in Little Clacton), SSX (Rice Bridge in Bolney), SUR (Ridgebridge Hill in Wonersh, Ricebridge Fm in Reigate), DRH (Risebridge near Durham), YOW (Risebrigg Hill in Martons Both), NTT (Roy's Bridge Lane in Rufford), and in field-names in various counties. Risingbridge in Geddington NTP contains an adjective *hrīsen* 'made of brushwood'.

The cognate ON *bryggja* means 'jetty, quay', not 'bridge'. This occurs in Brig Stones near Braystones CMB and Filey Brigg YOE. *brycg* sometimes becomes Brig- in northern counties, however, and Brigg in Glandford Brigg LIN refers to a bridge over the river Ancholme. The Fel-, Fell- names listed in the reference section have an ON qualifier but in spite of this -brigg clearly means 'bridge'.

Reference section for **brycg**

As simplex name: Bridge KNT, Bridgerule DEV, Bridge Sollers HRE, Bridgnorth SHR, Bridgwater SOM, Handbridge CHE, Swimbridge DEV (the affixes are explained in the article).

As first element: Bridgend LIN (a fen causeway); Bridgford NTT, STF; Bridgehampton SOM; Bridgham NFK, Brigham CMB, YOE; Brighouse YOW; Brigsley LIN; Brigstock NTP; Bristol GLO (*stōw* 'assembly place'); Brushford DEV, SOM.

As generic, with references to materials or methods of construction: Budbridge IOW (probably **butt* 'log'); Elbridge KNT (*thel* 'plank'); Felbrigg NFK, Fell Beck

ROADS AND TRACKS

Bridge (earlier *Felebrigbec*) YOW and Fell Briggs YON (ON *fjol* 'plank'); Rumbridge HMP (*hruna* 'tree-trunk'); Stalbridge DOR (*stapul* 'post'); Stambridge ESX and Stanbridge BDF, HMP ('stone'); Stockbridge HMP; Trobridge DEV and Trowbridge WLT ('tree'). Names of ME origin are Trussel Bridge CNW ('trestle') and Turn Bridge YOW ('swing bridge'). OE *hrīsbrycg* 'brushwood causeway' is the source of a number of minor names listed in the article.

With personal names: Abridge ESX (*Æffa*); Attlebridge NFK (*Ætla*); Curbridge OXF (*Creoda*); Dudbridge GLO (*Dudda*); Edenbridge KNT (*Ēadhelm*); Harbridge HMP (**Hearda*); Oughtibridge YOW (fem. **Ūhtgifu*).

With references to vegetation: Brambridge HMP ('broom'); Ivybridge DEV; Piercebridge DRH (**persc* 'osier'); Redbridge HMP ('reed'); Sawbridge WAR ('sallow willow').

With river-names: Axbridge SOM; Bainbridge YON; Cambridge CAM; Doveridge DRB, Weybridge SUR.

With references to domestic creatures: Bulbridge WLT ('bull'); Cowbridge ESX, WLT; Gad Bridge BRK ('goat'); Rotherbridge Hundred SSX ('cattle'). Ironbridge Fm in Shalford ESX is a corruption of *eowenabrycg* 'ewes' bridge'.

With reference to a wild creature: Beobridge SHR ('bees').

With references to topography or soil: Dunbridge HMP (**denu**); Fambridge ESX (**fenn**); Felbridge SUR (**feld**); Heronbridge CHE (*hyrne*, here 'river-bend'); Hubbridge ESX (**hōh**); Slimbridge GLO ('slime').

With references to categories of people: Groombridge KNT, Isombridge SHR and Knightsbridge GTL contain words meaning 'boy' or 'servant' (ME *grome*, OE *esne*, *cniht*). Kingsbridge DEV was presumably once in royal ownership. Uxbridge GTL is believed to contain the tribal name *Wixan*. Fordingbridge HMP contains an *-ingas* name meaning 'ford people', *Ford* being an earlier and alternative name for the settlement.

With references to position: Eastbridge KNT. Harrowbridge CNW and Horrabridge DEV may belong here, see discussion below. Bembridge IOW means '(place) within the bridge', from OE *binnan*.

With an earlier place-name: Corbridge NTB, from the Roman fort called *Corstopitum*.

With reference to an activity: Wakebridge (in Crich) DRB (ME *wake* 'festival').

Names with obscure or ambiguous qualifiers:
Bracebridge LIN. The Fosse Way crosses the river Witham here, before turning north into Lincoln. The first element is obscure. Ekwall 1960 suggested that it was an OE ancestor of modern dialect *brash* 'small branches, twigs' which might give a meaning identical to that of *hrīsbrycg*, 'brushwood causeway'.

Such an etymology is, however, at odds with the importance and antiquity of the crossing.

Heybridge (near Maldon) ESX, Highbridge SOM. The first element is 'high', but the significance is uncertain. The village of Highbridge occupies a site which is slightly higher than those of surrounding settlements in this part of the Somerset Levels. It is near the mouth of the river Brue, so 'higher up the river' is not an option, and in any case the usual term for that is *upp*. Heybridge near Maldon is by the estuary of the river Blackwater. It is possible that 'high' in these names means 'most important', as in High Street.

Horrabridge DEV, Harrowbridge CNW. Both places are at the junction of three parishes. The use of *hār* 'hoar' in the sense 'boundary' is disputed, but the colour/texture sense does not seem obviously appropriate with *brycg*.

Pembridge HRE, earlier *Penebruge*. The early spellings, which are quite numerous, indicate a first element with -*n*-, not -*nn*-, and this makes OE *penn* 'enclosure' unlikely. The frequent *Pene-* forms are also inappropriate for PrW **penn**. *Pen-* is an OE personal-name theme, albeit a rarely evidenced one, and a personal name **Pena* might be conjectured for this place-name.

Tonbridge KNT. This is generally considered to be one of a very small group of names which have *tūn* as first element, but an etymology 'bridge of the estate' is not altogether convincing.

Wombridge SHR. The first element is *wamb* 'womb'. This has been interpreted as 'womb-shaped' for Wombwell in Chapter 1, but the reference is not clear in a compound with *brycg*.

Woodbridge SFK. It is not clear whether 'wood' refers to topography or to building material.

fær 'passage', probably 'difficult passage', occurs in Denver NFK, Laver ESX, and the minor name Walter Hall in Boreham ESX. The word probably went out of use as a place-name-forming term at a very early date, and it is unlikely to be found in areas of late English settlement. It has been supposed to occur in Hollinfare LNC (in Warrington), but see under **ferja** for an alternative suggestion. It is not likely to occur as far west as Farway DEV, and an alternative suggestion for that name is offered under **weg**.

Denver NFK is on the east bank of the river Ouse, near the southern limit of the great fen which lies to the west of the river. There is a Roman road from Peterborough which runs past Denver and on towards the east coast. Denver is 'Danes' crossing', obviously referring to traffic with Denmark much earlier than the Viking period. Laver (High, Little and Magdalen) ESX appears to mean 'water' or 'flood passage', OE *lagu-fær*. The places are situated by the Roman road from London to Great Dunmow, at a point where its course has disappeared for several miles, perhaps because the ground was marshy. Another village in the gap is called Moreton. Denver and Laver may contain *fær* used in somewhat the same way as French *malpas* in Malpas CHE and a number of minor names in England and Wales.

Walter Hall ESX is *Walhfare* in its earliest reference. The Roman road from Chelmsford to Colchester must have been known in this stretch as 'Welsh Passage'. It crosses a small stream.

ferja ON, **ferrye** ME, 'ferry' occurs in Ferriby LIN, YOE, Ferrybridge YOW, and Kinnard's Ferry LIN. The last is *Kinerdefere* 1185, from a personal name which is ultimately OE *Cyneheard*. Apart from these names ME *ferrye* is occasionally used as an affix, as in Ferry Fryston YOW, Stoke Ferry NFK.

Hollinfare in Warrington LNC is *Le Fery del Holyns* 1352, *Holynfeyr'* 1504, *Helingfare* 1550, *Hollynfayre* 1556, and these forms probably reflect confusion of *ferry* with *fare* 'excursion for which a price is paid'. Hollin- is OE *holegn* 'holly tree'.

ford 'ford'. This is the commonest topographical generic in English place-names apart from **lēah** (which, like *tūn* among habitative generics, occurs with a different order of frequency). About 530 names are listed in the reference section *infra*, and there are many more in minor names. The DEV place-name survey (p. 663) lists over 350 in that county. The term is less frequent in other counties, but it is common in most areas.

In view of its frequency in the whole corpus of English place-names the occurrence of nine instances of *ford* in names recorded before AD 730 (Cox 1976, p. 59) is not especially remarkable. The element does, however, occur with marked frequency in some areas where there is early Anglo-Saxon archaeological material, such as the Oxford region and the valley of the Warwickshire Avon.

Places with *-ford* names which are on major rivers and have become towns or cities, such as Bedford, Bradford, Chelmsford, Dartford, Guildford, Hereford, Hertford, Oxford, Stafford, Stratford-upon-Avon, are vastly outnumbered by those which refer to river-crossings of local, rather than national, significance. Where they cluster thickly, as for instance in the Thames Valley west of Oxford, some of the crossings, like Oxford, Frilford, Garford, Wallingford, are on long-distance medieval routes, but most, like Duxford, Shellingford, Stanford, Hatford, Appleford, south of the Thames, and Shifford and Yelford north of the river, refer to routes by which villages communicated with neighbouring villages. It is possible that most *-ford* names arose in the latter context, and that the patterns of travel and transport which caused some of the places to become military, trading and administrative centres emerged later than the coining of the names. This consideration may affect the interpretation of some names, particularly Oxford. The significance of Oxford is discussed in Gelling 1984, p. 71, and that of Stafford in Gelling 1992, p. 153.

Some simplex names and some compounds with *ford* as qualifier are listed in the reference section, but these are very few compared to the mass of compounds in which *ford* is the generic.

Descriptive terms form much the largest category of qualifiers in *ford* compounds, perhaps because *ford* was not a word with a highly specialised meaning. Some compounds recur, fords which were 'broad', 'long', 'red', 'sandy' and 'stony' being particularly frequent. Some of the 'long' and 'stone' fords probably refer to causeways over marshy ground on either side of a stream crossing.

No other category of qualifiers in *-ford* names approaches that of descriptive terms, but personal names are very well represented. With only two exceptions, these personal names are masculine, three being ON. Monothematic and hypocoristic names predominate heavily over dithematic ones.

Names in which the qualifier refers to a topographical feature are almost as numerous as those with personal names. Some of these compounds recur, the most numerous being those with **strǣt**. References to vegetation form the next largest group. It could be argued that descriptive terms, together with references to vegetation and to topographical features, have as one of their functions that of providing guidance for travellers. This could apply also to the class of qualifiers which consists of words for buildings, a group in which compounds with 'mill' are particularly frequent. Travellers (occasionally picturesque) are directly referred to in some of the compounds listed under the heading 'groups or classes of people'. A recent paper, Hough 1995b, queries the association of Birdforth and Bridford with brides, but this name fits well enough into the category.

References to animals, birds and insects are fairly numerous. The most frequently recurring wild creature is the hart. With the exception of cocks, domesticated creatures in the *-ford* compounds noted here are all such as could be herded, driven or led. There are not many references to the goods transported over fords: 'barley' is for some reason the most frequently mentioned.

The analysis attempted in the reference section includes a substantial class of compounds the meaning of which is obscure or uncertain. This includes some 50 names out of over 500 compounds which are listed.

Reference section for **ford**

As simplex name: Ford DOR, HRE, NTB, SHR, SOM, SSX.

As first element: Forcett YON (*geset*, usually taken to mean 'fold for animals'); Fordham CAM, ESX, NFK (*hām* 'village'); Fordley SFK (*lēah*); Fordwich KNT (*wīc* in the Middle Saxon sense 'trading centre'); Forton HMP, LNC, SHR, STF (*tūn* 'settlement, estate'); Furtho NTB (**hōh**).

ROADS AND TRACKS

As generic with descriptive terms:

(a) *referring to surface*: Chalford GLO, OXF(2) ('chalk'); Chillesford SFK ('gravel'); Chingford ESX ('shingle'); Cinderford GLO; Clatterford IOW ('loose stones'); Flawforth NTT ('flagstones'); Greatford LIN ('gravel'); Mudford SOM; Rufford LNC, NTT and Rufforth YOW ('rough'); Sampford DEV(3), ESX, SOM(2) and Sandford BRK, DEV, DOR, OXF(2), SHR, WML ('sand'); Slaggyford NTB ('muddy'); Stafford DEV, DOR, Stainforth YOW(2), Stamford LIN, NTB, YOE, Stamfordham NTB (*hām* added), Stanford BDF, BRK(2), ESX(2), HRE(2), KNT, NFK, NTP, NTT, WOR, Staverton DEV (*tūn* added), Stoford SOM(2), WLT, Stowford DEV (several), OXF, WLT (all 'stone').

(b) *referring to length, breadth, depth or shape*: Bradford CHE, DEV, DOR(3), LNC, NTB(2), SOM, WLT, YOW(2), Bradiford DEV, Brafferton DRH, YON (*tūn* added), Broadward HRE (all 'broad'); Cromford DRB ('crooked'); Defford WOR, Deptford GTL, WLT, Diptford DEV ('deep'); Gainford DRH ('direct'); Holford SOM ('hollow'); Langford BDF, DEV, ESX, NFK, OXF, SOM(2), WLT, Longford DRB, GLO, GTL, HRE, SHR, WLT (all 'long'); Romford GTL ('spacious'); Scalford LEI, Shadforth DRH, Shawford HMP ('shallow'); Shelford CAM, NTT, Shilford NTB (**sceldu* 'shallow place').

(c) *referring to appearance*: Blackford SOM(2), Blackfordby LEI (ON *bý* added); Greenford GTL; Guildford SUR ('golden'); Radford DEV, NTT(2), WAR(2), WOR(2), Retford NTT ('red'); Shereford NFK, Sherford DEV, SOM ('bright'); Whitford DEV ('white').

(d) *referring to qualities of the water*: Freshford SOM; Washford DEV (probably *gewæsc* 'washing', referring to the action of the water).

(e) *referring to constructional features*: Bamford DRB, LNC ('beam'); Bretford WAR, Bretforton WOR ('plank', *tūn* added to WOR example); Sparkford SOM (**spearca* 'brushwood'); Stapleford CAM, CHE, ESX, HRT, LEI, LIN, NTT, WLT ('post'). Stockingford WAR ('tree-stump') and Treyford SSX ('tree') may belong here.

(f) *referring to periods of use*: Efford CNW, DEV(4), HMP ('ebb', these are fords passable at low tide); Somerford CHE, GLO, WLT ('summer').

(g) *'fair' and 'foul' fords*: Fairford GLO; Fulford DEV, SOM, STF, YOE.

(h) *'double' fords (twī-)*: Tiverton DEV (*tūn* added), Twyford BRK, BUC, DRB, GTL, HMP, LEI, LIN, NFK. These are crossings over two adjacent streams.

(i) *'hidden' fords (derne)*: Darnford SFK, Dornford OXF, Durnford WLT.

(j) *referring to other characteristics*: Aldford CHE ('old'); Halliford MDX ('holy'); Leeford DEV (*hlēow* 'shelter'); Offord HNT ('up', i.e. higher up the river); Radford OXF ('riding'); Sharnford LEI ('dung').

With personal names: Aberford YOW (fem. *Ēadburg*); Alford SOM (fem. *Ealdgȳth*); Alresford ESX (**Ægel*); Ansford SOM (*Ealhmund*); Aylesford KNT

THE LANDSCAPE OF PLACE-NAMES

(*Ægel*); Basford CHE (ON *Barkr*); Basford STF (*Beorcol*); Bastonford WOR (*Beorhstān* or *Beornstān*); Battisford SFK (*Bætti*); Bayford HRT (*Bǣga*); Beckford GLO (*Becca*); Bedford LNC (*Bēda*); Besford SHR, WOR (*Betti*); Brinsford STF (*Brūn*); Brinsworth YOW (*Brȳni*); Canford DOR (*Cana*); Chatford SHR (*Ceatta*); Chelmsford ESX (*Cēolmǣr*); Codford WLT (*Coda*); Copford ESX (*Coppa*); Cosford WAR (*Cossa*); Cottisford OXF (*Cott*); Culford SFK (*Cūla*); Dadford BUC (*Dodda*); Daylesford WOR (*Dægel*); Dodford NTP, WOR (*Dodda*); Dunsford DEV and Dunsforth YOW (*Dunn*); Duxford BRK (*Duduc*); Egford SOM (*Ecga*); Eggesford DEV (*Ecgen*); Elford NTB, STF (*Ella*); Eynsford KNT (*Ægen*); Fiddleford DOR (*Fitela*); Frilford BRK (*Frithela*); Gusford Hall SFK (*Gūthwulf*); Hankford DEV (*Haneca*); Hapsford CHE (*Hæp*); Hednesford STF (*Heddin*); Higford SHR (*Hugga*); Ickford BUC (*Ica*); Ideford DEV (*Giedda*); Kempsford GLO (*Cynemǣr*); Kirdford SSX (*Cynerēd*); Kittisford SOM (*Cyddi*); Knutsford CHE (ON *Knut*); Langford (earlier *Landeford*) NTT (*Landa*); Lawford ESX, WAR (*Lealla*); Lilford NTP (*Lilla*); Mainsforth DRH (*Mægen*); Moulsford BRK (*Mūl*); Mundford NFK (*Munda*); Otford KNT (*Otta*); Pendeford STF (*Penda*); Pitsford NTP (*Peoht*); Pontesford SHR (*Pant*); Putford DEV (*Putta*); Rainford LNC (*Regna*); Rofford OXF (*Hroppa*); Sedgeford NFK (*Secci*); Sibford OXF (*Sibba*); Sigford DEV (*Sicga*); Snarford LIN (ON *Snortr*); Tatterford NFK (*Tāthere*); Tugford SHR (*Tucga*); Ugford WLT (*Ugga*); Whittlesford CAM (*Wittel*); Winsford CHE (*Wine*); Wixford WAR (*Wihtlāc*); Woodsford DOR (*Wīgheard*); Wormingford ESX (*Withermund*); Yelford OXF (*Ægel*); Yelverton (earlier *Elleford*) DEV (*Ella*).

With reference to topographical features: Backford CHE (**bæc**); Barrasford NTB and Barrowford LNC (**bearu**); Bransford WOR and Brentford BUC ('brain', i.e. 'hill'); Byford HRE ('river-bend'); Castleford YOW and Chesterford ESX (*ceaster* 'Roman ruin'); Cringleford NFK (ON *kringla* 'circle', the precise reference has not been ascertained); Dishforth YON and Ditchford NTP, WAR ('ditch'); Garford BRK and Garforth YOW ('gore', i.e. triangular piece of land); Hackford NFK(2) and Hackforth YON (*haca* 'hook', probably referring to bends in roads); Halford DEV, SHR, WAR (**halh**); Harpford DEV, SOM (*herepæth* 'main road'); Hatford BRK (**hēafod**); Latchford CHE, OXF (**læcc**); Leckford HMP (*leaht* 'side-channel'); Ludford SHR (*hlȳde* 'loud stream'); Midford SOM and Mitford NTB ('river-junction'); Parford DEV (**pæth**); Peckforton CHE (**pēac**, *tūn* added); Pitchford SHR (there is a pitch well); Pulford CHE (**pōl**); Sandleford BRK ('sand spring' or 'sand hill'); Seaford SSX (a crossing over the old course of the river Ouse); Startforth YON, Stratford BDF, BUC(3), ESX, GTL, NTP, SFK(2), WAR, WLT(2), Strefford SHR and Stretford HRE, LNC (all **strǣt**); Stortford HRT ('tongues of land'); Styford NTB (**stīg**); Trafford LNC (**strǣt**); Walford HRE, SHR (**wælle**); Wansford NTP (*wylm*, perhaps 'whirlpool'); Wayford SOM (**weg**); Welford GLO, NTP (**well**); Wonford (near Exeter) DEV (*wynn* 'meadow'); Woodford CHE(2), CNW, ESX, NTP(2), WLT; Woodlesford YOW (perhaps *wrīdels* 'thicket').

ROADS AND TRACKS

With references to vegetation: Alderford NFK, Allerford SOM(2) and Alresford HMP ('alder'); Ampleforth YON ('sorrel'); Appleford BRK, IOW; Ashford DEV(2), DRB, SHR, Axford WLT and Ayshford DEV ('ash'); Ashford KNT (*æscet* 'ash copse'); Little Barford BDF ('birch'); Bramford SFK ('broom'); Bulford WLT (probably a derivative of *bulut* 'ragged robin'); Chagford DEV ('gorse'); Clatford HMP, WLT and Cloford SOM ('burdock'); Lackford SFK ('leek'); Linford BUC and Lynford NFK ('flax'); Mayford SUR ('mayweed'); Nutford DOR; Oakford DEV and Okeford DOR; Pyrford SUR ('pear'); Ribbesford WOR ('ribwort bed'); Rudford GLO ('reed'); Rushford NFK; Salford BDF, LNC ('sallow willow'); Slaughterford WLT ('sloe tree'); Stifford ESX (*stithe*, ?'lamb's cress'); Thornford DOR and Thorverton DEV ('thorn', *tūn* added to latter); Welford BRK (*welig* 'willow'); Wichenford WOR ('wych-elm'); Widford ESX, HRT, OXF (*wīthig* 'willow'); Wilford NTT, SFK (*wilig* 'willow'); Wishford WLT and Witchford CAM ('wych-elm'); Wytheford SHR (*wīthig* 'willow').

With references to groups or classes of people: Bayford SOM (probably 'boy'); Billingford NFK ('followers of Billa'); Birdforth YON and Bridford DEV (?'brides'); Conisford NFK (ON *kunungr* 'king'); Coppingford HNT ('traders', from a late OE derivate of ON *kaupmathr*); Hemingford HNT ('followers of Hemma'); Horringford IOW ('dwellers in horn-shaped land' or 'followers of Horn'); Huntingford DOR, GLO ('huntsmen'); Kingsford WAR ('king'); Kingsford WOR ('followers of Cēna'); Lattiford SOM (*loddere* 'beggar'); Maidford NTP; Manningford WLT ('followers of Mann(a)'); Orford LIN ('Irish'); Poundisford SOM ('pinder'); Salterford NTT and Salterforth YOW ('salter'); Scotforth LNC and Shotford NFK/SFK ('Scots'); Shellingford BRK ('dwellers on a ploughshare-shaped piece of land' or 'followers of *Scear'); Sleningford YON (unexplained folk-name); Tetford LIN and Thetford CAM, NFK (*thēod* 'folk'); Thenford NTP ('thegns'); Thursford NFK ('giant'); Walford (near Ross) HRE ('Welsh'); Wallingford BRK ('followers of Wealh'); Walshford YOW ('Welshman'); Whichford WAR (*Hwicce*).

With references to activities: Glandford NFK, Glanford Brigg LIN, Glemsford SFK (*glēam* 'merriment'); Mutford SFK ('meeting'); Plaitford HMP and Playford SFK ('play'); Shutford OXF (*scytta* 'archer'); Trafford NTP ('trap'); Watford HRT, NTP ('hunting').

With references to wild creatures: Beaford DEV (*bēaw* 'gadfly'); Beeford YOE (*bēo* 'bee'); Blandford DOR ('gudgeon'); Boresford HRE ('boar'); Brockford SFK (probably 'badger'); Buntingford HRT; Carnforth LNC ('crane'); Catford GTL and Catforth LNC; Cornforth DRH, Cranford GTL, NTP and Cransford SFK (all 'crane'); Desford LEI, Durford SSX ('deer'); Enford WLT ('duck'); Harford GLO, Hartford CHE, NTB, Hartforth YON and Hertford HRT (all 'hart'); Lostford SHR (?*lox* 'lynx'); Rockford HMP ('rooks'); Stinsford DOR (*stint* 'sandpiper'); Taddiford HMP ('toad'); Wansford YOE (*wand* 'mole'); Wilsford LIN, WLT(2) (*wifel* 'beetle').

With references to domesticated creatures: Gateford NTT and Gateforth YOW ('goat'); Gosford DEV, OXF, WAR and Gosforth CMB, NTB ('goose'); Handforth CHE and Hanford STF (*hana* 'cock'); Horsford NFK and Horsforth YOW; Oxford OXF; Rochford ESX, WOR (*ræcc* 'hunting dog'); Shefford BDF, BRK and Shifford OXF ('sheep'); Swinford BRK, LEI, STF, WOR ('swine'); Tickford BUC ('kid').

With river-names: Blithford STF (Blythe); Blyford SFK (Blyth); Brentford GTL (Brent); Camblesforth YOW (probably a British name corresponding to Welsh Camlais 'crooked stream'); Crayford GTL (Cray); Dartford KNT (Darent); Exford SOM (Exe); Ilford SOM (Isle); Ilford GTL (river Roden was earlier *Hyle*); Kentford SFK (Kennett); Lydford DEV (Lyd); Sidford DEV (Sid); Sleaford LIN (Slea); Tempsford BDF (*Thames*: Tempsford is by the confluence of the rivers Ivel and Ouse, but *Thames* may have been the name of one of the small streams in the neighbourhood); Tideford CNW (Tiddy); Warenford NTB (Warren).

With earlier place-names: Lintzford NTB (Lintz); Ludford LIN (Louth, here the sense appears to be 'on the way to ...'); Washford SOM (OE *Wecetford*, Watchet).

With references to buildings or other structures: Bottesford LEI, LIN (*botl* 'hall'); Bridgford NTT(2), STF and Brushford DEV, SOM (**brycg**); Burford OXF, SHR (*burh* 'fort'); Melford SFK, Milford DRB, HMP, WLT, YOW(2) and Milverton SOM, WAR ('mill', *tūn* added to last two examples); Quarnford STF (*cweorn* 'mill-stone', assumed to be used in the sense 'mill'); Stafford STF (**stæth**); Warford CHE ('weir'); Weeford STF ('heathen temple'); Wickford ESX and Wigford LIN (*wīc*); Wyfordby LEI (probably 'heathen temple', ON *bý* added); Yafford IOW ('hatch').

With references to the transport of goods: Barford NFK, NTP, OXF, WAR, WLT(2), Great Barford BDF ('barley'); Coleford GLO, SOM(2) ('charcoal'); Eyford GLO (probably 'hay'); Heyford NTP, OXF ('hay'); Salford OXF, WAR and Saltford SOM ('salt'). Wanford (near Thetford) SFK ('waggon') and Yoxford SFK (?'yoke of oxen') may refer to the means of transport.

Names with obscure or ambiguous qualifiers:

Alford LIN. The likeliest suggestion is 'eel', but some early spellings with *E*- are needed for confirmation.

Arnford YOW. A compound appellative *ærne-ford has been suggested on the analogy of the recorded *ærne-weg*, and the meaning 'riding ford' has been assumed with Radford OXF as parallel. The recorded instances of *ærne-weg*, however, and those of *ærningweg*, are all in the context of racing, not of ordinary travel. See the note on Rackenford *infra*.

Ashford MDX, *Ecelesford* 969; the brook is *Eclesbroc* 962. Echinswell HMP is *Eccleswelle* in DB, and the stream by which it stands is *ec(e)lesburna* in a boundary survey of 931. The later development of both names points to a palatalised *c*, which rules out derivation from *ecles* 'Christian community' as in Eccles and Eccles- names. There may have been a pre-English river-name **Ecel*.

Bedford BDF, Bidford WAR, Bideford DEV. Bidford and Bideford belong together, and Bedford is likely to have the same first element, though with only a few ME spellings available it is impossible to be certain. Bedford BDF has one OE spelling *Bydanford* which, like the spellings for Bidford and Bideford, suggests possible derivation from OE *byden* 'tub', used literally or in a transferred topographical sense. Names containing *byden* are listed in Chapter 4. One of these, Bidna DEV, is by the Torridge estuary, 2 miles north of Bideford.

Bedford LNC has been listed above as containing the personal name *Bēda* because it has a good run of ME spellings with *-e-*.

Belchford LIN, earlier *Beltesford*. The same first element occurs with **hlāw** in a hundred-name, Beltisloe LIN. Ekwall 1960 suggests a personal name **Belt* derived from the word *belt*, but this word may have been used in a transferred topographical sense.

Belford NTB. Possibly *bēl* 'funeral pyre', but there are other possibilities for Bel- in place-names.

Bickford STF. The choice lies, as usually with Bick- names, between **bica* 'something pointed' and *Bica* personal name.

Brampford DEV. Charter evidence shows that *Brenteford* was the OE form. *Brent* is evidenced as a pre-English river-name in Brentford GTL, and probably as a hill-name in Brent Knoll SOM, but neither seems suitable for Brampford, where the crossing was over the river Exe.

Briddlesford IOW. 'Ford of the bridle', perhaps indicating that it was only passable on horseback.

Britford WLT. This contains the same first element, *bryt-*, as is found in Britwell, discussed under **well**.

Chelford CHE. The first element could be *ceole* 'throat' in a topographical sense, or it could be the personal name *Ceola*.

Coxford NFK. This name presents the same problems as Coxwell BRK, discussed under **well**.

Doniford SOM. An OE **dunige*, of obscure meaning, has been postulated to explain the minor name Dunmore Pond in Brightwalton BRK (*dunian mere* 939) and a surname *atte Dunye*, 1382, in SUR. Such a word would suit Doniford.

THE LANDSCAPE OF PLACE-NAMES

Droxford HMP (*Drocenesford* 939). Ekwall 1960 suggested an OE **drocen* 'dry place'; this is not otherwise evidenced.

Farforth LIN. There is an interchange of *Far-* and *For-* in early spellings and this makes interpretation difficult.

Freeford STF. The meaning is 'free ford', but the significance is not apparent. It is unlikely that other fords were liable to tolls.

Hainford NFK. It is difficult to reconcile the earliest spellings – *Homfordham* c.1060, *Han-*, *Hamforda* DB – with 12th-century *Heinford* and the modern form.

Harford DEV, Hartford HNT, Hereford HRE(2). Harvington (near Evesham) WOR is *Herefordtūn*. The first element is *here* 'army', but the significance is open to debate. The names are discussed in Torvell 1992.

Hungerford BRK, SHR. The significance of 'hunger' in this name has not been established.

Ickleford HRT. This name and Ickleton CAM have a first element *Iceling*. It is probably a coincidence that they are both on the ancient route called Icknield (OE *Icenhilde*). The personal name *Icel* was borne by an ancestor of the kings of Mercia who is considered to be a historical, rather than a legendary, character, and to have led a migration from East Anglia into the midlands. Formally Ickleton and Ickleford could have been 'settlement/ford associated with Icel', using *-ing-* as a connective particle, but it is doubtful whether there is a reference to the migration, which would have been somewhat off course if it passed through Ickleford. The personal name was probably borne by non-royal personages, as it occurs also in Icklesham SSX, where it is not likely to refer to the founder of the Mercian dynasty. On the whole it seems likely that Ickleford is 'ford associated with a man named Icel', without reference to the historical character or to the Iceni who are referred to in the Icknield Way.

Keresforth YOW. Perhaps 'Cēnfrith's ford', though none of the early spellings has *-n-*. Another possibility is a collective genitive of *ceafor* 'beetle'.

Landford WLT. Early spellings show variation between *Lange-*, *Land(e)-* and *Lane-*, the last being the most frequent. The *Lane-* spellings suggest *lanu* 'lane', but this is a very rare term in major place-names.

Lapford DEV. The word *læppa* (modern *lap*) has varied senses in OE: it means 'skirt of a garment', 'lobe' (as in *dew-lap*), and, in one instance 'district, detached land'. There may be a topographical or administrative significance in Lapford and in Lapal WOR (with *hol* 'hollow') and Lapley STF (*lēah*), but there is not enough evidence to identify it.

Marlesford SFK. This name appears to have the same first element as Marlston CHE, for which EPNS CHE survey, 4 p. 163, suggests a personal name **Mǣrel*.

Ekwall 1960 prefers to derive Marlesford from a reduced form of Martley, a nearby settlement.

Marlingford NFK. Early spellings suggest that the OE form was *Mearthlingford*. Possibly a personal name **Mearthel* with connective *-ing-*.

Meaford STF. *(ge)mȳthe* 'confluence' would suit topographically, but the ME spelling *Medford* suggests **mæd** 'meadow'.

Montford SHR. Possibly OE *gemana* 'fellowship, society', with reference to the use of Montford Bridge as a meeting-place for negotiations between English and Welsh.

Mordiford HRE. The first part of this name is unexplained. Ekwall 1960 tentatively suggested Welsh *mor dy* 'great house', but this is not a convincing early Welsh name.

Orford SFK. Possibly **ōra**. That element has not otherwise been noted in East Anglia, but there is a low ridge of the appropriate shape here.

Orford LNC. Possibly 'upper ford', but available spellings are late and inconclusive.

Panxworth, earlier *Pankesford*, NFK. First element a word or personal name *panc*, not otherwise noted.

Pensford SOM. Either OE *penn* 'enclosure' or a British hill-name *Pen*.

Quatford SHR. There appears to have been a district-name *Cwat(t)*, surviving in this name and in Quatt, which is two miles south-east. No convincing etymology has been suggested.

Quemerford WLT. Ekwall 1960 suggested the personal name *Cynemǣr*. EPNS WLT survey suggested a British word meaning 'confluence' related to Welsh *cymmer*. The evidence for original *Cw-* is, however, stronger than admitted by either authority. A stream-name from OE *cwēman* 'to please' might be considered.

Rackenford DEV. This has been explained as a doublet of Arnford with a distinguishing prefix, perhaps *racu* 'stream-bed'. It is not certain, however, that such spellings as *Racarneford*, *Rakerneford* reflect the original form. Many of the early spellings have *Rachene-*, *Rakene-*, and the first *-r-* may be inorganic. The name could be **hracenaford* 'ford of the gullies' from *hraca* 'throat'. It is in broken country.

Seighford STF. On the basis of a single undated form *Cesterford* Ekwall 1960 says 'a Normanized form of Chesterford'. Nothing Roman is known here, however, and the other available forms suggest comparison with Chesford Bridge in Willenhall WAR, which is *Chesford* 1279, *Chestford* 1291, *Chesterford* 1370, and for which EPNS WAR survey suggests *ceast* 'strife'. At all events, these are not certain examples of *ceaster*.

Shillingford DEV(2), OXF. It is difficult to find an explanation which suits all three examples. OE *sciell* 'resounding' would be a suitable base for a stream-name, but the rivulets which join the Thames at Shillingford OXF are not likely to have deserved such an appellation. There was a *scillinges broc* in Little Wittenham BRK, across the Thames from Shillingford.

Snowford WAR. The etymology is not in doubt, but the significance is uncertain. Local observers say that snow does not lie exceptionally long at this crossing of the river Itchen.

Spofforth YOW. The first element is the word *spot*, but the meaning of this word in its rare occurrences in place-names has not been ascertained.

Stechford WAR. The spellings indicate *sticce*. This may be the word found in field-names in BRK, OXF, WLT for which a meaning 'hurdle' was tentatively suggested in EPNS BRK survey, p. 907.

Stickford LIN. Perhaps 'ford by a long strip of land': see the note on Stickney in Chapter 2.

Tilford SUR. The first element could be either the personal name *Tila* or the adjective *til* 'convenient'.

Trafford CHE. First element 'trough', but (as with Bidford) it is uncertain whether a literal or a transferred sense is intended.

Tuxford NTT. Spellings for this name and for Tuxwell SOM indicate OE *Tuces-*. No etymology is available for this.

Walford, earlier *Walteford*, DOR. OE *wealt* 'unsteady' has been suggested, but it is not clear in what way this would apply to a ford.

Wangford (near Southwold) SFK. Generally explained as containing the OE poetic word *wang* 'plain', but this does appear to be otherwise evidenced in settlement-names. ME *wong, wang* is frequent in field-names in some counties, particularly NTT, where derivation from the cognate ON *vangr* is likely. The ME word had specialised senses relating to open-field systems. Neither the OE nor the ON word offers an obvious answer for the etymology of Wangford.

Warnford HMP. The first element could be 'wren' or an unrecorded *$wr\bar{æ}na$ 'stallion'.

Wentford SFK. The only spelling available is *Wanteford* 1315. Ekwall 1960 suggests ME *went* 'passage', but this is not a very meaningful compound.

Yafforth YON. Generally explained as 'river ford' from *ēa*. Early forms are, however, different from those of other names which have *ēa* as qualifier. The initial *Y-* appears early and may represent an OE *G-*. The name is unsolved.

gelād 'difficult river-crossing'. A strenuous discussion of this term in Gelling 1984 has stood the test of time quite well, most of the conclusions being supported by Ann Cole's subsequent field-work.

The related word **lād** is discussed in Chapter 1. The two terms cannot always be distinguished from each other in place-name spellings. Certainty depends on the availability of OE spellings, like *Crecca gelad* for Cricklade WLT, *Euulangeladæ* for Evenlode WOR, *Hlincgelad* for Linslade BUC, or ME and modern spellings in which *ge-* is represented by *-i-*, as in *Framilade*, the Domesday Book spelling for Framilode GLO. Aqualate STF, contrary to what was said in Gelling 1984, probably qualifies on the basis of spellings like *Aguilade* 1227, *Akilote* 1282. Post-Conquest spellings for *gelād* names do not always show evidence of the diagnostic *ge-*, however, and there are some instances for which no final choice can be made between *gelād* and *lād*: Clevelode WOR (*infra*) is one of these.

Major names for which the spellings are not conclusive but which are likely to contain *gelād* are Lechlade GLO, Portslade SSX and Shiplate SOM. There are two minor names by the Severn north of Gloucester: Abloads Court (*Abbilade* 1210), which is certainly a *gelād*, and Wainlode, for which the early spellings, starting in 1378, do not demonstrate *gelād* though the topography is strongly in favour of it.

It will be seen that this is a small corpus, and since the distribution precludes early obsolescence, the term is likely to have been a highly specialised one, only used when particular circumstances demanded it. In the 1984 discussion topographical details were assembled which suggested liability to flooding at places with *gelād* names, and investigations by Ann Cole in the winters of 1989, 1990 and 1992 produced photographs of dramatic flooding at Cricklade, Evenlode, Lechlade and Abloads. At Wainlode, on the Severn near Abloads, the flooding was very impressive in the winter of 1990. At Aqualate, on the STF/SHR border near Newport, there is marshy ground where a Roman road crosses a stream, and oaks are still in evidence, so *āc-gelād* is the likely OE source, though spellings like *Aguilade* and *Akilot* show distortions respectively in the first and second elements.

The first element of Cricklade is obscure, but firm etymologies can be offered for all other certain or possible examples. Evenlode and Abloads are considered to contain personal names. Framilode refers to the river Frome and Lechlade to the river Leach. Shiplate refers to sheep. The first element of Portslade is here assumed to be the genitive of *port*, perhaps in the sense 'Roman harbour'. Wainlode contains OE *wægn* 'waggon' (modern dialect *wain*). Linslade BUC is OE *hlinces gelad* 'difficult river-crossing by a terrace road' – see **hlinc** in Chapter 5.

There are three names containing *gelād* in OE charter boundaries, *dyrnan gelad* (which was 'hidden') and *eanflæde gelade* (which belonged to a woman named *Ēanflǣd*) on the BRK bank of the Thames by Appleford and Wytham respectively, and *hafoc* ('hawk') *gelad* in Haseley OXF. In addition to these three OE place-names, an agreement concerning land at Worcester defines some meadow by reference to a *gelād* on the river Severn.

All discussions of *gelād* are complicated by the necessity to account for the use of an early Modern English word *lode* for ferries across the river Severn in the counties of GLO, WOR and SHR. The earliest unequivocal evidence noted for *lode* 'ferry' occurs in records dealing with property at Apley in Stockton SHR. In 1480 this property is described as the weir of Apley and *the loode* with a house and close, and in 1494 the ferry is 'the fery other whyles called the loode of Apley with the were to the same fery or lode belongyng'. Further down the Severn, Hampton and Hampton Loade occur on either side of the river at a point where there is a ferry, and Hampton Loade is first recorded as *Hemptons Lood* in 1594.

Possible earlier evidence is found in forms for Lower and Upper Lode near Tewkesbury. EPNS GLO survey 2, 65, cites references to *passag' de Wulmereslode* 1248, *passage de Overlode* 1300. This shows that there was a ferry (French *passage*) in the vicinity, but it does not prove that -*lode* meant 'ferry' in the 13th century. There are small streams, possibly canalised, here, which could have been referred to by the term lād. For Clevelode WOR, higher up the river, the spellings (*Clivelade* c.1086, *Clivelode* 1275) could derive from lād or *gelād*. Here also there was a ferry. In an early discussion (1956, Part 2, p. 9) A. H. Smith suggested that the meaning 'ferry' arose because there were coincidentally ferries at several points on the Severn where there were names containing either *gelād* or lād, but in the GLO place-name survey, in his discussion of Lower and Upper Lode near Tewkesbury, he put forward the contrasting view that both terms had the meaning 'ferry' in OE. There is no clear solution to this problem, but it should be noted that -*lode* names referring to ferries only occur on the river Severn, and that 'ferry' does not appear to be a possible sense of lād in the eastern fens or in SOM. The proven *gelād* names are in some cases open to the 'ferry' interpretation, but as Kitson (forthcoming, commenting on the charter occurrences) points out, this is not likely for *dyrnan gelad*. There is no question of it, either, for *hafoc gelad*, where the crossing is over Haseley Brook. Also, it seems unlikely, despite the evidence on the Severn, that ferries would have been consistently established at points on rivers which were exceptionally liable to flooding.

The exact nature of the 'crossing' in some *gelād* names which are by major rivers is uncertain; it may not always have been a crossing of the river. At Framilode GLO, the crossing may have been one which allowed the north-south road to negotiate the outfall of the Frome and continue along the east bank of the Severn estuary. Abloads (like Wainlode) is in an area where a road leading north out of Gloucester on the east bank of the Severn has to cross very wet ground.

Two of the literary occurrences of *gelād* are found in the poem *Beowulf*, where a *fengelad* is listed as one of the desolate features in the territory of the monsters, and an *uncuth* ('unknown') *gelad* as one of the grim places passed by the hero on his way to Grendel's mere. Brief mention was made of these in Gelling 1984, but the linking of poetic to place-name usage has now been explored in detail in a paper by D. Cronan (1987).

Cronan writes without knowledge of Gelling 1984, but he reaches much the same conclusion. He notes that words which have a specifically 'poetic' sense in OE poetry have a high rate of alliteration, and that the absence of this in the six occurrences of *gelād* in poetry points to the word having the same meaning as it had in non-poetic usage. He notes that the phrase *uncuth gelad* occurs not only in *Beowulf*, but also, twice, in the poem *Exodus*, the first occurrence referring to the passage taken by the Israelites across the Red Sea. He concludes that translation of this phrase as 'an unknown water-passage' rather than (as by editors) 'an unknown way' 'transforms the verse from the general comment on the journey itself to a specific danger faced by the travellers, a danger that was surely present in the terrain envisioned by the poet'.

The linking of *gelād* in place-names to particularly difficult river-crossings strengthens Cronan's perception of the sense of danger intended by the *Beowulf* poet. This is one of a number of instances in which landscape terms used in *Beowulf* can be better understood by a study of their use in place-names: see **pæth, hop** and **hlith**, in which terms, also, a sense of menace may be detected.

As regards the second *Beowulf* reference, it is likely that *fengelad* means 'difficult water-crossing in marsh' rather than 'fen-stream' as preferred by Cronan. The definition of **fenn** as 'linear marsh' suggested in Chapter 2 also contributes to the general picture of the landscape in which the monsters live.

hȳth 'landing-place on a river, inland port'. The limited distribution of this element was noted in Gelling 1984 and is shown on Fig. 13. The earlier study was based only on major names, but when minor names of proven antiquity are taken into account the distribution pattern is not altered. Special interest lies in the possibility that clusters of *hȳth* names in certain areas and along certain rivers preserve the outlines of systems of water transport which were operating in the early post-Roman period.

It will be seen from the items referred to *infra* that *hȳth* is particularly liable to confusion with commoner elements in ME and modern forms.

The largest concentrations of *hȳth* names are in the fenland of eastern England and along the river Thames. There is a cluster on the lower course of the river Trent, and a thin scatter along the east and south coasts, and there are two instances in the northwest. Southwestern outliers are Hyde by the Torridge estuary on the north Devon coast, a lost *Prattshide* by Exmouth on the south Devon coast, and, most importantly, Bleadney in the Somerset

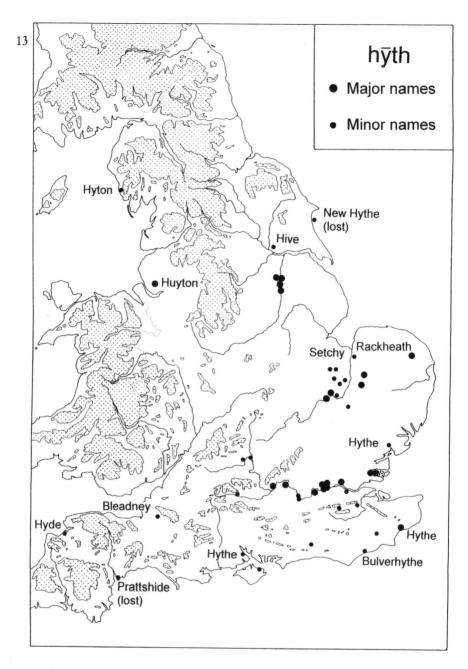

Levels. This last name affords a good starting point for a discussion of the category of *hȳth* names which occur at the junction of marsh and firm ground.

Bleadney (Fig. 14) is named in a charter of the early 8th century. The fragment of this text which survives in a cartulary of Glastonbury Abbey refers to *portam quae dicitur Bledenithe*, establishing that *hȳth* is the generic. It

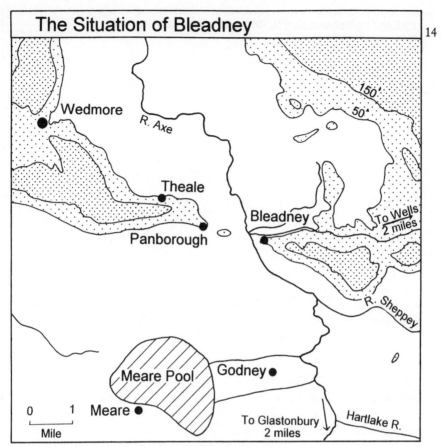

14 The Situation of Bleadney

has been suggested that *portam* is a miscopying of *pontem*, but bridges are not usually referred to by such phrases as *quae dicitur*, and the possible existence here of something deserving the term 'port' should be considered, though *portam* would, of course, be grammatically incorrect. Bleadney is situated in a narrow gap in the chain of hills running from Wedmore to Wells, which separates two areas of the Levels. The river Axe, which was an important water-route to the Severn estuary, flows through this gap. The 'hithe' is likely to have catered for the distribution of goods brought either along the river or over flooded moors to the settlements which are crowded along the narrow strips of raised ground on either side of Bleadney. There was at least one more *hȳth* in the Levels: the charter bounds of North Wootton mention one which must have been close to the north edge of Glastonbury's hill.

Distribution of goods to settlements on raised ground adjoining fen can be postulated also for Aldreth CAM. This place is situated at the western end of a strip of raised ground in the Fens, south-west of Ely (Fig. 15). There are five settlements along this strip, the eastern end of which abuts the Isle of Ely. The raised ground bifurcates at its west end, and Aldreth occupies the tip of the

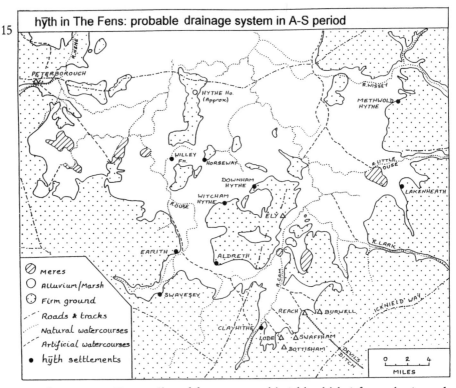

15 hȳth in The Fens: probable drainage system in A-S period

south-west arm. Four miles of fen separate this 'alder hithe' from the 'gravel hithe' at Earith HNT, and both hithes were surely responsible for the movement of cargoes from fenland channels to dry ground. The nearest firm ground to Aldreth is at Willingham, two miles south west, and there is a causeway between the two places which is first mentioned in 1172. A 'passage for boats' at Aldreth is referred to in 1169.

Swavesey CAM is another 'hithe' in this area which is situated at the junction of firm ground with fen, and Lakenheath SFK and Methwold Hythe NFK occupy similar situations near the eastern edge of the Fens.

There are water basins at Swavesey, but ironically the place where this function of handling movement of cargoes from fen to firm land is still most clearly visible does not have a *hȳth* name. Reach CAM was built on the northern tip of the Dark-Age earthwork called the Devil's Dyke, which stops at the edge of Burwell Fen. The installations of its hithe, which functioned into modern times, are carefully preserved, but at the time when the name was coined the relationship of the settlement to the Devil's Dyke must have seemed more important than its function as a port. This is discussed under ræc in Chapter 5.

There are four well-documented minor names in the CAM fens: Horseway and Willey Fm in Chatteris parish, and Downham Hithe and Witcham Hithe. The 19th-century 1" map, with its sensitive hachuring, shows that these places also are on the edge of slightly raised ground. The hinterlands they served would be small compared to that of Aldreth.

It is difficult in the eastern fenlands to ascertain whether places with *hȳth* names were once situated on navigable rivers. Setchey NFK, however, is clearly related to the river Nar, and Clayhithe CAM to the river Cam; and it is possible that some of the other places mentioned in the preceding paragraphs once handled river-trade, rather than goods punted along lodes or over flooded fen. The complicated topography of this region is mapped in Fig. 15.

Rackheath NFK was treated as a *hȳth* name in Gelling 1984, and is shown here on Fig. 13, though the spellings are not conclusive, and a devil's advocate could argue that the generic is equally likely to be **hæth**. The justification for *hȳth* is that the church, 1½ miles south of the river Bure, overlooks a tributary which rises from copious springs and flows to the Bure through a wide but deep valley – the *hraca* 'throat' of the qualifier. In earlier times there may well have been sufficient water here to enable transport from the Bure. Hive YOE must have been linked by a navigable channel to the river Foulness.

River-side *hȳth* names are to be found by the Trent and the Thames. There are three on the Trent near Gainsborough: Knaith, Walkerith and Stockwith (four if East and West Stockwith are counted separately). The Thames has a magnificent series, running from east of London to above Oxford. In a high proportion of these Thames names the modern form disguises the etymology. From east to west the series is: Greenhithe, Erith, Stepney, Rotherhithe, Queenhithe, Lambeth, Chelsea, Putney, Hythe in Egham, Maidenhead, Bolney, Hythe Bridge in Oxford, Bablock Hythe. Two KNT names, Riverhead near the source of the Darent and New Hythe on the Medway, are on Thames tributaries. In BRK there must have been a *hȳth* near Kintbury, on the river Kennet. A long curving valley runs down to the Kennet from the north, and along its course are farms called Great, Little and North Hidden. Hidden is OE *hȳth-denu* 'valley leading to the hithe'.

There is a scatter of *hȳth* names along the south-east and south coast. Hythe ESX, near Colchester, is by the Colne estuary, and Creeksea and Pudsey ESX are on either side of the Crouch estuary. Hythe KNT and Bulverhythe SSX refer to riverside landing-places very near the coast. Hythe HMP is on Southampton Water.

The two northwestern names which are considered to contain *hȳth* are Hyton CMB and Huyton LNC. Old Hyton CMB stands by the river Annas, which was presumably navigable for the short distance to the coast. At Huyton near Liverpool there was presumably a navigable channel leading to the Mersey.

There is no type of qualifier which is predominant in this small corpus. Domestic animals are well represented, and these could have been cargo, in which case the proximity of Bolney to Rotherfield ('cattle pasture') would be significant. As regards the function of the maidens at Maidenhead anybody's guess is as good as anybody else's.

Reference section for hȳth

As simplex name: Old Heath ESX (neighbouring Hythe was *La Newhethe*), Hive YOE, Hyde DEV, Hythe HMP, KNT, OXF(2), SUR, New Hythe KNT (early references are to an 'old' and a 'new' hithe), Downham Hithe and Witcham Hythe CAM.

As first element: Hidden BRK (**denu**), Huyton LNC, Hyton CMB.

As generic with references to classes of people: Bulverhythe SSX (*burhware* 'town dwellers'); Lakenheath SFK (*Lacingas*, probably 'people of a district characterised by drainage channels'); Maidenhead BRK ('maidens'); Swavesey CAM ('Swabian'); Walkerith LIN ('fullers').

With references to domestic animals: Bolney OXF ('bulls'); Horseway CAM; Lambeth GTL; Riverhead KNT and Rotherhithe GTL (*hrȳther* 'cattle').

With personal names: Ætheredeshyd (later Queenhithe) GTL (*Æthelrēd*); Bleadney SOM (**Bledda*); Prattshide DEV (**Prætt*); Pudsey ESX (**Pud*).

With descriptive terms: Earith HNt and Erith GTL ('gravel'); Greenhithe GTL.

With references to topography: Knaith LIN ('knee', i.e. 'river-bend'); Rackheath NFK (*hraca* 'throat', i.e. 'gulley').

With references to vegetation: Aldreth CAM ('alder'); Willey CAM ('willow').

With references to construction: Stepney GTL (probably 'stump'); Stockwith LIN ('tree-stump').

With an earlier place-name: Bablock Hythe OXF; Clayhithe CAM.

With reference to a wild creature: Putney GTL ('hawk').

With obscure or ambiguous qualifiers:
Chelsea GTL. Cole 1987 (pp. 52–3) points out that 'chalk landing-place' would give excellent sense as this could be a distribution point for chalk brought by boats for use on the clayey fields round London. The spelling *Cealchythe* in the Anglo-Saxon Chronicle suggests that the 'chalk' etymology was current in the late Anglo-Saxon period. Earlier references, however, have *Cælichyth*, *Celichyth*, and one of these, dated 801, occurs in a contemporary copy. There must be a presumption that this earlier form was rationalised to *Cealc-* in the 10th century, but later forms, and the modern one, are more easily derived from *Cælichyth*. OE *cælic* means 'cup, chalice'. A transferred topographical sense is perhaps not totally impossible.

Creeksea ESX. The word which suits the spellings best is **crūg** 'abrupt hill', discussed in Chapter 5. This would be admissable if the small hill by Creeksea Place, or any other feature above the 50ft contour, presents a striking

ROADS AND TRACKS

appearance from the Crouch estuary. The word *creek* probably had no OE antecedent so should not be adduced as a possibility (*pace* Ekwall 1960 and Gelling 1984).

Oteringhithe (later Methwold Hythe) NFK. OE *oter* 'otter' is perhaps used as a personal name here, since the *-ing-* construction is best evidenced with personal names.

Setchey NFK. The only spellings available are those in Ekwall 1960, and these show no sign of the genitival *-s-* which would be appropriate to the personal name *Secci* (postulated for Sedgeford). Neither do they suggest *secg* 'sedge'. A fuller collection of forms might help.

lād 'track'. The usual meaning of *lād* is 'artificial watercourse', and it is therefore discussed in Chapter 1. The 'track' meaning (noted in NED s.v. *lode* as no. 1) is rare, and is only found in minor names, but a few instances deserve mention in order to complete the treatment of the word. In Saddleworth YOW there is a farm called Load Hill which overlooks a deep holloway. Longload Lane in Middleton by Wirksworth DRB is recorded in 1309. *Milleslhade*, mentioned 1301 in Wallingford BRK, was the name of a road. Some of the CHE names listed in EPNS CHE Survey 5 (p. 256) probably contain *lād* in the sense 'track'.

16

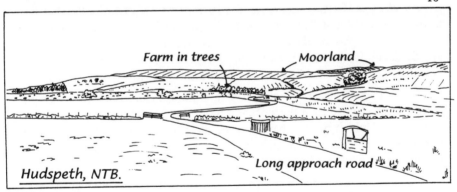

Hudspeth, NTB. Farm in trees. Moorland. Long approach road.

pæth 'path' is more frequent in minor names and field-names than in major settlement-names, but it occurs as the generic in at least six major names and is the probable first element in several others. It is a fairly frequent term in charter boundaries. Pave Lane, south of Newport SHR, preserves part of the name *diowuces pæth* in the OE bounds of Church Aston, and Pathe southwest of Othery SOM may represent *Wilbrittis pathe* in the bounds of Zoy.

Study of the situations of places with names containing *pæth* establishes that the commonest use is for an upland track. This definition suits Bagpath GLO, Brancepath DRH, Hudspeth NTB (a spectacular example, Fig. 16), Morpeth NTB, Painley YOW, Parford DEV, Pateley Bridge YOW, Roppa YON, Smythapark

DEV, Soppit NTB, Sticklepath DEV(2) and Urpeth DRH. Other *pæth* names are near heathland which is not particularly high: here belong Alspath WAR, Gappah DEV, Horspath OXF, Monkspath WAR. A few instances – Panborough and Pathe SOM, Tolpits HRT – are in low marshy ground.

The use of *pæth* in charter boundaries is examined in Kitson (forthcoming) with the conclusion that "the meaning is not so much 'path' in the modern sense, which in charter language would be *smalan weg*, but 'upland path'". He adds that a high proportion of charter names containing *pæth* and **stīg** occur on land over 400ft, or if there is no such land in the vicinity on the highest that there is.

Further comments on the special significance of some *pæth* names are offered in the reference section, where the small corpus is analysed according to qualifiers.

It is clear from charter boundaries that *herepæth* (literally 'army path') was the regular term for a main road in areas of West Saxon speech. Harepath occurs threee times as a minor name in DEV, and the compound appellative is the first element of Harpford DEV, SOM, and of two instances of Harford (in Crediton and Landkey) DEV.

Reference section for **pæth**:
As simplex name: Pathe SOM.

As first element. Certain examples are Parford DEV (*Pathford* 739), Pateley Bridge YOW (**lēah**) and Pathlow WAR, north west of Stratford upon Avon. At Pathlow there was a **hlāw** which marked the meeting-place of Pathlow Hundred, and further north, on the outskirts of Birmingham where it crosses Hockley and Shirley Heaths, the road is called Monkspath Street.

Possible examples are Painley YOW (**halh**) and Panborough SOM (**beorg**), but these can only contain the word if it be accepted that there was a feminine form **pathu** which would have a gen. pl. **pathena**. In Panborough the reference might be to the prehistoric log causeways of the Somerset Levels: Panborough appears on Fig. 14.

As generic with animal names: Bagpath GLO (probably 'badger'); Doepath NTB; Gappah DEV ('goats'); Horspath OXF; Urpeth DRH ('wild cattle').

With descriptive names: Roppa YON (ON *rauthr* 'red'); Smythapark DEV (*Smethepath* 1270, 'smooth'); Soppit NTB (*soc*, probably 'drain'); Sticklepath DEV(2) (*sticol* 'steep').

With personal names: Alspath WAR (*Ælle*); Brancepeth DRH (probably ON *Brandr*); Hudspeth NTB (**Hod*).

With references to danger: Dupath CNW ('thief'); Morpeth NTB (*morth* 'murder'). The town of Morpeth seems anomolous both as regards its status, which is unusually high for a place with a *pæth* name, and etymology, since the latter implies a lonely track over empty moorland. But Morpeth only attained its

90

high status after the Norman castle was built, when it displaced its neighbour, Mitford.

Unclassified: Tolpits in Watford HRT (*Tolpade* 1365, *Tollepathe* 1529, first element 'toll'). Perhaps a name of ME origin. The place is in low, marshy ground by the river Colne.

ritu- British 'ford' is the ancestor of Welsh *rhyd* and is cognate with Latin *portus* and English **ford**. It is well evidenced among place-names recorded from Roman Britain, e.g. *Anderitum* (Pevensey SSX), *Camboritum* (Lackford SFK) and *Durolitum* (Chigwell ESX). Rivet and Smith 1979 (p. 251) discuss its use in Continental names. British *ritu-* occurs in Leatherhead SUR, Penrith CMB, and probably in Ridware STF. The etymology of Leatherhead was satisfactorily elucidated for the first time in Coates 1980; it is a British name meaning 'grey ford', an important addition to the corpus of names which indicate substantial British survival south-west of London. Penrith means 'hill ford' or 'chief ford'. The crossing referred to was probably about a mile south-east of the modern town, where the Roman road from Brougham to Carlisle crossed the river Eamont, and the high ground east of Penrith might be felt to overlook this.

Ridware (Hamstall, Pipe and Mavesyn) STF has OE *-ware* 'dwellers' as second element, and was probably a district-name. If the first element is the Primitive Welsh reflex of *ritu-* that word may have been functioning as a simplex name when English-speaking people settled in the area. Or it may have been in frequent use as an appellative by Welsh speakers and mistakenly apprehended as a proper name by the English. In either case the feature must have been an important one. The 'ford' was perhaps a road through wet ground between the rivers Blithe and Trent.

Tretire HRE (*Rythir* 1212) is a Welsh name meaning 'long ford', probably, in view of the order of the elements, a post-Roman coinage.

stæth 'landing-place' occurs as a simplex name in Stathe SOM (east of Athelney) and Statham CHE (near Lymm), which is from the dative plural. It is the first element of Stafford STF, and the second element of Bickerstaffe LNC, Birstwith YOW, Brimstage CHE and Croxteth (near Liverpool) LNC.

The cognate ON **stǫth** is considered to occur in Toxteth (Liverpool) because the first element is the ON personal name *Tóki*. The plural of *stǫth*, which is *stǫthvar*, occurs in Burton Stather and Flixborough Stather on the river Trent in LIN. Other places in eastern England have such names associated with them, e.g. Brancaster Staithe and Overy Staithe on the Norfolk coast, but there is no way of knowing whether the element is OE or ON when it is singular. Staithes YON (*Setonstathes* 1451, from Seton Hall) is presumably from *stæth* as it has an English plural.

Smith 1956 gives more of these names under *stoþ*, and considers that OE *stæth* (only recorded with the meaning 'river-bank') acquired the sense 'landing-place' when it was influenced by the ON word. Stafford, however, seems likely to be an early OE name, and 'ford by a landing-place' makes more sense than 'ford by a river-bank'. It seems best to assume, as in Ekwall 1960, that *stæth* had the sense 'landing-place' before the period of the Norse settlements. It is possible that this was an earlier meaning than 'bank' in OE, and that it became obsolete and was revived by the influence of ON.

The nature of the 'landing-place' requires careful consideration in some instances. Bickerstaffe LNC ('bee-keepers' *stæth*') is not on a river. The village occupies a slight ridge in what may, in early times, have been very wet ground. Possibly there was here a marsh settlement, of a type known to archaeologists, whose inhabitants depended on dug-out boats for their communications. A similar explanation might suit Brimstage (probably 'Brūna' landing-place') in Wirral. Dodgson (EPNS CHE survey 4, p. 235) says "the meaning of *stæth* 'a landing-place' is improbable here for the smallness of the water at Brimstage", but by the same token the use of a word meaning 'river-bank' seems pointless. Croxteth Hall near Liverpool, however, is lower down the river Allt than Huyton, which is considered to contain **hȳth**, so 'landing-place' seems feasible. The early topography of Toxteth could only be evaluated with the help of expert local knowledge. Croxteth and Toxteth are considered to contain personal names, OE *Crōc* and ON *Tóki*.

Birstwith YOW is included here in spite of the rejection by Smith (EPNS survey 5, p. 131) of Ekwall's derivation from OE *byrg-stæth*, 'landing-place of the fort'. Smith says that this is 'an improbable explanation topographically', but as Birtswith stands on a hill overlooking the river Nidd it does not seem unreasonable.

stæth and *stoþ* are rare elements in minor names and field-names but a few instances have been noted in CAM, DRB, GLO, NTT, YOE and YOW.

stīg OE, **stígr** ON 'upland path', related to *stīgan* 'to climb', and so probably felt to be specially appropriate to an ascending track. Kitson (forthcoming) classes *stīg* and **pæth** together as regards meaning in charter boundaries, but says that there is a dialectal distinction, *stīg* being Hwiccan. In settlement-names and in ME field-names, however, *stīg* is not restricted to the southwest midlands. In some northern counties instances may be ascribed to ON influence, and this is valid for the fairly frequent appearances in Lincoln street-names, but there are several minor names in SSX and one in WAR (*Hawkestye*, later Hawkes End in Allesley) which must derive from the OE word.

As regards ancient settlement-names, OE *stīg* or (where appropriate) ON *stígr* is the generic in Bransty CMB, Bringsty HRE, Corpusty NFK, Gresty CHE and Wolsty CMB, and a possible first element in Styal CHE and Styford NTB. *stīg* cannot be distinguished on phonological grounds from *stigu* 'stye'. Compounds with *hogg* and *swīn* are usually translated 'pig-stye'.

Bransty CMB, now in Whitehaven, means 'steep path': the old road to Cockermouth ran along the present Bransty Road, and it is steep. Bringsty HRE, now the name of a common east of Bromyard, means 'brink path', and the road here follows the curving edge of a hill-spur. Corpusty NFK is explained in Ekwall 1960 and Smith 1956 as 'Corp's path', from an Anglo-Scandinavian personal name, but it may be an ON name from *stígr* and *korpr* 'raven'. Ravensty, perhaps another track from which ravens were seen, occurs as a minor name in Colton LNC. Gresty CHE is translated 'badger-run' in EPNS survey, but see **sol(h)** for the suggestion that *græg* means 'wolf'. Wolsty CMB is probably 'wolf path', but the name is not well recorded.

Smith 1956 cites some minor names in addition to the examples given above. He says of the group Corpusty NFK, Hardisty YOW, Ravensty LNC, Spruisty YOW, Thorfinsty LNC and *Wolmersty* LIN 'all with personal names', but this is not necessarily true of Corpusty and Ravensty, and in EPNS YOW survey 5 (p. 100) Smith himself offers a different origin, from *sprota* 'sprout, shoot', for Spruisty. Personal names do, however, occur in Hardisty (*Hardulf*), Thorfinsty (ON *Thorfin*), and probably *Wolmersty* (*Wulfmǣr*), and in this, as in the use of animal names as qualifiers, the compounds with *stīg*, *stígr* are comparable to those with **pæth**.

strǣt 'Roman road' is a West Germanic loan from the second component of the Latin term *via strata* 'paved road'. It occurs as a simplex name in Streat SSX, Street HRE, KNT, SOM(2), Strete DEV(2) and Stroat (in Tidenham) GLO, but it is more frequently used as a qualifying element in a compound name.

Habitative terms compounded with *strǣt* are *hām* (Streatham GTL, Stretham CAM), and, much more frequently, *tūn* (Stratton BDF, GLO, HMP, NFK(2), OXF, SFK, SOM(3), SUR, WLT, Stretton CHE(2), DRB, HRE(2), LEI(2), RUT, SHR, STF(2), WAR(4), Sturton LIN(3), NTB, NTT, YOW). Some instances of *strǣt-tūn* may belong in the newly-recognised category of 'functional' *tūn* names. The settlements may have performed a special function in relation to the road in a manner similar to that suggested in Chapter 1 for *ēa-tūn* and *mere-tūn*.

Topographical terms compounded with *strǣt* include **feld** (Stratfield BRK/HMP), **ford** (Startforth YON, Stratford BDF, BUC(3), ESX, MDX, NTP, SFK(2), WAR(2), WLT(2), Strefford SHR, Stretford HRE, LNC, Trafford LNC), **halh** (Strethall ESX) and **lēah** (Streatlam DRH, Streatley BDF, BRK, Stretley ESX, Streetly CAM, WAR, Strelley NTT).

The translation 'Roman road' is sound when *strǣt* occurs in major settlement-names (Gelling 1978, p. 153), but it does not always hold good for boundary-marks in Anglo-Saxon charters, and it cannot automatically be applied to field-names. In charter boundaries 'main road' is sometimes an acceptable rendering. In field-names and minor settlement-names it is necessary to allow for the modern dialect sense 'straggling village'.

The modern sense 'urban road' is found in late OE, and in this sense *strǣt* is the commonest term in the names of more important streets in towns.

vath ON 'ford' occurs as a simplex name in Wath YON(2), YOW, and Waithe LIN. It is the first element of Wassand YOE ('sandbank by a ford'), and the final element in about 15 major settlement-names. As an element in minor names and field-names it is present in some counties where there was Norse settlement, but not in all of them. It is probably commonest in CMB, and is the second element of Solway in Solway Firth. Modern forms often have -with or -worth, due to confusion with ON **vithr** and OE *worth*.

The small corpus of compound names in -**vath** resembles the much larger corpus in -**ford** in that the commonest type of first element is a word which describes the ford. Here belong: Brawith YON ('broad'); Helwith YON ('flat stone'); Langwathby CMB, Langwith DRB, NTT, YOE, Langworth LIN(2) (all 'long'); Sandwith CMB; Stenwith LIN ('stone'). Personal names may occur in Mulwith YOW, Ravensworth YON and Snilesworth YON, but these may be references to mules, ravens and snails. There is certainly a personal name (Old Swedish *Vinandr*) in Winderwath WML. For Flawith YON and Rainworth NTT alternative etymologies are available. Flawith may contain ON *flatha* 'flat meadow' or OE *fleathe* 'water lily' (Ekwall 1960 has another suggestion, ON *flagth* 'female troll, witch'). Rainworth may be 'clean ford' or 'boundary ford'.

gewæd 'ford' is cognate with ON **vath**. Most of the settlement-names in which it occurs are situated in BDF, IOW, KNT, NFK and SFK, and it is likely that in most of England *gewæd* went out of use at an early date, being superseded by the ubiquitous **ford**. Langwathby CMB is sometimes cited under this heading, but it seems more satisfactory to interpret it as containing ON **vath**, with some interchange of -*th* and -*d* in early spellings. *gewæd* is probably not found north of NFK, but it must have remained in occasional use in the south-west until a relatively late date. Wadebridge CNW is *Wade* 1382, and there was a lost *Ayleswade* in Harnham near Salisbury WLT, *Aylyswade Brigge* being the early name of Harnham Bridge. No early forms are available for Wade Hill Fm and Wade Park Fm, by the river Blackwater, west of Southampton.

The settlement-names which certainly contain *gewæd* are Biggleswade BDF, Cattawade SFK, Iwade KNT, Lenwade NFK, *Wathe*, now St Lawrence, IOW, Wade in North Cove SFK and Wadebridge CNW. There is a KNT parish called St Nicholas at Wade, and *Wade* was probably the name of the settlement here before the church was built, though this does not emerge so clearly from the early references as the comparable process does for St Lawrence IOW.

Cattawade, Iwade and Wadebridge are places where the crossing is over tidal water. St Nicholas at Wade was approached from the west by a long causeway over the marshes bordering the Wantsum Channel. It is possible that 'causeway' as well as 'ford' was implied by the use of *gewæd*.

As regards qualifiers, Biggleswade contains a personal name **Biccel*, Cattawade refers to cats and Iwade to yew trees. Lenwade may contain *lane* in the sense 'slow-moving stream', conjectured by Ekwall from a Scottish dialect use.

Landwade CAM may contain *gewæd*, but the early spellings show an unusual degree of variety.

weg 'way' is the general OE term for a road, and much the commonest one in charter boundaries. Unlike the other 'road' terms discussed in this chapter, *weg* appears to have had no specialised sense. It is most common as a qualified generic in names of roads, such as Icknield Way, Fosse Way and the ubiquitous Ridgeway. It is not common in ancient settlement-names.

Way occurs frequently as a simplex name in DEV, but only one instance, Way in Bridestow which is a DB estate, qualifies as a major settlement-name. As a first element *weg* occurs in Wayford SOM, Whaley CHE and (according to Gelling 1973, postulating a compound of *weg* with *hōh*, to which *hill* was added) in Weyhill HMP; but a stream-name seems more likely in Weybread SFK.

As a generic, *weg* is most frequently qualified by a descriptive term, a bias which arises naturally from the unspecialised nature of its meaning. Major settlement-names in this category are Bradway DRB, Broadway SOM and WOR ('broad'); Holloway GTL and Holway SOM ('hollow'); Stantway GLO (*stānihtan* 'stony') and Stanway ESX, GLO, HRE, SHR, Stowey SOM (2) (all 'stone'). A descripton of some 'stone ways' is given in Cole 1994b. In charter boundaries there are numerous 'green' ways and some 'white' and 'red'. DEV has a Greenway and a Whiteway which are recorded in DB, and three DEV settlement-names – Reddaway in Sampford Courtenay, Roadway in Morthoe and Rudway in Rewe – are almost certainly 'red'. Radway WAR is probably 'red way', referring to a track up the clay slope of Edge Hill, but an etymology 'riding way' is sometimes preferred. This latter seems appropriate for Rodway SOM.

In Gelling 1984 it is stated that three major names in *weg* have personal names as qualifiers: Garmondsway DRH, Hanwell OXF and Thoresway LIN. For Thoresway an alternative etymology has more recently been suggested, based on the rarity of personal names with *weg* and the DB spelling *Toreswe*. In the EPNS LIN survey Part 3, p. 151, John Insley is quoted as suggesting Old Danish *Thōrswǣ* 'shrine dedicated to the god Thor'.

Halsway SOM is 'pass way' (OE *hals*, literally 'neck'). Roundway WLT has been explained as 'cleared way', with *rȳmed*, past participle of *rȳman*, as qualifier. Two references have been noted to vegetation, Barkway HRT (*beorc*

THE LANDSCAPE OF PLACE-NAMES

'birch') and Spurway DEV (*sprǣg* 'brushwood'). References to goods transported occur in minor names and charter boundaries, where salt ways and hay ways are particularly common, but the only instance noted in an ancient settlement-name is Highway WLT, where the qualifier is probably the West Saxon form, *hig*, of *heg* 'hay'.

Farway DEV has been considered to contain the word **fær** discussed *supra*, but this is not likely, as that word seems to have become obsolete at an early date. A possible alternative is *fǣr* 'danger', modern *fear*. Farway is approached from several directions by roads which make very steep descents to the village.

In Flotterton and Hartington NTB (*Flotweyton* c.1160, *Hertweitun* 1171), *tūn* has been added to compounds in -*weg*. Hartington refers to harts. At Flotterton a road crossing a stream might have been supported on floats.

*werpels. This is the conjectured OE form of a dialect word *warple*, which is well-evidenced from 1565 onwards. In the first NED reference (from Wimbledon court rolls) the word is used for access ways between units in common fields. The second reference, dated 1658, is to 'one whaple or bridle way', and *whapple way* and *warple way* occur in the 19th century as terms for a bridleway or a cart track. *warple* probably derives from the verb *warp* 'to throw', OE *weorpan*.

The sparse occurrences of *werpels* in place-names establish the word's existence, but do not provide a base for conjecture as to the precise nature of the tracks so called. *werpels* is the first element of Warpsgrove OXF, Worplesdon SUR and Wapsbourn Fm in Chailey SSX. It occurs in two LNC names, Wirples Moss in Ormskirk and a lost *Wirpeslid* in Tatham. There are a few occurrences in ME field-names.

CHAPTER 4

Valleys, Hollows and Remote Places

Shelter from the elements is an obviously desirable characteristic in a settlement-site, and the category of place-name-forming terms for which this is the main connotation is a large one.

The main Old English terms for valleys are **cumb, denu** and **hop**, and the distribution of these in settlement-names is shown on Figs 20, 23 and 25. It can be seen that they cluster in upland regions, showing the inverse relationship to the landscape which was commented on in the discussion of water terms in the Introduction to Chapter 1. The single word 'valley' is an inadequate translation for these words. Modern English still has the words *coomb* and *dean*, but these are now used for empty hollows in hillsides and for steep, wooded stream-valleys. Probably no modern observer of a Compton would say that it was in a coomb, and dean in its modern sense does not suit the long, open valleys of the Chilterns and the South Downs. The third word, **hop** (modern *hope*), has kept its main place-name sense, but only as a dialect term in northern England. All three terms conveyed a great deal more to speakers of Old English.

It is possible to translate **cumb, denu** and **hop** with a few sentences, if not with a single word. The term **halh** is more difficult to explain, but since it is one of the commonest topographical place-name elements every effort has to be made to understand its significance. It is hoped that the discussion in the text, combined with the sketches in Fig. 24 and the map which is Fig. 25, goes some way to meeting this challenge. The distribution of the names in which 'hollow' is the appropriate translation shows that, in contrast to the other 'valley' words, this is not a term found mainly in upland country. It is most characteristic of areas where the relief is shallow; and careful scrutiny is sometimes needed in order to see that a **halh** site is slightly recessed. In areas like the flat lands of eastern England and the central part of Staffordshire a very slight recess can provide welcome shelter.

Names of topographical significance coined in England by speakers of Norse languages exhibit much less subtlety and variety than Old English names coined in earlier centuries. The only Norse 'valley' term which is common is **dalr**. This is widely used in the northern half of England for long valleys, and it is clear from the evidence cited in the article on **dalr** that this word and English **denu** were interchangeable in the topographical vocabulary of the late Anglo-Saxon period.

Germanic immigrants to this country must have been impressed by the variety of sheltered settlement-sites available, and the terms discussed here bear witness to their appreciation of this aspect of the English landscape. Twenty words are included, of which two are British and three Old Norse.

botm, bothm OE, **botn** ON, **bytme, bytne, bythme** OE 'bottom'. A thorough study of names containing these words was presented in Cole 1988, and a revised distribution map is reproduced in Fig. 17. In most instances in major place-names, and in early-recorded minor names, the terms refer to "a flat alluvial area, restricted in size, moist and often easily flooded and with a marked change of slope at the valley sides". There is a cluster of *botm* names near Manchester (giving rise to several locally-biased surnames), and in the majority of these the *botm* is "a locally wider area of very flat, wet land adjacent to a river ... The flood plain or river terrace narrows almost to vanishing point upstream and downstream of the 'bottom' ... In each case the valley sides rise abruptly and quite steeply from the valley floor, and the settlement, if any, is on rising ground – not on the valley floor".

The land-form is so distinctive that it may sometimes be decisive in cases where an etymology is uncertain. Beamsley YOW overlooks a perfect 'bottom' as defined here. The -*s*- genitive, however, and the early spellings would require a neuter side-form *bethme rather than any of the recorded forms. Bitteswell LEI, on the other hand, for which Ekwall 1960 proposed a similar solution of a neuter *bythme*, fails the topographical test as its valley is a wide v-shape.

Leaving aside disputed cases, there are some 30 instances of names recorded before 1500 which contain either one of the OE words or ON *botn*. It is important to note the date of the first record, as in names of later origin Bottom is used all over southern England with the more general sense 'lowest part of a valley'.

The conclusion in Cole 1988 is that in and near the Pennines both OE *botm* and ON *botn* are used in the sense described above, with only three minor names failing to fit the pattern. South of the Pennines, where the valleys of lowland Britain are less steep-sided and deep, there are examples like Bottoms Fm, east of Bristol between Abson and Doynton, where the feature is "like the majority of the Pennine 'bottoms', but on a gentler scale". The sparse southern instances are all named on Fig. 17, and comments on the situation of each can be found in Cole 1988.

The CMB name Wythburn (earlier *Wythbottom*) looks anomalous on the modern map because the level of Thirlmere has risen since it was dammed in the late 19th century. The village was originally on a small fan overlooking an area of very damp flat land where streams flowing into the lake had silted up

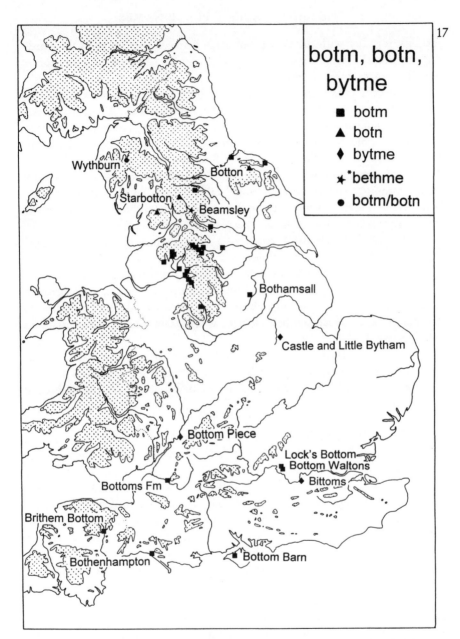

its southern end. Several examples overlook silted-up pro-glacial lakes; a notable example is Starbotton YOW. Bothamsall NTT (also discussed under **scelf** in Chapter 5) overlooks a large expanse of alluvium between the rivers Meden and Maun.

Cole points out that the use of *botm, bythme, botn* in the specialised sense which she describes is best evidenced in northern and south-western England,

and this, together with the identity of usage between the OE and ON words, suggests that this sense developed at a relatively late stage in the evolution of the OE topographical vocabulary.

In the corpus assembled in Cole 1988 there are eight simplex names and several in which the 'bottom' term is the first element. Where there is a qualifying element it may refer to the shape of the feature, as in Broadbottom, Longbottom, Sidebottom (the last from *sīd* 'wide'), or to vegetation. In the second category are Oakenbottom LNC, Ramsbottom LNC ('wild garlic') and Wythburn CMB ('willow'). Keldbottom YON and Starbotton YOW are ON names meaning 'spring bottom' and 'stake bottom' (ON *stafir*). Shipperbottom LNC, which is believed to be the origin of the surname Shufflebottom, was *Schyppewellesbothem* in 1285, 'sheep spring bottom'. Ekwall 1922 (p. 64) suggests that Shillingbottom contains a stream-name meaning 'resonant'. Winterbottom occurs several times as a minor name in LNC and CHE, and Brook Bottom in Saddleworth YOW was earlier *Winterbottom*.

Some qualifying elements are of undetermined derivation, as in Brithem Bottom DEV. The surname Higginbottom has been traced to a field-name in Bredbury CHE, of which the first element is uncertain; there is a discussion in EPNS survey Part 1, 266-7.

More minor names than usual have been included in this discussion, as few places named from these words grew into major settlements. The main exception is Ramsbottom, which owes its development to the Industrial Revolution.

byden, beden 'vessel; tub' is sometimes used of deep valleys. Simplex names are Beedon BRK, Bennah and Betham DEV (both earlier *Bydene*), Bidden HMP, Bidna CNW, DEV. As a first element, *byden* in its 'valley' sense occurs in Bedlested SUR, a lost DB settlement called *Bedenesteda* in Sandon ESX, and Bibbern in Stalbridge DOR (*Bydeburnan* 933). When combined with terms for springs, however, *byden* has its literal sense and refers either to a drinking vessel or to a container which holds water from the spring. Bidwell (or Bedwell) is a recurring name, and this was perhaps a compound appellative: *v.* Rumble 1989 and cf. *supra* under **funta**. Some possible occurrences with **ford** (Bedford, Bidford, Bideford) are discussed in Chapter 3.

canne 'vessel for holding liquid' is used of a steep-sided valley in Cann DOR, and of a less dramatic feature in Canney (in Cold Norton) ESX. It also occurs in a few minor names in DEV, and in the minor name Howcans ('hollow valley') in Morley YOW. In names like Canford DOR and Canwell STF, where there is no trace of *-nn-* in the spellings, a personal name *Cana* is to be preferred.

VALLEYS, HOLLOWS AND REMOTE PLACES

cetel, WSax ci(e)tel, cytel, 'kettle, cauldron', is used of valleys in Chettle DOR, Cheddleton STF and Chittlehampton DEV. In some other names, e.g. Chattlehope Burn NTB, Chettiscombe DEV, Kettlewell YOW, it could alternatively refer to bubbling or swirling water. Wherwell HMP, from *hwer* 'cauldron', may be compared; here the reference is to eddies in the river Test.

*clof, clofa 'something cut'. Only the weak form *clofa* is on record, meaning 'a cut-off section of a chirograph', but it is believed that both forms existed, and that the strong form *clof* had a topographical meaning similar to that of the related ON *klofi* 'crevice'. The meaning may have been 'triangular indentation', like one of the cuts in the jagged edges of chirograph sections.

If the place called *Clofesho, Clofeshoas*, where the rulers of Mercia held synods during the period 716–825, is correctly identified with Brixworth NTP (Davis 1962), the *clof* will be the valley of the stream which rises south of the church and bisects the western edge of the spur on which the settlement stands. Professor Davis points to the topographical appropriateness of the name as one of the factors in favour of the identification, and this is discussed further under hōh, *infra*.

The gen. sg. of *clof* is considered to be the first element of Closworth, south of Yeovil SOM, which is on a hill-spur with narrow stream-valleys on either side.

The third name which has been considered to contain *clof* is Clovelly DEV, where the village occupies a spectacular cleft in the sea-cliff for which this word would be highly appropriate. The village is a late 16th-century creation, but the harbour at the foot of the cleft must always have been a very important feature on this dangerous coast. In Gelling 1984 Clovelly was explained as a compound of *clof* and a hypothetical hill-name *Felg*, literally 'wheel-rim', referring to the large curved hill which overlooks the harbour. This has been challenged by Professor Richard Coates (1996, 1997). It is indeed open to serious objections, the weightiest being perhaps the occurrence of the second component by itself as the name of farms called Higher and Lower Velly which are inland, well out of sight of the coastal hill. Professor Coates suggests that the *clof* element in Clovelly derives from an early form of Cornish *cleath* 'dyke', referring to Clovelly Dykes, a great earthwork a mile from the coast, and he explains Velly as a personal name, perhaps that of a saint.

With Clovelly deleted, or at least rendered doubtful, *clof* is a very rare place-name term, but it deserves mention because of the special importance of the name *Clofesho*. It is related to *clēofan* 'to split', modern *cleave*. The past participle of *clēofan* is the first element of Clannaborough, north-west of Crediton DEV ('cloven hill'), and a derivative, *clēofung*, is thought to be the source of Cleaving near Londesborough YOE, referring to a steep valley in the Wolds.

*clōh OE, **clough** ME, only found in the north and the north midlands. Duignan 1902 says: 'Clough. A common name in the N. Staffordshire moorlands, but unknown S. of Stone ... It means a ravine or narrow valley, with steep sides, usually forming the bed of a stream'. This suits ME *clough*, which is common in moorland country in the north of England, sometimes becoming -cleugh. But the situation of the two ancient settlement-names in which *clōh* is considered to occur, Clotton CHE and Cloughton YON, suggests that the OE word (which is not on independent record) was used for much less dramatic valleys. Clotton north-west of Tarporley has a low-lying site enclosed by the 150ft contour, and Dodgson, EPNS survey 3, 272, translates it 'farm at a dell'. Cloughton north of Scarborough does have a small ravine, but is in undramatic country. It is not clear why the ME reflex of *clōh had the sense 'ravine'.

Apart from the two compounds with *tūn*, *clōh occurs occasionally as a first element, as in Cloffocks near Workington CMB and Clougha, a hill southeast of Lancaster, both with -hōh; but it is more frequently the second element, as in Deadwin (i.e. 'dead woman') Clough LNC and Wildboarclough CHE, or used as a simplex name. Clougha, which is *Clochhoch* 1199, may be one of the earliest instances of ME *clough* used of a dramatic feature.

Haltcliff CMB (southeast of Hesket Newmarket) was *Halteclo, Hautecloch* in the 13th century: this is probably a ME name in which *clough* 'ravine' is qualified by French *haut* 'high'.

corf 'valley, pass'. This is a derivative of *ceorfan* 'to cut', modern *carve*. In SHR it was the term applied to the long, broad valley between Wenlock Edge and the Clee Hills. The name was transferred to the river Corve, and the valley then became known as Corve Dale. Corfham and Corfton are in this valley.

In the southwest, *corf* acquired the more specialised sense 'pass'. In Corfe Castle DOR it is used of the pass which divides the ridge of the Purbeck Hills. A less dramatic gap between hills is referred to in Corfe Mullen DOR, but Corfe south of Taunton SOM refers to a pass in a steep ridge, as does Corton Denham north of Sherborne. Coryates and Corton DOR refer to passes through an escarpment east of Abbotsbury. See Fig. 18.

Ekwall 1960 explained Corscombe DOR and Croscombe SOM as OE *corfweges cumb* 'valley of the pass road', and this was accepted in Gelling 1984. Closer consideration of the earliest spellings for both names, however, reveals that this is hardly tenable. Corscombe DOR is *Corigescumb* in charters of 1014 and 1035, and Croscombe SOM is *Correges cumb* in a charter of 705. This looks like a double occurrence of a meaningful compound, but it is not likely that *corfweges* would have been reduced to *Coriges-, Correges-* in Old English spellings. The names are unsolved. The bounds of Corscombe in the 1014 charter mention a *miclan corf*, but this is probably a coincidence rather than

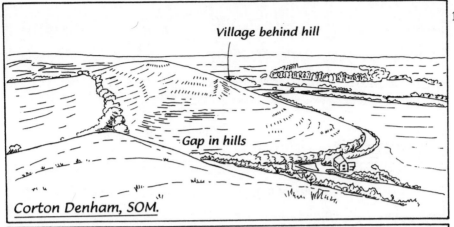

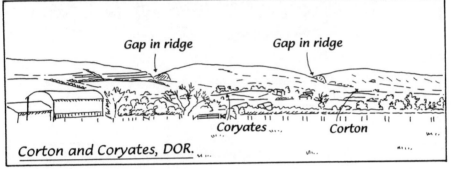

an indication that Corscombe contains *corf*. (Corscombe DEV is a different name, of uncertain etymology).

cumb 'valley', more specifically 'short, broad valley, usually bowl- or trough-shaped with three fairly steeply rising sides'. The definition of *cumb* in Cole 1982 has been widely quoted in recent landscape studies and in works on the Old English language. This study of the most frequently used valley terms, *cumb* and **denu**, in the chalklands of southern England is rightly regarded as a classic.

cumb was used in this highly specialised sense all over England, except in DEV. It is commonest in chalk and limestone escarpments, where the relevant landform is most likely to occur, but examples are found with other types of geology. As Cole comments 'it is the relative proportions and resulting overall shape that are important'. See Fig. 19, overleaf.

A *cumb*-shaped valley offered a distinctive type of settlement-site. It was sheltered, but it lacked the facilities which would encourage growth. Because of the steepness of the enclosing hillsides many *cumbs* are reached by dead-end roads, so they cannot be on main travel routes. Also there is no room for indefinite expansion of the settlement or for the plantation of daughter settlements. The only *cumb* settlement which achieved outstanding adminis-

19

Telscombe, SSX.

Tidcombe, WLT.

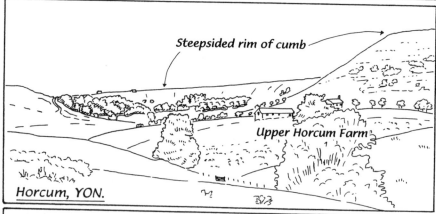

Horcum, YON.

Coombe, IOW.

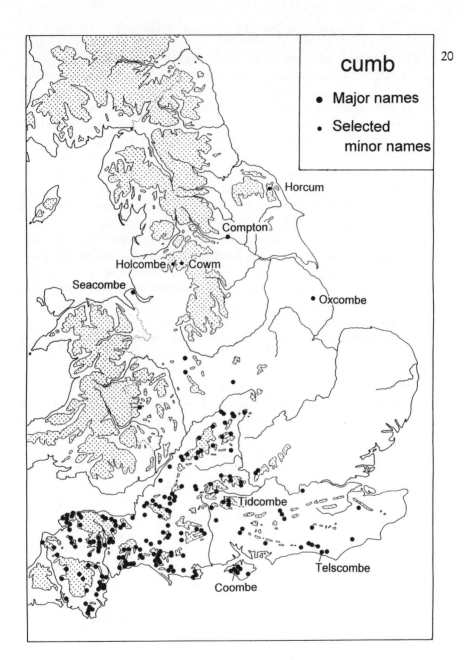

trative status before modern times is Winchcombe GLO, which occupies an atypically large valley. Other settlements in relatively large *cumbs*, such as Castle Combe WLT, were parish centres, however. Some of the Comptons have less tightly enclosed sites than most names in which *cumb* is the generic, and these have sometimes become large villages.

Distribution is governed mainly by geology. The term is rare in counties like STF, where the landform is poorly represented, and the few *cumb* names which do occur in clay country in the midlands denote sites which have only a rough approximation to that of the sharply delineated chalkland *cumbs*. No major name in *cumb* has been noted in East Anglia, LEI or NTT, and LIN has only Oxcombe, in a round-ended valley south of Louth. See Fig. 20.

The rarity of *cumb* in SHR and HRE is due not so much to geological factors as to the adoption of the term **hop** in those areas to denote features which were frequently *cumb*-shaped, but which had an extra quality of extreme remoteness. *cumb* probably does not occur in ancient settlement-names north of Yorkshire, and this, also, may be partly due to the use of **hop**.

In the CMB names Cumcatch, Cumcrook, Cumdivock, Cumrew, Cumwhinton, the generic is not OE *cumb* but the British word *cumbo-*. These names are wholly British, though of post-Roman origin. The British word is also used as an affix to an English name in Cumwhitton. The relationship between the British and English words is discussed *infra*. Pencombe HRE may be a British name meaning 'place at the end of the coomb'.

cumb is much more common in south-west England than in any other region. Ekwall 1960 includes 42 settlement-names containing *cumb* in SOM, 36 in DEV and 26 in DOR, so these counties contain more than half of the instances included in the dictionary. Fig. 20 gives the overall distribution.

It seems possible that the frequency of genuine *cumb* landforms in DOR and SOM caused other 'valley' terms to be largely forgotten, so that when English names were established, at a somewhat later date, in DEV, any valley, whatever its shape, was liable to be called a *cumb*. The statistic of 36 major names given above and the analyses in the reference section *infra* do not give an adequate impression of the use of *cumb* in DEV: if minor names are taken into account it is the commonest place-name element in the county apart from *tūn*. But a great many of the *cumb* names in DEV do not refer to *cumb*-shaped valleys, and this may account for the frequency of qualifiers referring to shape. Widdecombe in the Moor occupies a large basin, appropriately described as 'wide', and there are several other 'wide' coombs in DEV. Smallacomb(e)(5), Smallicombe(2) and Smallcombe(3) are 'narrow'. There are several 'long' and 'short' coombs and numerous compounds with *hol* 'hollow'. There are two Deancombes, perhaps describing a *cumb*-shaped feature in a longer valley. Some DEV valleys are accurately denoted by other terms, such as **denu** and **slæd**, and some of the *cumb* names are in *cumb*-shaped valleys, but the precision of the valley terminology may have been breaking down when English names were being coined in this county.

Details of the distribution pattern, and of the precise topography of many *cumb* names, can be found in Cole 1982 and Gelling 1984.

cumb has generally been regarded as one of the handful of OE words which are adaptations of Primitive Welsh terms. Primitive Welsh had a word

VALLEYS, HOLLOWS AND REMOTE PLACES

represented by philologists as *cumbo-*, which has become Modern Welsh *cwm*. Gelling 1984 expressed doubts about this being the whole story. There is an Old English word *cumb* 'cup, vessel' which could have been used in a transferred sense, like **byden, canne, cetel** and **trog**. This may be at least partly the origin, with the resemblance to the Welsh word helping *cumb* to take precedence over other 'vessel' terms used to denote valleys. This suggestion was made much earlier in NED, but was ignored in Smith 1956 and Ekwall 1960.

cumb is frequently used as a simplex name, though many examples acquired affixes after the Norman Conquest. As a first element it is found mainly with *tūn*. When used as a generic it has qualifiers which show a fairly even spread over the several categories of personal names, vegetation names, topographical features and descriptive words. The last category occurs mainly in the south-west, though the compound with *hol* is widespread. Other categories of qualifier are sparsely respresented, and there is a rather high incidence of names in which the first element defies firm identification.

Reference section for cumb

As simplex name: North and South Coombe, Coombe Martin, Combpyne, Combe Raleigh and Combeinteignhead DEV; Coombe Almer and Coombe Keynes DOR; Combe GLO(2), HRE, OXF, SOM, WAR; Coombe HMP, SUR; Abbas Combe, English Combe, Combe Florey, Combe Hay, Monkton Combe, Combe St Nicholas, Combe Sydenham and Temple Combe SOM; Coombe Bissett and Castle Combe WLT; Coombes SSX; Cowm LNC.

As first element: Combwich SOM; Comhampton WOR; Compton BRK(2), DEV, DOR(4), GLO(3), HMP, IOW, SOM(7), SSX, STF, SUR, WAR(5), WLT(3), YOW; Congreve STF (**græfe**).

As generic with personal names: Babbacombe DEV (*Babba*); Bettiscombe DOR (*Betti*); Branscombe DEV (Welsh *Branoc*); Burlescombe DEV (*Burgweald*); Butcombe SOM (*Budda*); Catcomb WLT (*Cada*); Chacombe NTP (**Ceawa*); Elcombe WLT (*Ella*); Faccombe HMP (**Facca*); Harescombe GLO (**Hersa*); Icomb GLO (**Ica*); Ilfracombe DEV (*Ælfrēd*, WSaxon *Ielfrēd*); Letcombe BRK (*Lēoda*); Odcombe SOM (*Uda*); Snorscomb NTP (**Snoc*); Telscombe SSX (**Titel*); Tidcombe WLT (**Titta*); Winscombe SOM (*Wine*).

With references to vegetation: Appuldurcomb IOW ('apple-tree'); Ashcombe DEV; Boscombe WLT (a conjectural OE **bors*, probably 'spiky plants'); Farncombe SUR ('fern'); Kingcombe DOR (*Chimedecome* DB, from *cymed* 'wall-germander'); Lyscombe DOR (a conjectural OE **lisc* 'reed'); Nettlecombe DOR, IOW, SOM; Thorncombe DOR(2); Widdicombe (in Stokenham) DEV, Witcombe SOM, Withycombe DEV, SOM (all 'willow').

With references to topographical features: Awliscombe DEV (*āwel* 'fork', here 'river-fork'); Chettiscombe DEV ('kettle', referring either to a spring or to a deep hollow); Pyecombe SSX (**pīc** 'pointed hill'); Seacombe CHE; Watercombe DOR; Welcombe DEV, WAR, Woolacombe DEV and Woolcombe DOR (all **well**).

With earlier hill- or river-names: Chilcombe DOR, HMP (an element *cilt-*, found in a number of names in southern England, believed to be a pre-English hill-name); Rendcomb GLO (*Hrinde*, a stream-name meaning 'thruster'); Yarcombe DEV (river Yarty).

With descriptive terms: Challacombe (near Lynton) DEV ('cold'); Drascombe DEV (*drosn* 'dirt'); Holcombe DEV(3), DOR, GLO, LNC, OXF, SOM ('hollow'); Horcum YON ('dirt'); Pitcombe SOM (probably an OE *pide* 'marsh', only recorded in place-names); Shalcombe IOW ('shallow'); Smallacombe, Smallcombe, Smallicombe (a recurrent name in DEV, *smæl* 'narrow'); Watcombe OXF, Whatcombe BRK, DOR ('wet'); Whitcombe DOR, IOW, WLT, Widcombe SOM(2), Widdecombe in the Moor DEV, Witcombe GLO (all 'wide'); Winchcombe GLO (*wincel* 'bend').

With references to wild creatures: Bittiscombe SOM ('beetle'); Creacombe DEV and Crowcombe SOM ('crow'); Renscombe DOR ('raven'); Stinchcombe GLO (*stint*, a bird-name); Stitchcombe WLT (*stūt* 'gnat'); Ulcombe KNT ('owl'); Yarnscombe DEV ('eagle').

With references to domestic creatures: Challacombe (in Combe Martin) DEV ('calves'); Gatcombe IOW ('goats'); Moulsecoombe SSX ('mule'); Staddiscombe DEV ('herd of horses'); Swyncombe OXF ('swine').

With references to crops or products: Melcombe DOR ('milk'); Salcombe DEV ('salt'); Timbercombe SOM; Watcombe DEV ('wheat'); Wexcombe WLT ('wax').

With references to classes of people: Charlcombe SOM ('farmers'); Hascombe SUR and Hescombe SOM (*hætse* 'witch'); Hestercombe SOM (*Hegsteldescumb* 854, the significance of *hægsteald* 'warrior, bachelor' is unclear); Huntercombe OXF; Parracombe DEV ('pedlars').

With references to structures: Haccombe DEV ('hatch'); Hippenscombe WLT ('stepping-stones'); Liscombe SOM and Loscombe DOR (*hlōse* 'pig-stye'); Pencombe HRE ('enclosure'); Seddlescombe SSX (*setl* 'dwelling').

With references to position: Milcombe OXF ('middle'). Bowcombe IOW is a phrase, '(place) above the coomb'.

With references to activities: Motcombe DOR ('meeting'); Syndercombe SOM ('cinder', presumably an indication of industrial activity).

With obscure or ambiguous first elements:
Balcombe SSX. Two spellings, *Baldecombe* 1279 and *Baldcomb* 1284, are consistent with an etymology 'Bealda's coomb', but earlier and more numerous spellings lack the *-d-*. No convincing etymology has been suggested for the *Balecumba, Belecumbe* forms.

Batcombe DOR, SOM(2). 'Bata's coomb' is formally acceptable, but the triple occurrence of the name in the southwest suggests a meaningful compound. No recorded OE word suits.

Bickham DEV. There are five instances of this name in DEV. These are in Buckland Monachorum (*Bykecombe* 1291), in North Huish (*Bykecomb* 1330), in Okeford (*Bichecoma* DB), in Kenn (*Bikecumb* 1249) and in Branscombe (*Bykcomb* 1495). This five-fold occurrence rules out the personal name *Bica*. The qualifier may be *bica 'point', cited for Bicknor under **ofer** and **ōra**, in which case Bickham belongs with the group of DEV names in which an imprecise use of *cumb* is qualified by a descriptive term. Fieldwork is required to check this hypothesis.

Bincombe DOR. The post-Conquest spellings are compatible with *bēan* 'bean', but an Anglo-Saxon charter has *Beuncumbe*, which is unexplained.

Burcombe WLT. The spelling *Brydancumb* 937 rules out 'valley of the bride', which would be OE *brydecumb*. This is one of several place-names which require an element *brȳda*.

Chaffcombe DEV, SOM. A personal name *Ceaffa* has been conjectured, but there is little evidence for it. The word chaff (OE *ceaf*) might be considered, though the persistent *-ff-* in the forms for both places tells against it.

Corscombe DOR, Croscombe SOM. The problems presented by this name are discussed *supra* under **corf**.

Daccombe DEV. Neither of the etymologies which have been suggested – personal name *Dæcca* or *dā* 'doe' – accounts for all the early spellings. These include *Doccuma* 1185 and *Daggecumba* 1193.

Luccombe IOW, SOM. Either the personal name *Lufa*, or *lufu* 'love' referring to amorous activities.

Morecombelake DOR (*Mortecumb* 1240). This is one of a group of names containing an element *mort* which has not been satisfactorily explained.

Pitchcombe GLO (*Pichenecumbe* 1221). Apparently an adjective from *pic* 'pitch', but EPNS survey 2, p. 169, expresses doubt about the presence of pitch here.

Riddlecombe DEV. A derivative of Cornish *rid* 'ford' has been suggested for Riddle-. OE *hridel* 'sieve' would suit formally, but it is not clear what the reference would be.

Sutcombe DEV. Ekwall's derivation from *sūth* 'south' does not suit either the earliest spellings or the modern form. The same first element may be found in Sotwell BRK, but there is no other evidence for a personal name *Sutta.

dalr ON, **dæl** late OE, **dale** ME 'main valley'. If the suggestion put forward in the next article be accepted, ON names containing *dalr* may be as early as those in which OE *dæl*, ME *dale*, is used in the sense 'main valley'. It is not possible to make a firm distinction between the OE and ON terms; they are treated together here and in the reference section. In the lists of names, however, note is taken of ON, as opposed to OE, qualifying elements.

ME *dale* (or OE *dæl* in the sense 'main valley') is only common north of a line from the Mersey to the Humber, excluding NTB and DRH. The comparative rarity in these two counties is significant, supporting the hypothesis that the degree of ON influence in an area is the crucial factor in the use of the word in place-names. The common name Dalton is only found in northern counties. South of LNC and Yorkshire, the county with most major names in -dale is DRB, where Cowdale, Edale, Longdendale and Kingsterndale are in the northwest, and Dale Abbey and Ravensdale are respectively east and north-west of Derby.

The incidence of *dæl* 'valley' and ON *dalr* is difficult to assess because when there are abundant early spellings it is frequently clear that **denu** has either been replaced by or has alternated with the other term. Saxondale NTT, for instance, is *Saxeden* 1086, *Saxenden* 1316, but has -*dala*, -*dale* spellings starting *c*.1130, and Debdale Lodge in Rothwell NTP is named from a valley which is called *deopandene* in the OE bounds of Kettering.

Edale DRB is *Aidele* in DB. There is then a long gap in the available spellings, *Eydal(e)* being the usual form from 1275 on. The first element is ēg, and Edale is one of a group of ēg names in north DRB in which the precise meaning is difficult to determine. This problem is discussed in Chapter 2. The settlement lies in the valley of a tributary of the river Noe, and the name may originally have referred to this short feature rather than to the main valley known as the Vale of Edale. This may be an instance of *dæl* in the sense 'valley', in which case the compound with ēg suggests a relatively early currency for that meaning. The absence of 12th-century spellings, however, makes it impossible to be certain that there was no alternation with -*den* from **denu**.

Reference section for **dalr** ON, **dæl** OE, **dale** ME

Some names which are likely to be early OE coinages using **dæl** in its earlier sense 'pit, hollow' are omitted from these lists. The YON group – Dalton, Givendale, Cundall – has been included, but see the discussion of **dæl** for reservation about whether the generic means 'valley' in these names. Some items listed here are district-names rather than referring primarily to settlements.

As simplex name: Dale Town YON.

As first element: Dalby IOM, LEI(2), LIN, YON(3) (ON *bȳ*); Dalton DRH(2), LNC(2), NTB(2), WML, YOE(3), YON(3), YOW(2) (*tūn*). Dawdon in Dalton-le-Dale DRH (*Daldene c*.1050) is a compound of *dæl* and **denu**.

VALLEYS, HOLLOWS AND REMOTE PLACES

As generic with personal names: Ainsdale LNC (*Ægenwulf*); Bilsdale YON (ON *Bildr*); Botesdale SFK (*Bōtwulf*); Bransdale YON (ON *Brandr*); Cuerdale LNC (*Cynferth*); Ennerdale CMB (ON *Anundr*); Garsdale YOW (ON *Garthr*); Givendale YOW (**Gȳthla*); Martindale WML (*Martin*); Patterdale WML (*Patric*); Thixendale YOE (ON *Sigstein*).

With descriptive terms: Bleasdale LNC (ON *blesi* 'light spot', perhaps used of a bare place on a hillside); Deepdale LNC, YON (Dale Abbey DRB was earlier *Depedale*); Kildale YON (ON *kíll* 'narrow bay' used in a transferred sense); Silverdale LNC; Sterndale DRB (*stæner* 'stony ground').

With references to vegetation: Bannerdale WML (ON *beinvithr* 'holly'); Birkdale LNC (ON *birki* 'birch copse'); Farndale YON ('fern'); Lindal and Lindale LNC ('lime-tree'); Matterdale CMB ('madder'); Ramsdale YON ('wild garlic'); Yewdale LNC.

With references to wild creatures: Cadwell (earlier *Cattedale*) LIN ('cat'); Ravendale LIN and Ravensdale DRB; Woodale YON, Wooldale YOW and Uldale CMB (all 'wolf', Uldale from ON *ulfr*).

With references to domestic creatures: Cowdale DRB and Kiddal YOW (both 'cow'); Grisedale CMB, WML, Grizedale LNC(2) (ON *gríss* 'young pig'); Rosedale YON (ON *hross* 'horse'); Withersdale SFK ('wether').

With references to topographical features: Knaresdale NTB (the district was probably called *Cnearr* 'rugged rock'); Mardale WML (**mere**); Mosedale CMB (ON **mosi**); Ragdale LEI (*hraca* 'throat' used of a straight, narrow stream-valley); Ravenstonedale WML; Wasdale CMB, WML (ON *vatn* 'lake'); Wheldale YOE ('wheel', referring to the curves of the river Aire).

With river- or stream-names: Allerdale CMB (river Ellen); Borrowdale CMB, WML ('fort-river', ON *borgar á*); Clapdale YOW (an OE stream-name *Clæpe* 'noisy'); Givendale YOE (*Gifl*, a British river-name meaning 'forked'); Glaisdale YON (a British stream-name *Glas* 'clear'); Kendal WML (river Kent); Lonsdale LNC/WML (river Lune); Rochdale LNC (river Roch); Swaledale YON (river Swale).

With references to man-made structures: Baysdale YON (ON *báss* 'cow-shed'); Bowderdale CMB (ON *búthar* 'booths'); Crossdale CMB; Kirkdale LNC, YON (ON *kirkja* 'church'); Scawdale CMB (ON *skáli*; 'hut').

With references to owners or inhabitants: Bishopdale YON; Bretherdale WML ('brothers'); Colsterdale YON ('charcoal burner'); Cummersdale CMB ('Britons'); Saxondale NTT; Scugdale YON(2) (probably *scucca* 'goblin').

With reference to a crop: Rydal WML ('rye').

With reference to position: Westerdale YON.

With obscure or ambiguous first elements:
Bannisdale WML, earlier *Banandesdale*. An ON personal name *Bannandi* or *Bannandr* has been conjectured, as a derivative of *banna* 'forbid, hinder, curse'. Fellows-Jensen 1985 (p. 102) observes that this could have been the name of the beck.

Cundall YON. Ekwall 1960, followed by Fellows-Jensen 1972, explains this name as OE *cumb* with an 'explanatory' ON *dalr*. The topography is against this; there are no features to which either the OE or the ON terms are clearly appropriate. The early spellings regularly have *Cun-* or *Kun-*, and the DB spellings have *-del*. It is possible that Cundall is an early OE name, *cūnadæl*, 'hollow of the cows'.

Edale DRB. The qualifier is ēg, but the precise sense is uncertain. See the discussion under ēg in Chapter 3.

Smardale WML. The qualifier could be ON *smjor* or OE *smeoru* 'grease, fat, butter', referring to rich pasture, or ON *smári*, *smæra* 'white clover', which survives in Lakeland dialect as *smere*.

dæl 'pit, hollow'. This is the same word as '**dæl** late OE' in the preceding article. It is treated separately here on the hypothesis that there are some place-names in which it has an earlier significance, more akin to the 'pit, gulf' senses which are evidenced in literary sources in addition to the meaning 'valley'.

This hypothesis is tentative, and it is rejected in Kitson (forthcoming). The occurrence of *dæl* in Doverdale WOR and Dalham SFK, and perhaps in Edale DRB, may be seen as evidence against it, but probably not (*pace* Kitson) the occurrence of *eardel* in the OE bounds of Madeley STF, as the valleys in this vicinity are not major features such as are described by ME *dale*. At Dalham SFK the river Kennet runs in a long valley, but the *dæl* of the name may refer specifically to the wider embayment in which the settlement lies. Viewed from the church the settlement is certainly in a hollow.

Two ancient settlement-names in KNT contain *dæl*. Dalham near High Halstow is on a low hill overlooking a broad recess called Buck Hole. The early topography of the other example, Deal, is difficult to evaluate, but there can hardly have been a valley of the sort for which OE **denu** and ME **dale** are used. A well-recorded minor name in KNT, Wormdale west of Sittingbourne, refers to a farm in a hollow, and this situation is clearly distinct from the **denu** of nearby Danaway.

There are a few minor names in SSX: Daleham (a **hamm**) in Fletching, Dale Park in Madehurst, Summersdale in New Fishbourne, and two instances of a compound which has 13th-century spellings *Hindedal(e)*, *Hyndedale*. These last are Hendall in Buxted and Hendal Fm in Withyham. EPNS SSX survey says 'deer valley', but the meaning could be 'gully behind the settlement'. This certainly suits Hendal in Withyham, where there is a deep ravine behind the farm.

The single instance of *-dale* in DEV, Dalwood, is first recorded in 1195, and may be of ME origin. In Stavordale SOM it is clear from the early spellings that *dale* replaced **denu** in the mid-13th century.

VALLEYS, HOLLOWS AND REMOTE PLACES

North of the Thames, the only ancient settlement-name in the south-east midlands which may contain *dæl* is Dawley GTL. The EPNS survey of MDX objects on the grounds that the area is flat, but the word may have been used of a pit or hollow. There are no settlement-names containing *dæl* in BUC, OXF, GLO, or in the counties immediately north of these, with the single exception of Doverdale WOR.

Probably, although *dæl* existed in the language, it was not widely used in place-names until the introduction of ON **dalr** influenced its meaning, after which it replaced **denu** in some areas.

Some north country names in -dale might, if their sites were examined carefully, appear more likely to contain *dæl* in the sense 'pit' than in the sense 'valley'. Givendale south-east of Ripon YON is a possible instance. Cundall and Dalton, in the same area of YON, have -*del*, *Del*- spellings in DB, and EPNS survey gives *dæl* for these names.

dell 'dell'. This word is realted to **dæl**. It has a limited provenance in place-names in southern England, occurring mostly in minor names and field-names. Two examples, Arundel SSX and Ramsdell HMP, have, however, become major settlements. At both these places there are tiny valleys, and the only instance in SUR, Farthingdale in Lingfield, overlooks a little stream-valley. Arundel and Farthingdale, and probably Ramsdell, have plant-names as qualifiers: *hārhūne* 'horehound' (river Arun is a back-formation), 'fern', and probably 'wild garlic', though for names with ME *Rames*- there are always alternative possibilities.

Besides being used for natural features *dell* was applied to artificial hollows. *Chalkdell* is common in HRT field-names, referring to pits from which chalk was dug. Kitson (forthcoming) observes that in charter boundaries *dell* is frequently used for ancient quarries.

Crondall HMP, *Crundellan* c.880, derives from a term *crundel* which is probably a compound of *dell* with *crumb* 'crooked'. *crundel* occurs frequently in charter boundaries in a limited area, two-thirds of the instances being in WLT and west BRK. Location of some *crundels* on the ground suggests that the term was applied especially to large quarries which were overgrown with trees. Kitson says that in the charter material *dell*, **dæl** and *crundel* have complementary distributions and are all used of steep-sided valleys or hollows whether natural or artificial.

denu 'valley'. This was the standard OE term for a main valley. It occurs in about 185 major settlement-names, and the distribution is widespread (Fig. 23). In counties where there are no examples in major names there are usually a few in minor names. There was probably no region in which *denu* was not part of the place-name-forming vocabulary, though in northern counties it was to some extent replaced by **dæl** or **dalr** in the late OE period.

The incidence of *denu* is heavily obscured on modern maps by confusion with **dūn** and *tūn*. It is also necessary to bear in mind that in south-east England many names now ending in -den are from **denn**, not *denu*.

The type of valley to which *denu* was applied can be seen on Figs 21, 22. The definition in Cole 1982 is '*denu* is mostly used of long, narrow valleys with two moderately steep sides and a gentle gradient along most of their length'. Such features can be found in parallel along the dip-slopes of escarpments (as in Fig. 21), or in isolation, as at Croydon GTL, Helmdon NTP, Sheldon DEV. The characteristic relationship with dip-slopes contrasts with the frequent occurrence of **cumb** features on scarp-slopes, but **cumb** and *denu* intermingle in the South Downs because the long *denu* valleys have **cumb**-shaped features opening off them. Most valleys designated *denu* are sinuous, but those of the South Downs are sometimes straight, Houndean on the western edge of Lewes being a good example.

Sometimes where there is a fine series of such valleys there are no *denu* place-names because the settlements are named from the rivers which flow in the valleys. This can be seen in DOR, where the villages called Tarrant, Winterborne, Milborne, Dewlish, Cerne, and those named from the river Piddle, might have had *denu* names.

The great majority of settlements with *denu* names have sites which conform to the pattern outlined here. The most striking anomaly is Howden near Goole YOE. Howden is recorded as *Heafuddene* in a charter of A.D. 959, when it was the centre of a large, composite estate. Smith (EPNS survey p. 251) argues for a meaning 'valley by the spit of land', saying that the *denu* is the old course of the river Derwent, and the **hēafod** ('headland') the stretch of land between the Ouse and the old Derwent. This is very forced, however, and it seems wiser to concede that nothing in the topography of the area deserves either of the terms *denu* or **hēafod** in the senses in which they are evidenced elsewhere. It is worth noting that Onions 1966 says under *dean* (the modern form of *denu*) that the word is related to *den* 'animals' lair', and that under *den* he says 'the basic meaning may be 'open or flat place''. This sense (though difficult to reconcile with the later meanings of both words) would be well suited to the Vale of York. An alternative translation to Smith's 'valley by the spit of land' might be 'chief plain', if it could be assumed that *denu* is here used in an archaic sense.

The type of valley for which *denu* is used offers settlement-sites with greater potential for growth than a **cumb**, hence the frequency of Great and Little prefixed to *denu* names on Fig. 21. Assendon, perhaps the most splendid of the Chiltern *denu*s, comprises three settlements, Upper, Middle and Lower. Because of the easy gradients and the effortless emergence onto higher ground, *denu* valleys frequently have through roads running along them, so that these settlements would have more frequent contacts with the outside world than those in valleys called **cumb** or **hop**.

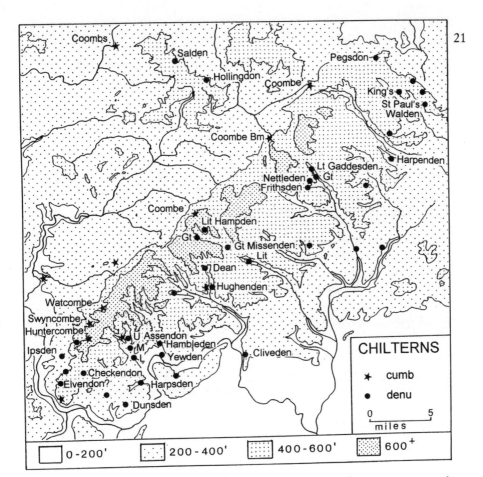

denu is frequently used as a simplex name. As a qualifier it occurs mostly with *tūn*. As the generic in compound names it is most frequently qualified by personal names or descriptive terms. Group-names in *-ingas* are well represented. The number of compounds in which the first element presents philological problems is not unduly large.

Reference section for **denu**

As simplex name: Dean BDF, CMB, DEV, GLO, HMP(2), OXF, SSX(2), WLT, Deane HMP, LNC, Deene NTP, Denes NFK. Deanham NTB and Denholme YOW are from the dative plural.

As first element: Denford BRK, NTP; Denham BUC (near Oxbridge), SFK(2) (*hām*); Denton CMB, DRH, HNT, KNT(2), LIN, LNC(2), NFK, NTB, OXF, SSX, YOW (*tūn*); Denwick NTB (*wīc*); Dunbridge HMP.

As generic with personal names: Addlestone SUR (**Ættel*); Agden HNT (*Acca*); Arkesden ESX (probably late OE *Arcel* from ON *Arnkell*); Balsdean SSX (**Beald*); Basildon BRK (**Bæssel*); Bevendean SSX (**Bēofa*); Biddesden (near

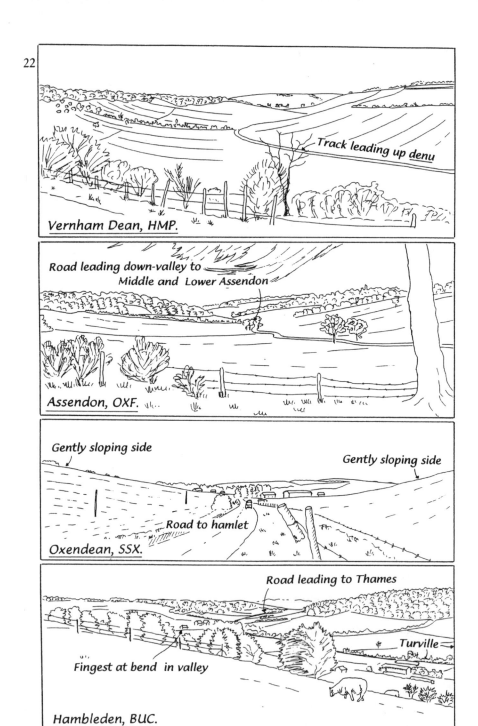

VALLEYS, HOLLOWS AND REMOTE PLACES

Chute) WLT (probably *Bīede*); Bittadon DEV (**Beotta*); Buckden HNT (fem. *Bucge*); Calmsden GLO (**Calemund*); Chaddesden DRB (*Ceadd*); Chillenden KNT (*Ceolla*); Colesden BDF (**Col*); Croxden STF (*Crōc*); Duddoe (earlier *Dudden*) NTB (*Dudda*); Dunsden OXF (*Dynne*); Egdean SSX (*Ecga*); Elsdon NTB (*Elli*); Erringden YOW (ON *Eiríkr*); Essendine RUT (*Esa*); Figheldean WLT (**Fygla*); Framsden SFK (*Fram*); Gangsdown OXF (*Gangwulf*); Gransden CAM, HNT (**Grante*); Helmdon NTP (**Helma*); Hoddlesden LNC (**Hodd*); Hundon SFK (*Hūna*); Kelvedon ESX (*Cynelāf*); Lavendon BUC (*Lāfa*); Lexden ESX (**Leaxa*); Lyveden NTP (*Lēofa*); Minsden HRT (**Myndel*); Ovenden YOW (*Ofa*); Paddington (DB *Padendene*) SUR (*Pada*); Pledgdon ESX (**Plycca*); Riddlesden YOW (*Hrēthel*); Scammonden YOW (ON *Skammbeinn*); Shoddesdon HMP (*Scot*); Silsden YOW (ON *Sigulfr*); Togstone (earlier *Toggesden*) NTB (**Tocg*); Wandon End HRT (**Wafa*); Wantisden SFK (*Want*); Wilsden YOW (*Wilsige*).

With descriptive terms: Ballidon DRB (*belg* 'bag'); Blackden CHE and Blagdon NTB ('black'); Bradden NTP ('broad'); Chaldean's Fm (in Much Hadham) HRT ('chalk'); Chisledon WLT ('gravel'); Debden ESX, Depden SFK, Dibden HMP, KNT, Dipton DRH ('deep'); Flaunden HRT ('flagstone'); Gordano SOM ('gore'); Grendon HRE ('green'); Hambleden BUC ('maimed', i.e. misshapen, probably referring to the right-angle bend in the valley, and to its broken sides); Holden YOW ('hollow'); Horden DRH ('dirty'); Howden NTB ('hollow'); Howden YOE ('head' perhaps in the sense 'most important'); Langdale (earlier *Langedene*) WML ('long'); Longdendale CHE/DRB; Meriden WAR and Munden (in Watford) HRT ('merry', i.e. 'pleasant'); Nevenden ESX ('level'); Salden BUC and Shalden HMP ('shallow'); Standean SSX and Standen BRK, LNC, WLT ('stone'); Trawden LNC ('trough-shaped'); Waterden NFK and Waterdine SHR; Wendens ESX ('bend').

With references to vegetation: Agden YOW ('oak'); Ashington NTB ('growing with ash', OE *æscen*); Aspenden HRT ('aspen'); Berkesden HRT ('birch'); Bramdean HMP ('broom'); Croydon SUR ('saffron'); Embleton DRH ('elm'); Haslingden LNC ('hazel'); Hatherden HMP ('hawthorn'); Hebden YOW(2) ('hip'); Hesledon DRH ('hazel'); Nettleden HRT (*netelig*, 'growing with nettles'); Ramsden ESX, OXF (probably *hramsa* 'wild garlic'); Rushden HRT, NTP; Sabden LNC ('spruce'); Sawdon YOW ('willow').

With references to wild creatures: Amberden ESX (a bird-name *omer*, found in ModE *yellowhammer*); Croydon CAM ('crows'); Elveden SFK ('swan'); Foxton DRH, NTB; Harding WLT ('hare'); Hendon DRH ('hinds'); Horndean HMP (OE *hearma*, which may be 'shrew' or 'weasel'); Ousden SFK ('owl'); Ravensden BDF ('raven'); Rodden SOM ('roe').

With references to domestic creatures: Assendon OXF ('ass'); Buckden YOW ('bucks'); Cogdean DOR ('cock'); Houndean SSX ('hounds'); Shipden (lost in the sea near Cromer) NFK ('sheep'); Swinden YOW ('swine'); Wetherden SFK ('ram').

With references to crops: Barden YOW ('barley'); Haydon Bridge NTB ('hay'); Whaddon WLT and Wheddon SOM ('wheat'). The reference to 'thatch' in Theydon ESX puts it loosely in this group.

THE LANDSCAPE OF PLACE-NAMES

With river-names: Luddenden YOW (a stream-name from *hlūd* 'loud'); Ripponden YOW (river Ryburn); Skelding YOW (river Skell); Turkdean GLO (*Turce*); Yelden BDF (*Gifl*, believed to be an earlier name of river Til).

With topographical terms: Brogden YOW (**brōc**); Campden GLO and Compton DRB (gen.pl. of *camp*, perhaps a district-name meaning 'open spaces'); Cheddon SOM (probably **cęd**); Cliveden BUC (**clif**); Dundon SOM (**dūn**); Frithsden HRT (**fyrhth**); Ipsden OXF (*yppe* 'upland'); Pegsden BDF (**pēac**); Sheldon DEV (**scelf**).

With a group-name in -ingas: Bagendon GLO ('Bæcga's people'); Manuden ESX ('Manna's people'); Monewden SFK ('Munda's people'); Ovingdean SSX ('Ofa's people'); Pangdean SSX ('Pǣga's people'); Rottingdean SSX ('Rota's people'); Yattendon BRK ('Geat's or 'Eata's people').

With classes of people: Lothersdale (earlier *Lodersden*) YOW ('beggar'); Quendon ESX ('women'); Saxondale (earlier *Saxonden*) NTT ('Saxons'); Walden ESX, HRT, YON ('Britons'); Whissendine RUT (*Hwicce*).

With references to structures: Berden ESX and Burdon (south of Sunderland) DRH ('byre'); Haydon CAM (probably *(ge)hæg* 'enclosure'); Hidden BRK (**hȳth**); Pinden KNT (possibly **pinning* 'enclosure'); Stagsden BDF ('stake'); Stavordale (earlier Stavorden) SOM ('post'); Thedden HMP (*thēote* 'water conduit'); Tincleton DOR (possibly **tȳnincel* 'small farm').

With references to boundaries: Marsden LNC, YOW (*mercels* 'sign, mark', here probably 'boundary'); Todmorden YOW ('Totta's boundary-valley', with *(ge)mǣre*).

With references to activities: Finedon (earlier *Thingdene*) NTP (*thing* 'assembly'); Plowden SHR ('play').

Names with obscure or ambiguous first elements:

Arkendale (earlier *Arkenden*) YOW. There are several place-names which presuppose OE terms *earc* or *earca*. A possibility which seems not to have been considered is the word *earc, earce* 'ark, chest', which could have been used in a topographical sense. In Arkendale the word would have the weak feminine genitive, *earcan*, a form which is recorded. As with Howden the site barely qualifies as a **denu**. The settlement is on a hill between two very small streams which run in slight depressions.

Arlecdon CMB. The suggestion that the first part of the name is a compound *earn-lacu* 'eagle-stream' was made in Ekwall 1960 and accepted in EPNS survey, but this is not compatible either with the early spellings or with the distribution and meaning of **lacu**. Something like **arlāc* seems to be required. The name is unsolved.

Barden YON. The DB form has *Berne-*, but the *-n-* does not appear again. If *Berne-* rather than the later *Ber-* is to be trusted the first element could be *berern* 'barn', *beren* 'growing with barley', or the personal name *Beorna*. If the

VALLEYS, HOLLOWS AND REMOTE PLACES

DB form is an aberration this name will be the same as Barden YOW, from *bere* 'barley'.

Biddlesden BUC and Biddlestone (*Bitlesden* 1181) NTB. Ekwall 1960 suggested an OE **bythle* 'building' for these names and for Betteshanger KNT, but the triple occurrence as first element in the genitive, with no other manifestations, raises doubt about this.

Bozen Green (in Braughing) HRT, earlier *Bordesden*. This is one of a group of names in *Bord(es)-*. The word *board* probably appears in some, but there was a personal name *Brorda* from which dissimilation of *-r-* could have produced **Borda* and a strong form **Bord*.

Checkendon OXF. An element *ceacca* is indicated by the early spellings for this name and for Chackmore BUC and Checkley HRE, STF. A hill-name **Ceacce* or a personal name **Ceacca* have been suggested. An onomatopoeic bird-name might be considered.

Chidden HMP. An etymology '*Citta's valley' would suit the spelling *Cittandene* 956, but it is invalidated by the occurrence in the bounds of the same charter of the phrase *to cittanware becun*. Another charter of the same date calls this landmark *citwara becon*, and in a third boundary survey it is *cytwara bæcce*. The boundary mark is 'ridge(s) of the *citta* dwellers'. OE *ware* 'dwellers' can be added to a significant word or to a place-name, but there appears to be no instance of its being added to the first element of a compound name, so *cittanware* is not likely to be an abbreviation for 'dwellers at Chidden'. It seems necessary to postulate a toponym and to interpret Chidden as 'valley of the feature called *Cit, Cyt* or *Citta*'. Kitson (forthcoming) takes *citwara, cytwara* as presenting the correct form and follows Jackson 1953 (p. 327) in taking this to be a WSaxon development of PrW *cẹ̄d* 'wood'. He explains the inappropriate *Cittan-* spellings as due to folk-etymology bringing in the personal name *Citta*.

Cliddesden HMP. The forms indicate OE *clȳde*. Ekwall 1960 suggested a derivative of *clūd* 'rock', and Coates 1989 suggests that **Clȳde* is an oblique case of a hill-name **Clūd*, referring to the adjacent Farleigh Hill.

Cutsdean GLO. The earliest forms are *Cottesdene, Cotesdene* from the 12th century. This is considered to be the same estate as *Codestun* in charters of 977, 987, and the first element of both names is to be associated with that of Cotswold, discussed under **wald** in Chapter 6.

Droylsden LNC. Early spellings indicate ME *drīles-*, for which no convincing etymology has yet been suggested.

Essendon HRT. A document of 11th-century date refers to *Eslingadene* in a context of association with Hatfield which makes it virtually certain that the place is Essendon. It is, however, difficult to reconcile the spelling with the other early forms for Essendon. These begin in the late 11th century with *Essendene*, and none of them has *-l-*. The first part of Essendon has been generally considered to be *Eslinga* ('of Esla's people') in accordance with *Eslingadene*, but it is difficult to account for the total disappearance of *-l-* and the early reduction of *-inga-* to *-en-*. Without the *Eslingadene* form the name would have been interpreted as 'Esa's valley', a doublet of Essendine RUT.

Frostenden SFK. Generally explained as 'frog valley', OE *froxadenu*, which suits the DB form *Froxedene*, but not the later *Frostedene*, *Frostendene* spellings.

Gaddesden HRT. Gaddesden is OE *Gætesdene*, and the river Gade, which runs down the valley, is *Gatesee* 1242, with the same first element and *ēa*. *Gates-* occurs also in Gatesbury near Braughing, also in HRT but a considerable distance from Gaddesden. Ekwall 1960 suggests a personal name **Gǣte*, a derivative of *gāt* 'goat'. A small stream joins the river Rib at Gatesbury, and a stream-name, applied to both river Gade and this stream, would have seemed appropriate if it had not been for the apparent genitival composition in *Gatesee*. *ēa* is sometimes added to earlier river-names, but not with an *-es* genitive.

Hampden BUC. The qualifier is **hamm** in one of its senses.

Harpenden HRT, Harpsden OXF. Names containing 'harp' have not yet been satisfactorily explained. Earliest forms for the HRT name indicate derivation from uninflected *hearpe*, but some 13th-century ones have *Herpen-*, perhaps indicating the currency of an alternative form with the genitive *hearpan*. The *-s-* of Harpsden is not original; the early spellings are identical with the *-n-* forms for Harpenden, so here also 'valley of the harp' is at least partly the etymology. The name of the long valley which has Harpsden at its mouth occurs again in a different form, *hearp dene*, in a charter boundary of A.D. 966. Kitson (forthcoming) suggests that *hearp(e)* in these names is an abbreviation of *herepæth* 'main road', but neither the Harpsden nor the Harpenden valley is followed by a road of outstanding importance. The Harpsden valley is curved, but the Harpenden one is straight, which seems to rule out a reference to shape.

Hellidon NTP. The spellings indicate *hælig-*. The recorded word *hǣlig*, which is a side-form of *hālig* 'holy', would suit, but this is a Northumbrian form, perhaps unlikely to be manifested in a NTP name. In addition to the uncertainty about the first element there is some doubt as to whether Hellidon is a *denu* or a *dūn*. Spellings with *-den* and *-don* both occur among the earliest forms, which are late 12th century, and the village is on a hill overlooking a *denu*-type valley.

Hughenden BUC. A personal name **Huhha* has been conjectured to suit this and some other place-names. The same first element is likely to occur in

VALLEYS, HOLLOWS AND REMOTE PLACES

Howfield KNT, discussed in Chapter 7, but the two instances of Houghton in DEV, cited in this connection in Ekwall 1960, may be of later origin with ME *Hugh*.

Lydden KNT(3). For Lydden near Dover there is an 11th-century spelling *Hleodæna* which suggests a compound of *denu* with *hlēo* 'shelter'. The post-Conquest spellings for Lydden in Thanet and Lydden near Sandwich are consistent with this. The triple occurrence is surprising, however, and if *hlēo* is the first element it is uncertain whether the 'shelter' is a building, a place of refuge from enemies, or a place which offers protection from harsh weather.

Marden WLT. Probably 'fertile valley' with OE *mærg*, *merg*, 'marrow, fat'. This word is not universally accepted as a place-name element, but (*pace* Smith 1956 and Ekwall 1960) it seems the likeliest source, used as a simplex name, for Merrow SUR. Marden has an OE form *Merh dæne* and ME forms *Mereghedene*, *Mereweden*.

Missenden BUC. The river Misbourne runs down the valley, and both names probably derive from a noun **mysse*, with an *-n* genitive in Missenden. This was probably not the name of the river, as *burna* does not seem to be added to earlier river-names. A derivative of *mēos* 'moss' might be considered, with a meaning such as 'mossy place'.

Munden, Great and Little, HRT. The first element could be *mund* 'protection' or a personal name *Munda*.

Polesden SUR. Two spellings, *Palesden(n)* from 1202, 1204, suggest 'valley of the pole' from OE *pāl*. This word could conceivably have been used as a hill-name. Poling SSX is an *-ingas* name with the same base, and a personal name *Pāl* has been conjectured to suit these two place-names.

Rattlesden SFK. The verb *rattle* is only recorded from the 14th century and the noun from the 16th, but it might perhaps have an OE origin. It occurs in the modern names of a number of plants.

Sarsden OXF. ME spellings are *Cerchedene*, *Cercendene*, with occasional *Cercbesdene*. The only suggestion yet offered is that in Ekwall 1960, where 'church valley', OE *circan denu*, was proposed. There is, however, no other instance of OE *cirice* giving spellings comparable to those for Sarsden.

Scackleton YON (earlier *Skakelden*). This is one of several place-names which contain the word *shackle*. The meaning of this word in place-names has not been determined.

Walkden LNC. Ekwall 1960 postulated as OE river-name **Wealce*, 'rolling', for this name and that of the river Walkham in DEV. A case can also be made for a personal name **Wealca*.

Walsden LNC. Perhaps a personal name **Walsa*, but this does not seem to be otherwise evidenced.

Wilden BDF. There are several possible first elements of which the likeliest are probably the personal name *Willa* or *wilig* 'willow'.

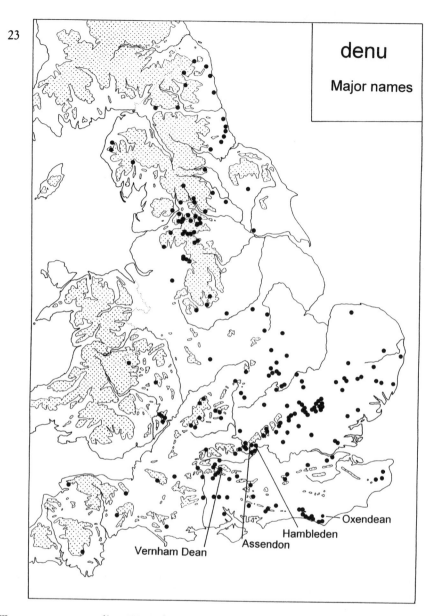

23 denu Major names

Wooperton NTB, earlier *Wepreden*. A term *wepre* has given modern Wepre in FLI. Wyn Owen 1988, pp. 105, 112, accepts the suggestion in Ekwall 1960 that Wooperton is OE **wēoh-beorg-denu*, 'valley by the hill with a heathen temple', and interprets Wepre as *wēoh-beorg*. This etymology is not wholly convincing, however, and in a later allusion to Wepre (Wyn Owen 1997, p. 273) he interprets the FLI name as *wearpe* 'silted land'. This might be possible for Wooperton, which overlooks tributaries of the river Till, but *wearpe* in this sense is not a firmly established OE term.

gil ON 'ravine, deep valley with a stream'. This term was used by Norwegian settlers in north-west England. It is commonest in minor names, but occurs in a few settlement-names, including Gaisgill (near Tebay) WML, Gazegill and Howgill (near Barnoldswick) YOW, Ivegill (near Dalston) CMB, Reagill, Rosgill and Sleagill WML and Scargill YON. The qualifiers are ON. Gaisgill and Gazegill refer to geese, Reagill to foxes, Rosgill to horses. Scargill contains ON *skraki*, a bird-name. Sleagill contains ON *slefa* 'saliva', used either as a nickname for a person or as a stream-name. Ivegill refers to the river Ive (probably an ON name meaning 'yew stream'). Howgill, which is a fairly common minor name in YOW, is 'hollow ravine' from ON *holr*.

In Gillcambon (near Greystoke) CMB, *gil* is used as the generic in a compound made in the Gaelic manner, with the qualifying element, an ON personal name of Irish origin, placed last.

glennos British 'valley'. This is the likeliest origin of Glen LEI, which is the name of two villages, Great Glen and Glen Parva, five miles apart on the river Sence, south of Leicester. A charter of A.D. 849 was issued from *Glenne*, assumed to be Great Glen, and Ekwall (1960) considered that it was the pre-English name of the river. The use of a British river-name for an important settlement on its course is evidenced elsewhere, as is the survival of such a name for settlements, but not for the river. Sence is probably an OE name, which could have replaced *Glenne*. Alternatively, *Glenne* could have been a British district-name which was adopted by the Anglo-Saxons and given to two settlements which lay in the district, only coincidentally by the river. (The river-name Glen, which occurs in LIN and NTB, has a different British origin).

Gaelic *gleann*, from the same root, is a very common place-name element in Scotland and the Isle of Man. Welsh *glyn* and Cornish *glin* which are the direct descendants of British *glennos*, are much less frequent in Welsh and Cornish names.

There are a few names in the most northerly English counties, such as Glencoyne CMB/WML, Glendue NTB, Glenridding WML, which are considered to contain the British word.

Smith 1956 postulates an OE word **glenn*, but there is very little evidence, if any, to support this. The modern English word is a borrowing from Gaelic.

halh (Kentish, W. Saxon **healh,** dative *hale, heale*) 'nook'. This word was treated exhaustively in Gelling 1984, and there is not a great deal to add to that discussion.

halh is related to *holh* 'hollow', and the use for sunken places brings it clearly within the scope of this chapter: but it does not always mean 'valley' or 'recess'. NED gives pertinent OE and ME quotations under the modern forms *haugh* and *hale*, and the general impression conveyed by these is more

24

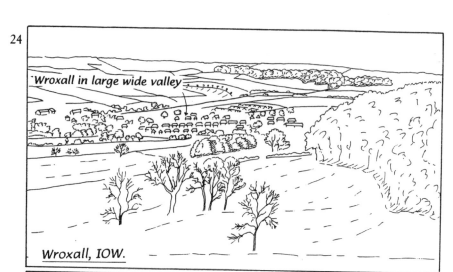

Wroxall, IOW.

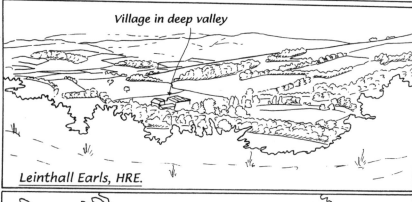

Leinthall Earls, HRE.

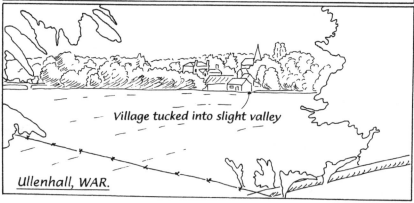

Ullenhall, WAR.

illuminating than any translation. No modern English words can convey the topographical uses adequately. The 'nook' is sometimes formed not by contours but by water. In some areas *halh* is used for land between rivers or in a river-bend. In others it appears to be used for slightly raised ground isolated by marsh, a sense which seems the only possible one for some names in

VALLEYS, HOLLOWS AND REMOTE PLACES

Lincolnshire and East Anglia. There is also a clearly-evidenced 'administrative' sense, in which *halh* denotes a piece of land projecting from, or detached from, the main area of its administrative unit. Instances of this include Bracknell and Broomhall BRK, and the same sense is found in Surrey, where Portnall is in a southern projection of Egham parish, and Michen Hall ('large nook') is in an angle of the parish of Godalming. There is a temptation to resort to this last sense when, as occasionally happens, all possible topographical explanations fail to suit.

In its 'valley' or 'hollow' sense, *halh* can be difficult to pin down in settlement-names. Wroxall IOW (Fig. 24) is a satisfying example, because although the settlement is quite large it has not obliterated the outline of the large, irregular depression in which it lies. Other instances of large settlements with *halh* names which occupy large, shallow basins without obliterating them can be found in Shropshire (Shifnal) and Staffordshire (Bednall). But when *halh* is used for less extensive hollows the original features cannot be identified with certainty in the area of a town or a large village. Also a slight shift of settlement site (harmless in the case of a **dūn** or **denu** name) could obscure the connection with a small *halh*.

Sometimes, if the settlement remains very small, however, the evidence can be visually very satisfying. A good instance is Blunt's Hall near Witham ESX, where the ancient house occupies a slight but clearly-defined hollow. The name is *Blunteshale* in early spellings; its modern development, like that of nearby Rivenhall, illustrates the misleading similarity to *hall* 'large building'. The latter word is in fact very rarely found in settlement-names of pre-Norman-Conquest origin.

The nearest one can come to a definition of the 'valley' and 'hollow' senses may be that *halh* is used for formations which are not sufficiently well-defined or firmly-shaped to merit the terms **cumb** or **denu**. Kitson (forthcoming) observes that in charter boundaries *halas* are never linear.

There is a marked clustering of *halh* names in some areas where the relief is shallow and surface features are not sharply defined. The part of Staffordshire west of Stafford may be instanced. The element is common in Essex and in Norfolk and Suffolk, and in most of this large region the usual meaning is 'hollow'. In East Anglia, however, there are some names in deep fenland, where *halh* cannot have this meaning. The four villages called Wiggenhall south of King's Lynn (– St Mary the Virgin, – St Germans, – St Peter, – St Mary Magdalen) presumably owe their existence to the presence of patches of firm ground in the fens traversed by the river Ouse. In Gelling 1984 attention was drawn to the contrasting use of ēg in the naming of three settlements called Tilney (– All Saints, – cum Islington, – St Lawrence) which lie immediately west of the Wiggenhalls, and it was tentatively suggested that *halh* was used for the lowest eminence which made settlement possible. Mildenhall SFK cannot be explained in terms of a hollow, and Mepal CAM has an island site in

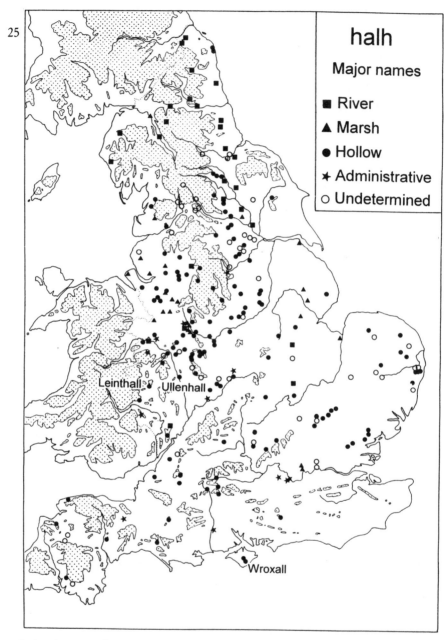

fenland, west of Ely. In Lincolnshire the villages of Great and Little Hale occupy raised gravel islands surrounded by boulder clay, and in Lancashire there is a series of *halh* names running from near Liverpool to near Garstang for most of which 'dry ground in marsh' seems the likeliest sense.

The meanings 'land between rivers' and 'land in a river-bend' are occasionally found in the midlands, Hallow WOR being a striking instance: but they are most frequent north of the Humber. In northern English speech, *haugh*, the modern form of *halh*, means 'low-lying land by a river'.

A more detailed analysis of the senses in which *halh* is used in settlement-names will be found in Gelling 1984. Fig. 25 endeavours to show these by the use of different symbols. No map and no discussion will ever be definitive, however, and readers will doubtless re-interpret some names in the light of local knowledge.

The use of *halh* in such varied senses causes the classification of qualifying elements to be less meaningful than it is for terms which have a more unitary significance. Nevertheless the analysis of *halh* compounds presented in the reference section brings out some interesting points.

Personal names form much the largest class of qualifiers with *halh*. Some of those adduced in the reference section, such as **Bracca* in Bracknell, **Scuffa* in Shifnal and **Cod* in Codsall, rest on slender foundations, but even when allowance has been made for some which more critical analysts may discard there are more than three times as many names in this category as in any other. Peter Kitson's analysis of *halh* names in charter boundaries also finds personal names the commonest qualifier, though not so far ahead of other categories as seems to be the case in settlement-names. Three of the personal names are ON, and five of the OE names are dithematic. There is only one possible feminine name.

This clear bias towards association with individuals is balanced by the occurence of seven *-inga-* compounds.

Other categories with transparent qualifiers are all small, but a marked feature of the corpus is the relatively large number of names in which the first element is ambiguous or obscure. This category would be even larger if a less charitable view were taken of some of the personal names allowed into the first category.

In so far as conclusions can be drawn from these observations, the analysis suggests that *halh* had a long life as a place-name generic. Early use is suggested by the *-inga-* and *ecles-* compounds and perhaps by the high degree of obscurity in qualifiers, late use by dithematic OE and occasional ON personal names.

Since this article was written Dr Patrick Stiles has published an important study of *halh* (Stiles 1997) based on the topographical information set out in Gelling 1984. He offers strong support for the meaning 'slightly raised ground in marsh', which was suggested for the first time in 1984, calling in evidence the Frisian use of the corresponding word. He also discusses the senses 'recess' and the 'administrative' meaning, setting out a scheme for a suggested relationship between the apparently disparate usages. He accepts 'low-lying land liable to flooding', which is evidenced in the northern dialect descendant *haugh*, but he rejects 'land in a river-bend' on the grounds that 'corner, angle' is not appropriate to the 'original' meaning of the ancestral **halhaz*. Examples

THE LANDSCAPE OF PLACE-NAMES

in which *halh* has been interpreted as 'land in a river-bend' or 'between rivers' could obviously be accommodated under other 'watery' meanings.

Reference section for **halh**

As simplex name: Hale CHE, CMB, HMP, LIN, LNC, SUR, WML, Hales NFK, SHR, STF, WOR, Halam NTT, Halefield (earlier *Hale*) NTP, Hallam DRB, Haulgh LNC, Hawne WOR, Hele SOM (and *freq.* as minor name in DEV). Hales is nominative plural, Hal(l)am and Hawne are dative plural.

As first element with tūn: Haighton LNC, Hallaton LEI, Halloughton NTT, WAR, Halton BUC, CHE, LIN(3), LNC, SHR, YOW(3), Haughton CHE, DRH, LNC, NTB, SHR(4), STF, Holton OXF, SOM, Houghton LNC(2), YOW(2).

With other generics: Halford DEV, SHR, WAR; Hallow WOR (*hagan*, dat. of *haga* 'enclosure'); Halston SHR ('stone').

As generic with personal names: Abenhall GLO (*Abba*); Badenhall STF (*Bada*); Beadnell NTB, Bedale YON and Bednall STF (*Bēda*); Bethnal Green GTL (*Blītha*); Blaxhall SFK (*Blæc*); Blunt's Hall ESX (*Blunt*); Boningale SHR (*Bola*+-*ing*-); Bossall YON (*Bōt*); Bothall NTB (*Bōta*); Bracknell BRK (*Bracca); Breadsall DRB (*Brægd*); Burnsall YOW (*Brȳni*); Campsall YOW (*Cammi*); Cattenhall CHE (*Ceatta*); Chepenhall and Chippenhall SFK (*Ceobba*); Chunal DRB (*Cēola*); Codsall STF (*Cod); Coppenhall CHE, STF (*Coppa); Cossall NTT (*Cott*); Croxall STF (*Crōc*); Dagnall BUC (*Dæcca or *Dægga); Dunkenhalgh LNC (Irish *Duncan*); Ellel LNC (*Ella*); Finghall YON (*Fin*+-*ing*-); Frizinghall YOW (*Frīsa*+-*ing*-, perhaps 'the Frisian' rather than a personal name); Gomersal YOW (*Gūthmǣr*); Hadnall SHR (*Headda*); Hempnall NFK (*Hemma*); Hensall YOW and Henshaw NTB (ON *Hethin*); Hucknall DRB, NTT (*Hucca*); Humshaugh NTB (*Hūn*); Iddinshall CHE (*Ida*+-*ing*-); Ilketshall SFK (ON *Ulfketill*); Kelsale SFK and Kelsall CHE (*Cēol*); Kelshall HRT and Kilsall SHR (*Cylli*); Kenninghall NFK (*Cēna*); Kersall NTT (probably *Cynehere*); Killinghall YON (*Cylla+-*ing*-); Loversall YOW (*Lēofhere*); Luffenhall HRT (*Luffa*); Mattishall NFK (*Mætt); Mildenhall WLT (probably feminine *Milde*); Moddershall STF (*Mōdrēd*); Monsal DRB (*Mōrwine*); Pertenhall BDF (*Pearta); Pethill DEV (*Pydda*); Portnall (earlier *Poddenhale*) SUR (*Podda*); Riccal YOE (*Rīca*); Rushall WLT (*Rust); Sedsall DRB (*Secg*); Shifnal SHR (possibly *Scuffa); Sicklinghall YOW (*Sicel+-*ing*-); Snelshall BUC (*Snell*); Symond's Hall GLO (*Sigemund*); Tattenhall CHE (*Tāta*); Tattershall LIN (*Tāthere*); Thremhall ESX (*Thrymma); Tittleshall NFK (*Tyttel*); Tottenham Court (earlier *Totenhale*) GTL (*Totta*); Tughall NTB (possibly *Tucga); Uggeshall SFK (*Uggeca); Walsall STF (*Walh*); Wappenshall SHR (*Hwætmund*); Whixall SHR (*Hwittuc*); Wiggenhall NFK (*Wicga*); Worsall YON (*Weorc*); Wrentnall SHR (*Wrenta*); Wribbenhall WOR (possibly *Wrybba).

With -inga- *compounds*: Arminghall NFK (possibly 'Ēanmǣr's people'); Edingale STF (possibly 'Ēadwine's people'); Killinghall YOW ('Cylla's people'); Rickinghall SFK ('Rīca's people'); Rippingale LIN (perhaps 'nook of the people

VALLEYS, HOLLOWS AND REMOTE PLACES

from the district called *Hreope*': an OE *Hreope*, which occurs in Ripon YOW and Repton DRB, is considered to be a tribal name which was sometimes used as a district name); Stanninghall NFK (probably 'Stān's people'); Willingale ESX ('Willa's people').

With references to vegetation: Aspall SFK ('aspen'); Benthall SHR ('bent-grass'); Bramhall CHE, Broomhall BRK, CHE and Broomhaugh NTB (all 'broom'); Brundall NFK (? oblique case of an adjective *brōmede* ('broomy'); Clothall HRT ('burdock'); Elmsall YOW ('elm' with collective use of the gen. sing.); Gowdall YOW ('marigold'); Hepple NTB ('hips'); Kersal LNC ('cress'); Maghull LNC ('mayweed'); Meppershall BDF ('maple'); Nuthall NTT; Peasenhall SFK ('peas'); Redenhall NFK ('reed'); Rushall STF; Saughall CHE(2) ('sallow willow'); Stivichall WAR ('tree stump'); Wirrall CHE and Worral YOW ('bog-myrtle'); Wyddial HRT ('willow').

With descriptive terms: Blakenhall CHE ('black'); Chisnall LNC ('gravelly'); Darnall YOW and Darnhall CHE ('hidden'); Dippenhall SUR ('deep'); Greenhaugh NTB; Gressenhall NFK ('gravel'); Langhale NFK and Langho LNC ('long'); Posenhall SHR (possibly 'bag-shaped', from OE *posa*); Pownall CHE ('pouch-shaped', from OE *pohha*); Rivenhall ESX (probably 'rough'); Sandal(l) YOW(2); Stonnal STF ('stone'); Wettenhall CHE ('wet').

With references to wild creatures: Arnold NTT, YOE ('eagle'); Beausale WAR ('gadfly'); Cattal YOW; Crakehall, Crakehill YON (?'raven'); Finchale DRH; Knettishall SFK ('gnat'); Midge Hall LNC and Midgehall WLT; Renhold BDF (probably 'roe deer'); Snaygill YOW ('snake'); Snydale YOW ('snipe'); Spexhall SFK ('woodpecker'); Tivetshall NFK ('lapwing'); Ullenhall WAR ('owl'); Wilsill YOW ('beetle'); Wolf Hall WLT; Worminghall BUC ('reptiles'); Wraxall DOR, SOM, WLT(2) and Wroxall IOW, WAR (a bird of prey, perhaps 'buzzard').

With references to domestic creatures: Chignall ESX (probably 'chicken'); Kidall YOW (possibly 'kid'); Oxenhall GLO; Rossall LNC, SHR ('horse'); Shephall HRT ('sheep'); Ticknall DRB and Tixall STF ('kid', OE *ticcen*); Wetheral CMB ('wether'); Yen Hall CAM (probably 'lamb').

With topographical terms: Beal YOW (*bēag* 'circle', referring to a river-bend); Calow (earlier *Calehale*) DRB (*calu* 'bare', used substantively of a bare hill); Cromhall GLO (*crumbe* 'crook', referring to a river-bend); Houghall DRH (hōh); Howtel NTB (holt); Knoddishall SFK ('knot', used of a hill); Painley YOW (pæth); Pannal YOW ('pan', used of a hollow); Pickhill YON (perhaps the gen.pl. of pīc 'pointed hill'); Strethall ESX (strǣt); Styal CHE (probably stīg).

With river-names and earlier place-names: Dinsdale YON ('*halh* belonging to Deighton'); Edenhall CMB (river Eden); Isel CMB (*Ise*, a pre-English river-name, may have been the name of the tributary which joins the river Derwent at Isel Hall); Leinthall HRE (British river-name *Lente*); Lyonshall HRE (British district-name *Lēon*); Seghill NTB (Ekwall 1960 suggests an OE stream-name *Siga*, 'slow-moving').

With references to crops: Barhaugh NTB ('barley'); Benhall SFK ('beans'); Ellenhall STF (possibly 'flax nook', with ēa prefixed); Ryhall RUT ('rye'); Wadden Hall KNT ('woad').

With references to buildings: Gornal STF ('mill'); Kirkhaugh NTB; Skellow YOW (probably **scela* 'shieling, hut'); Wighill YOW (*wīc*, in an undetermined sense); Worthele DEV (*worthig* 'enclosed settlement').

With references to direction: Northolt GTL; Southall GTL; Westhall SFK.

Chrishall ESX is 'Christ's nook'.

Hassall CHE is 'witch's nook'.

Ludgershall BUC, WLT, Luggershall GLO, Lurgashall SSX and a lost *Lotegareshale* in Arkesden ESX have as qualifier an OE **lūtegār* in the genitive singular. In Tengstrand 1940 it was suggested that *lūtegār* was a term for a trapping spear, and that a 'nook of the trapping spear' contained a device which would shoot when disturbed by an animal.

Eccleshall STF, Eccelsall YOW and Exhall WAR(2) are compounds of *halh* with the Old English loanword from Primitive Welsh (ultimately Latin *ecclesia*), which is believed to have been used by pagan Anglo-Saxons when referring to communities of British Christians. The fourfold occurrence with *halh* is perhaps not beyond the bounds of coincidence for a topographical term. All are in country where *halh* would be appropriate in the sense 'hollow'.

Strensall YON. This recurring compound of *halh* with what appears to be the genitive of *strēon* 'strain, descent' is of particular interest, because *Streoneshalh* was the name of the site of St Hilda's monastery. The group of names was discussed in Gelling 1988 (p. 189) and 1984 (pp. 109–10). There is a review of the problem in Coates 1981, and further suggestions are offered in Styles 1998.

With obscure or ambiguous first elements:

Balsall WAR. A personal name *Bælli* has been inferred from Balscott OXF, Balsdon CNW, Balsham CAM, *Bælles weg* in a charter boundary, and this name: but this would require a double consonant and only the boundary mark and Balsdon CNW clearly have -*ll*-. The others are more likely to have -*l*- (in spite of a single pre-Conquest -*ll*- spelling for Balsham CAM). The etymology is best left open.

Birdsall YOE. The qualifier could be either a personal name *Brid* or the word *bird* used in a collective gen. sing.

Bonsall (earlier *Bunteshale*) DRB. A personal name *Bunt(a)* has been conjectured from a small group of place-names, but a bird-name seems possible, related to ME *bunting*.

VALLEYS, HOLLOWS AND REMOTE PLACES

Brignall YON. The spellings show variation between *Bringen-* and *Brig(g)en-*, with the latter more frequent. A group-name *Brȳningas* has been suggested, but the development to *Brig(g)en-* would be surprising from this base.

Bucknall LIN, STF. The qualifier could equally well be *bucca* 'he-goat' or *Bucca* personal name.

Buxhall SFK. The qualifier could be *bucc* 'male deer' or a personal name of the same form.

Coggeshall ESX. The qualifier could be *cocc* 'cock' with the gen. sing. used collectively, or *cocc* 'heap' (as in *haycock*).

Coltishall NFK. The spellings available (in Ekwall 1960) are not consistent, and fuller material is needed.

Consall STF. The spellings indicate an element **cun*, not otherwise noted.

Courteenhall NTP. A noun *corta* is recorded in a charter boundary, and is probably the generic in Dovercourt ESX. This would suit the spellings for Courteenhall. The meaning, however, is uncertain; there is a discussion in Gelling 1988, pp. 79–80. See also Cullen 1998.

Emanuel Wood ESX (*Munehala, Monehala'* 1086, *Monehale* 1089–93, *Munehale* c.1145, *Monehalis* c.1170, *Manhale* 1254). The spellings suggest an OE *mon* or *mun*.

Etal NTB. This could be 'Ēata's nook', but another possibility is *ete* 'grazing', found in *etelond* 'pasture'.

Ettingshall STF. A personal name *Etting* is a reasonable conjecture, as is a verbal noun from OE *ettan* 'to pasture'.

Fundenhall NFK. The forms require an OE *funda*, but there is no such noun or personal name on record.

Gnosall STF. The forms indicate **Gnoweshalh*, with the exception of *Gnodweshal, Gnodeshall* in 1199.

Halsall LNC. A personal name **Hæle* from *hæle* 'hero' is a plausible suggestion, but the element has only been noted twice, the other instance being Halesworth SFK, and this is hardly sufficient to establish an unrecorded personal name.

Hothersall (earlier *Hudereshal*) LNC. Dodgson (1987) proposes an OE noun **hūder* 'shelter' for this and several other names. A personal name *Hūda* is recorded, however, and **Hūder* could be a derivative of this.

Kneesall NTT. The spellings suggest an OE *cnēs*, which is obscure. Such an element is also required for Kneesworth CAM.

Lyneal SHR. Both the modern form and early spellings such as *Lunyhal* 1221–2, *Linial* 1326 suggest an OE adjective with *-ig*. This is a type frequently formed from plant-names, but no recorded plant-name suits.

Markshall NFK. This name appears to share with Marksbury SOM, Maesbury SOM and Masbrough YOW a first element *Merces-*. *mearc* 'boundary' does not suit as it does not have an *-es* genitive. (NB Markshall ESX is a different name, not containing *halh*).

Mepal CAM. A personal name *Mēapa* has been conjectured to explain this name and Meopham KNT.

Mildenhall SFK. Despite the modern form this is not the same name as Mildenhall WLT. The earliest spelling, in a writ of *c*.1050, is *Mildenhale*, but this is followed by a series of forms which led Ekwall 1960 to suggest that the name was *middelan hale* 'middle nook'. It is difficult to see what 'middle' could mean here, and the etymology is best left open.

Ordsall LNC, NTT. A personal name *Ord* is a reasonable conjecture, since *Ord-* is found in compound names; but *ord* 'point' is a well-evidenced place-name element.

Pelsall STF. A personal name *Peola* has been inferred from this name and a charter boundary-mark *Pioles clifan*.

Rednal WOR. The name is *Wreodenhale* in three Anglo-Saxon charters. An adjective connected with *wridan, wridian* 'to grow, flourish' might be considered.

Runhall NFK. *hruna* 'fallen tree' is perhaps the likeliest of several possibilities.

Rushall NFK, earlier *Riveshale*. The only recorded OE word which suits is *hrif* (womb, belly). This could have been used in a transferred sense, as the synonymous *wamb* certainly was, but there is no other instance in place-names. Ekwall 1960 suggests as an alternative a personal name from *hrife* 'fierce'.

Ruthall SHR, earlier *Rothal, Routhale*. Derivation from *rōt* 'root' was tentatively suggested in EPNS SHR survey 1, p. 255. Hough 1996 subsequently advocated this as the first element of a number of names previously ascribed to *rōt* 'cheerful'. Ruthall is a particularly strong candidate for the 'root' etymology as the spellings show no sign of a medial *-e-* which would justify derivation from an adjective or from a personal name *Rōta*.

Somersal DRB. The qualifier could equally well be 'summer' or a personal name derived from it.

Taxal CHE. The spellings require an OE *tacc* or ME *tak*. Dodgson (EPNS CHE survey 1, 172–3) suggests that it is ME *tak* 'lease, tenure, revenue'. The name is not recorded until *c*.1251, so it could be of ME origin, but the word *tak* is not otherwise evidenced in place-names, and a genitival compound with the ME reflex of *halh* seems unconvincing.

Tessal WOR. The element *tess* is completely obscure.

VALLEYS, HOLLOWS AND REMOTE PLACES

Tettenhall STF. A personal name *Teot(t)a* has been inferred from this name and Teddington WOR.

Wainhill OXF, **Willenhall** STF, WAR, **Winnal** HRE. The qualifier could be the personal name *Willa* or an adjective *wilgen* 'growing with willows'.

Yoxall STF. *geoc* 'yoke' occurs in several place-names, but the significance is obscure.

hop 'remote enclosed place'. The origin of this word has not been established. The only certain occurrence in literary Old English is in *Beowulf*, where the monsters' lairs are called *fenhopu* and *morhopu*, 'marsh retreats'. There is, however, abundant evidence for a range of meanings in place-name usage.

Previous discussions (Smith 1956 and NED under *hope* sb²) give confused accounts of *hop*, failing to distinguish it from ON *hóp* 'a small land-locked bay or inlet'. The ON word is not certainly evidenced in pre-Conquest names in England, though it passed into ME and is seen in a few coastal names, including Stanford le Hope ESX, Middle Hope SOM and Hope (by Bolt Head) DEV. In earlier usage it may occur in a few coastal names in PEM, and it is certainly found in Scotland and in the names of harbour villages in Orkney. There is no reason to suppose that the ON word had any influence on the senses of OE *hop*. Most of the major names in which the OE word occurs must be supposed to date from before the time of the Scandinavian settlements in England, and Smith's discussion is seriously misleading in its suggestion that names containing *hop* in the sense 'small enclosed valley' may not have been coined before the Middle English period.

With very few exceptions OE names in *hop* fall easily into one of two main categories, 'enclosure in marsh or wasteland' and 'remote valley'. The sense of concealment is probably the link between the two.

In the Beowulfian sense 'piece of enclosed land in marsh' *hop* is clearly evidenced in Kent (as in Hope All Saints in Romney Marsh) and in Essex, where the EPNS survey notes that it is common in field-names in the coastal marshes. NED gives a quotation from *c*.1200 concerning *unam hopam marisci* in Essex. There are a few minor names in both these counties for which 'enclosure in waste' is more appropriate than 'enclosure in marsh': these include Hope House in Little Bursted ESX and Hope Fm in Hawkinge KNT. It is possible that (like **hamm** and **holmr**) *hop* developed a meaning 'promontory jutting into marsh'. This would be appropriate for Hopton on the north boundary of Suffolk, and it seems possible also for Hopton north of Lowestoft. Hope Fm in Beckley SSX is on a promontory jutting into the Rother Levels, and Hope Fm in Rudgwick SSX is on a promontory by the river Arun.

The senses 'enclosure in marsh or moor' are also found in the north of England. Hope Green, south of Stockport CHE, looks like 'enclosure in marsh'

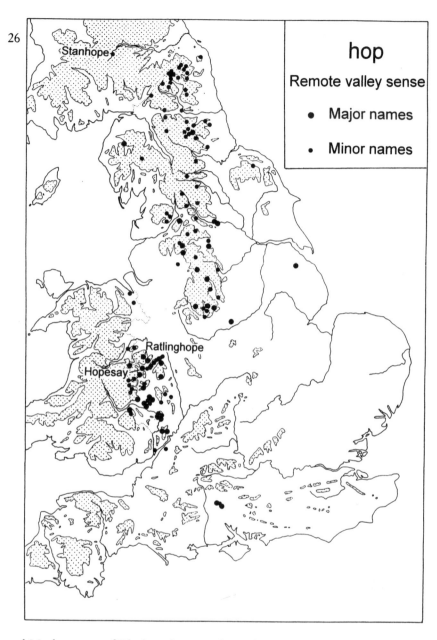

and Mythop, east of Blackpool LNC, and Meathop, north-east of Grange over Sands WML, are very good instances of this meaning. North-eastern examples are less certain, but Ryhope and Cassop DRH may refer to enclosures in moor. Ryhope, on the coast, is at the mouth of a very small valley, but this seems unlikely to be the referrent. Hoppen NTB, south of Bamburgh, seems much more likely to be 'at the enclosures' than 'at the valleys'.

VALLEYS, HOLLOWS AND REMOTE PLACES

In the sense 'remote valley' *hop* is characteristic of three regions: the central part of the Welsh Marches, the edges of the Pennine Chain, and the mountainous areas of Durham, Northumberland and southern Scotland. In Northumberland it was still used in this sense in 1542, when a Border Survey reported that the fertile valleys which lay among the boggy hills of the Cheviots could not be cultivated because of the thieves of Teviotdale. No-one would live there because 'the said valyes or hoopes of Kydland lyeth so dystante and devyded by mounteynes one from an other that suche as Inhabyte in one of those hoopes, valyes, or graynes can not heare the Fraye, outecrye, or exclamac'on of suche as dwell in an other hoope or valley upon the other syde of the said mountayne, nor come or assemble to theyr assystance in tyme of necessytie'. This reference was supplied by Mr V. Watts, who points out that the region in question, north-west of Rothbury, has a great many -hope names. These are in small print on the O.S. map, some referring to single farms, others to uninhabited valleys. In less rugged country further south the seclusion of *hop* sites was not a disincentive to settlement, though it obviously placed restrictions on growth. It was probably possible to think of a *hop* as a secure place provided the surrounding countryside was not too mountainous and the conditions of the time were not too lawless.

The use of *hop* for a valley does not depend on the criterion of shape. Valleys so designated are often *cumb*-shaped, but they exhibit considerable variety. The most commonly shared characteristic may be a constricted entrance. Fig. 26 shows a distribution pattern which suggests that the Anglo-Saxons felt the need of a special word when they encountered country where valley settlement-sites were particularly remote.

South of the river Thames the meaning 'valley' has only been noted in one name, Wallop HMP. Nether, Over and Middle Wallop are in the upper part of the valley of Wallop Brook, and if the first element is a form of **well** the reference is probably to the river-source at Over Wallop. Attention was drawn in Gelling 1984 to the anomaly of the use of the west-midland form **wælle**, and of the choice of the word *hop* for a valley in this area, and this remains unexplained. The compound occurs three times; there is another Wallop in Shropshire, west of Pontesbury, and Wallhope Fm in Tidenham GLO has the same origin. The three-fold occurrence probably tells against a pre-English origin for the Hampshire name, though it has been regarded as the place called *guoloppvm, id est cat guoloph*, named in a 10th-century Welsh text as the site of a 5th-century battle between two British leaders. *hop* is not common in any sense in southern England. It is not evidenced at all in Berkshire or Wiltshire, or in Devon (provided that Hope near Bolt head is ascribed to the ME reflex of ON *hóp*).

There are a few minor names in Gloucestershire (Cannop Brook, Hope Fm in Falfield and Wallhope Fm in Tidenham) in which *hop* means 'valley', but the common use of this term in major settlement-names starts on the

Gloucestershire/Herefordshire border, where Longhope and Hope Mansel are to be found, and the first concentration of such names lies a few miles to the north, where Sollers Hope, Fownhope, Littlehope and Woolhope are associated with the feature known to geographers as the Woolhope Dome. There is another group north-west of Hereford, where Hope under Dinmore, Gattertop, Westhope, Lawton's Hope and Burghope surround another massif. Brinsop and Yarsop are in valleys to the south-west of this, and there are a few examples north-west and north-east of Hereford, such as Cusop on the west boundary of the county, Hopton near Stoke Lacy, and Hopleys Green near Almeley (*Hope* in DB). Covenhope and Brinshope Fm occur in the hilly land near Wigmore, and Miles Hope is north-east of Leominster.

The densest concentration of settlement-names in which *hop* means 'valley' is probably in Shropshire, south of the river Severn. One of the most striking features of the toponymy here is the series of *hop* names between the parallel ridges of Wenlock Edge and the Aymestrey limestone escarpment, which extends for 14 miles south-west of Much Wenlock. Part of the valley between them is known as Hope Dale, and the settlements are Presthope, Easthope, Wilderhope, Millichope, Middlehope, Westhope and Dinchope. All but the first lie in funnel-shaped hollows where streams rise which run through the Aymestrey limestone to the river Corve, except for that at Dinchope which runs west through Wenlock Edge. Presthope, at the north end of the series, is not in its valley now, but it overlooks an identical feature.

In the hilly country to the east and west of Wenlock Edge *hop* is used for valleys of varied shape and size, with varying degrees of seclusion, the most extreme example being Ratlinghope, in a long, deep valley on the western side of the Long Mynd. There are also *hop* names in the hills which run up to the Severn on the Shropshire/Montgomeryshire border, a fine example being Hope near Minsterley, at the lowest point of a v-shaped valley.

There is a series of *hop* names in Wales, in a narrow belt close to the west boundaries of Herefordshire, Shropshire and Cheshire. In some of these – Burlingjobb, Evenjobb and Cascob (all RAD) – the modern form has been affected by Welsh speech. For most of them the 'valley' sense is probably the relevant one, but the sites are not as secluded as the Shropshire and Herefordshire ones. Hope north of Wrexham is, however, hidden from the road if approached from the southeast.

'Remote valley' names in the Pennine group include a concentration north of Manchester (Bacup, Cowpe, Hope in Eccles, Hopwood), several examples in narrow valleys running in from the western edge, and a line (Bradnop, Hope, Stanshope STF, Alsop, Hopton DRB) along the southern edge of the chain. The word is used in this sense in CMB, WML, NTB, DRH and Yorkshire, and in southern Scotland.

One isolated *hop* name in Yorkshire deserves comment. This is Fryup YON, near the head of a valley which runs off Esk Dale. *hop* has been explained as referring to this large side-valley. There are actually two valleys, Great and Little Fryup Dale, but they are broad features and do not deserve the term *hop*. The reference will be rather to the short, narrow valley which links the two Dales at their southern end. Fryup Hall stands at the opening of this in Great Fryup Dale. It is a very secluded spot, and additional interest is lent by the probability that the qualifier is a reference to the goddess *Frīg*. Pagan religious practices could have lingered here for a long time.

It will be apparent that in its 'remote valley' sense *hop* is a particularly rewarding subject for field-work, and there is a temptation to continue detailing particular instances; but this must be resisted if the discussion is to be kept in proportion. *hop* is less liable than many words to distortion in modern forms, and interested readers can pursue -hop(e) names and the simplex Hope with confidence. Some further observations on the distribution may be offered here, and some comments on the analysis of qualifiers in the reference section.

It is noted above that *hop* is rare or absent in counties south of the river Thames. It is similarly rare in the south-east and east midlands, and in East Anglia. Middlesex (including part of GTL) has Hope House in Edmonton and a few field-names, but in a number of counties for which there are EPNS surveys – BUC, OXF, HRT, BDF, HNT and CAM – *hop* has not been noted at all. There is as yet no survey for LEI, but *hop* certainly does not occur in major names in that county, and there is no sign of it in the recent very detailed survey of RUT. There are two major names in LIN, Swinhope, south-west of Grimsby, probably a 'valley' site, and Skellingthorpe (earlier *Scheldinghop*) west of Lincoln, an 'enclosure in marsh'. The two instances in NTT, Warsop and Worksop, are very difficult to explain. Both are in major river-valleys, and the sites do not give any impression of remoteness or seclusion.

hop is commonly used as a simplex name, and is frequent in the compound Hopton. When it is the generic in compound names the qualifiers are fairly evenly spread over a number of categories. Personal names are well represented, and there is a category in which *hop* is combined with a personal name plus either the singular connective -*ing*- or the genitive plural of -*ingas*. These names form a fairly compact geographical group in the Welsh Marches, and they should be considered together. They have been listed in the reference section with a heading which reflects the author's predilection for -*ing*- rather than -*inga*-. It is difficult to envisage six -*ingas* groups this far west, in regions where -*sǣte* was the normal way of forming group names. Apart from the WOR outlier Easinghope they are characterised either by assibilation (producing the -ch- of Dinchope, Millichope and the old form, *Ratchop*, of Ratlinghope) or palatalisation (producing the -j- of Burlingjobb and Evenjobb, and evidenced in the modern pronunciation of Bullingham). They may be a roughly contemporary group of names resulting from land-grants by the rulers of the short-lived kingdom of the Magonsæte. The phonological

27

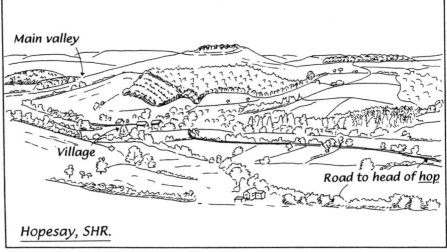

Ratlinghope, SHR.

Hopesay, SHR.

Stanhope, PEB.

VALLEYS, HOLLOWS AND REMOTE PLACES

problems which they present have not been resolved and their origin must be regarded as an open question; but they should certainly be considered as a discrete group. There is a fuller discussion in Gelling 1982.

Some minor names have been included in the analysis, as study of this element would be incomplete if confined to major names.

Reference section for hop

As a simplex name: Hope CHE, DEV, DRB, HRE(3), KNT, LNC, SHR(3), YON, Hoppen (dat.pl.) NTB. Some Hopes have manorial affixes (Hope Mansel, Sollers Hope HRE, Hope Bagot, Hope Bowdler, Hopesay SHR), but Hope under Dinmore HRE and Hope All Saints KNT have other distinguishing additions.

Three HRE names, Brinsop, Woolhope and Fownhope, were *Hope* in DB, as was Longhope GLO, but these were turned into compounds shortly after the Conquest by the prefixing of personal names (*Brȳni*, and fem. *Wulfgifu*) or adjectives ('multi-coloured', OE *fāgan*, and 'long').

As first element with tūn: Hopton DRB, HRE, MTG, SHR(6), SFK(2), STF, YOW.

With other generics: Hopwas STF (**wæsse**); Hopwell DRB; Hopwood LNC, WOR.

As generic with personal names: Alsop DRB (*Ælle*); Eccup YOW (*Ecca*); Edenhope SHR (*Ēada*); Glossop DRB (**Glott*); Warsop NTT (**Wǣr*); Wilderhope SHR (fem. *Wilthrȳth*); Yarsop HRE (*Ēadrēd*).

With personal names + -ing: Bullinghope HRE (*Bulla*); Burlingjobb RAD (*Berhtel*); Dinchope SHR (*Dodda*); Easinghope Fm (in Doddenham) WOR (*Ēsi*); Evenjobb RAD (*Emma*); Millichope SHR (?**Milla*); Ratlinghope SHR (**Rōtel*); Skellingthorpe LIN (*Scheldinchope* DB, *Scheldinghop* 1141, *Sceld*).

With references to vegetation: Ashop DRB; Bramhope YOW, Broomhope NTB; Kershope CMB ('cress'); Wythop CMB ('willow').

With topographical terms: Bacup LNC (**bæc**); Burghope HRE ('prehistoric fort'); Chesterhope NTB ('Roman fort'); Wallhope GLO, Wallop HMP, SHR ('spring').

With descriptive terms: Black Blakehop NTB and Blacup YOW ('black'); Bradnop STF ('broad'); Ryhope DRH ('rough'); Staindrop DRH (probably *stǣner* 'stony ground'); Stanhope DRB ('stone'); Sweethope NTB; Widdop YOW ('wide').

With domestic animals: Cowpe LNC ('cow'); Oxenhope YOW and Oxnop YON; Swinhope LIN.

With wild creatures: Emblehope NTB ('caterpillars'); Harehope NTB, Harrop (several in CHE and YOW); Hartsop WML; Rookhope DRB.

With river-names: Cusop HRE (British *cyw*, identical with Chew SOM); Dunshop YOW (British, 'black stream').

With references to direction: Easthope, Middlehope, Westhope (all in Hope Dale SHR).

With references to distinctive individuals: Fryup YON (the goddess *Frīg*); Hassop DRB ('witch'); Lurkenhope SHR ('wise woman'); Presthope SHR ('priests').

With obscure or ambiguous qualifiers:

Blenkinsopp NTB. Early forms are *Blencheneshopa* 1178, *Blencanhishop* 1236, *Blenkeneshop* 1256.

Brinsop LNC. There are two instances of this name in LNC. Early spellings suggest the personal name *Brȳni*, which was added in post-Conquest times to Brinsop HRE, but, as Ekwall (1922, p. 43) remarks, it would be a remarkable coincidence to find two examples of 'Brȳni's valley' in LNC. OE *bryne* 'burning' might be considered.

Cassop DRH. The available forms are not sufficient to support Ekwall's suggestion of OE *Catteshop* 'wild cat's valley'. Some spellings with -*t*- would be expected from this base.

Covenhope HRE. The spellings are hard to reconcile with each other. The name is *Camehop* DB, *Comenhop(e)* 1292, 1316, *Kovenhop* 1242.

Gattertop HRE. Here again there are two different types of spellings, one having *Gadred-*, *Gatred-*, the other *Gaterilde-*, *Gaderilde-*. The forms are set out in Coplestone Crow 1989.

Meathop WML, Mythop LNC, Middop YOW, Midhope YOW. These four identical names were discussed in Gelling 1984, p. 117. The translation given in Ekwall 1960 is 'middle *hop*', but this is not satisfactory because OE *mid* is not an adjective. V. Watts, reviewing Gelling 1984 in *Nomina* 9, pointed out that there is an OE adjective *midd*; but the dictionary examples given for this word suggest that if it were used in place-name formation it would occur in the superlative, *midemest* (which is occasionally found in early field-names). It seems necessary to accept the suggestion in Smith 1956 that these names contain the preposition *mid*, used elliptically to give the meaning '(land) among *hops*'. Other names cited by Smith as having this formation are Mid(d)ridge DRH, SOM, Midhurst SSX and Midsyke YON. The fourfold occurrence with *hop* is probably not outside the bounds of coincidence. The names are not topographically homogeneous, Meathop and Mythop being notable 'enclosure in marsh' names, while Middop and Midhope are in valley country.

Stanshope STF. Either '*Stān*'s valley' or 'valley of the stone'.

Worksop NTT. Either '*Weorc*'s remote place' or 'remote place with a building construction' (OE *geweorc*, modern *work*).

nant, nans Cornish, **nant** Welsh, 'valley'. This word is commonly found as *nance* in west Cornwall and as *nant* (the earlier form) in east Cornwall. It is often corrupted to *lan* or *lam* in Cornish names, leading to frequent confusion with *lan* 'sacred enclosure'. Trenance and Trenant mean 'valley farm' and Pennance and Pennant mean '(place at) the end of the valley'. Nanplough in Cury CNW has a doublet in Lamplugh CMB; both (as pointed out in Padel 1981) derive from *nant bluch* 'bare valley'. Another recurrent name is 'pleasant valley' from Welsh *hawdd* and *nant*. This has become Hoddnant in Wales, where there are several instances. In Cornwall the corresponding name has become Hennett, Henon and Huthnance. This compound is believed to be the origin of Hodnet SHR.

slæd 'valley'. It is difficult to give a meaningful translation for the OE ancestor of modern *slade*. The modern word has a variety of dialect meanings, but these are not certain to have belonged to the OE and ME forms. The soundest evidence for the meaning of the OE word is probably that of charter boundaries, where it is fairly common. Kitson (forthcoming) says that it denotes 'flat-bottomed, especially wet-bottomed valleys'. The watery connotation is reflected in the qualifiers of *slæd* names in charter bounds, 'rush' and 'withy' being the commonest. In Berkshire boundaries, where some dozen instances occur, there are a *morsled*, two instances of *riscslæd* and two of *wæterslæd*. Waterslade was noted 10 times in Cambridgeshire field-names. This aspect lies behind some modern dialect senses of *slade*. In Shropshire field-names, for instance, Slade is used for 'a patch of ground in a ploughed field too wet for grain and left for greensward' (Foxall 1980, p. 18). On the other hand *slæd* is sometimes used for dry valleys in charter boundaries and in surviving names.

The valleys for which *slæd* is used are probably never main valleys. Some of them are certainly very small, which does not accord with the occurrence in King Alfred's translation of Orosius which describes the decoying of King Cirus into a *micel slæd* where he and 2,000 men were slaughtered.

There are only a few well-documented settlement names containing *slæd*, and most of the settlements are very small. At Bagslate Moor ('badger slade'), west of Rochdale LNC, a reservoir obscures the topography, but the area is clearly wet and there is no main valley. At Castlett Fm, south of Temple Guiting GLO (earlier *Catteslade*), there is a deep, steep-sided little valley, with a flat, wet bottom. Chapmanslade WLT ('merchant's slade'), north-west of Warminster, is a linear settlement along a ridgeway; presumably the name once referred to one of the small valleys on the edge of the ridge. Weetslade NTB ('withy slade') is the name of two small settlements north of Newcastle. High Weetslade overlooks a tiny valley. Walderslade KNT, near Chatham, contains **weald**. The village is at the head of a long, narrow valley, and it is possible that here (as in some charter boundary marks) *slæd* is used of part of a valley. In Sleddale WML(2) and Sledmere YOW, *slæd* is used as a qualifier, probably referring to short side-valleys.

Slad(e) is a frequent minor name and field-name. Slad Valley between Stroud and Painswick GLO, has been made famous by Laurie Lee's *Cider with Rosie*. The name has not been traced back beyond 1779, however, so it is not certain that it embodies an OE use of *slæd*. The area has other examples: Stroud Slad Fm, Througham Slad Fm, and Slade House near Stroud. The term is not appropriate to the main valleys here, but it may be used of the gullies which run into them. A picture of Laurie Lee's valley published in *The Times* recently shows these gullies clearly.

slakki ON 'shallow valley' is fairly common in minor names and field-names in some areas of Scandinavian settlement, and occurs in a few settlement-names, including Hazelslack and Witherslack WML. Gelling 1984 suggested that *slakki* was more likely than **lacu** to be the generic in Elslack YOW. **lacu** certainly does not suit the mountain stream at Elslack, but the first element presents a problem if the name is assumed to contain *slakki*. There is no obvious ON qualifier which would fit the Elslack spellings.

trog 'trough' is occasionally used to denote valleys. Simplex names are Trough LNC, YON, YOW and Trow WLT. As first element *trog* occurs in Trawden LNC, Troughfoot CMB and Troughton LNC. In Trafford CHE, where the 'valley' sense is not appropriate, the reference may be to a literal trough, or to a characteristic of the ford or its approach roads.

CHAPTER 5

Hills, Slopes and Ridges

One of the new observations which was made in the course of collecting and arranging material for my 1984 book was that the largest category of topographical settlement-names was the one studied in this chapter: that in which the primary reference is to a raised landscape feature. Concavity and convexity are both very well represented in the Old English topographical vocabulary, with convexity having a substantial lead, and the consequence for the present book is that Chapter 5 is much the longest of the seven into which the material is divided. In fact Chapter 5 could be longer than it is, as a dozen or so terms which only occur in a few settlement-names have been omitted.

This category of generics supplies a few items to the tiny handful of words borrowed by Old English speakers from the British language. It is not always possible to distinguish firmly in place-names between adoptions of pre-English names and use of borrowed words, but the relatively high frequency of the three British words **crūg, mönith** and **penn** supports the latter interpretation. Eight of the words discussed here are British, seven are Old Norse and 28 are Old English. The Norse words include **berg, hlíth, nes** and **skjalf** which are not clearly differentiated from their English equivalents.

The hills and ridges referred to in settlement-names may be either the sites of settlements or adjacent features which serve as visual identifiers. In the first category the most important term is **dūn**, and it is hoped that the discussion of this word here and in the earlier book will elevate it from a lowly status as a simple topographical label to that of a quasi-habitative term of considerable significance. To labour this would be to repeat what is said in the **dūn** article, but the point is sufficiently important to deserve notice in more than one place.

The Old English topographical vocabulary is at its most discriminating in the classification of hills and ridges, and this aspect of the study can afford great pleasure to the informed observer. Here, most of all, we are seeing what the earliest Anglo-Saxon settlers saw. This claim sometimes provokes the objection that I am not allowing for greater tree-cover, but in fact it is doubtful that tree-cover was significantly greater. The damp oak forests of which historians spoke so frequently in the first half of this century have been largely demolished by landscape historians in the second half. And in any case, when hills and ridges are tree-covered, this may emphasise rather than obscure their shape, as can be seen from the sketches of View Edge SHR, and the hill-spur at Middle Tysoe WAR, on Fig. 36.

The shapes of the ridges called **hōh** and of those called **ofer** and **ōra**, and the angle of the slopes at those called **clif** and at those called **helde**, are so clearly and consistently differentiated that the application of these terms must be regarded as systematic. It is terms of the sort discussed in this chapter for which a function as travellers' guides can most convincingly be argued.

bæc 'back'. This anatomical term is used in place-names of ridges, varying in type from a low ridge in marshy ground to much more dramatic features. In Berkshire charter boundaries it is used several times for the banks of linear earthworks and once for a strip linchet. In major place-names *bæc* is best evidenced in the northern half of the country, but it also occurs in the southwest, and in Gelling 1984 it was suggested for the first time that it occurs in East Anglia. It is not common and is not evidenced in all counties.

Bashall YOW, earlier *Bacshelf*, refers to a narrow tongue of land between two brooks extending for half a mile from Backridge (also named from it) to the large farm called Bashall Town. This farm is situated on the **scelf** formed by a broadening of the ridge at its west end. The elevation is less than 100ft but it is a striking landscape feature. Backford CHE lies on a low hill which is the southern end of a broad ridge. Beckhampton WLT is probably named from the long, narrow ridge west of the village. This is a higher feature than those at Bashall and Backford, and high ridges are also referred to in Bacup LNC and Backwell SOM.

The meaning of this element has to be established from names in which it is the first element, because as a final element it is difficult to distinguish on formal grounds from **bæc, bece** 'stream-valley'. There are some names in which the topography is the deciding factor. As a first element *bæc*- gives *Bac*- in early spellings, but as a final element the dative inflection (giving *bæce*) could result in ME spellings like *-bache* and modern forms like *-bache, -bage*. There may also have been an early locative form *bæci*, in which the vowel was raised giving OE *bece*. This last form would produce ME and modern spellings indistinguishable from those derived from **bæc, bece** 'stream-valley'. In the three instances of Burbage LEI, WLT, DRB, *bæc* in the sense 'ridge' seems more probable than the 'stream' word. These names were discussed in Gelling 1984, where it was also suggested that Debach SFK is most likely to mean 'ridge overlooking the river Deben'.

In Gelling 1984 a case was also presented for considering the locative case of *bæc* 'ridge' to be the generic in some fenland names. These include Landbeach, Waterbeach and Wisbech CAM, and Pinchbeck and Holbeach LIN. This still appears a reasonable hypothesis, since slightly raised ground would be the most desirable factor in choice of settlement-sites in the areas where these places are. The article was, however, mistaken in asserting that **bæc**,

HILLS, SLOPES AND RIDGES

bece 'stream-valley' does not occur in the fens. As explained under that element in Chapter 1, there is a charter dated 1022 which refers to stretches of water near Whittlesey Mere CAM with names in -*bece*. These can only be flat watercourses, and this would seem to be a late, specialised use of the 'stream' term in an area where there are no valleys of the sort for which this term is used elsewhere. It seems unlikely that such a usage lies behind the major settlement-names listed above.

The first element of Burbage is *burh* 'fort'. At Burbage DRB there is an ancient fort called Carl Wark; *burh* could refer to this, but another possibility is that the long outcrop called Burbage Rock had a fortress-like appearance. At Burbage WLT the village is a short distance away from Easton Hill, a ridge which appears from charter boundaries to be the *burgbec*; the fort is probably Godsbury. Burbage LEI has a hill-top site, but no fort is shown on maps.

Landbeach and Waterbeach are adjacent parishes north of Cambridge. Both were originally called *Beche*. Land- and Water- appear in records in 1218 and 1235 respectively. Waterbeach is the nearer of the two to the edge of firm ground. For the qualifier of Wisbech CAM *wisse 'marshy meadow' seems a likelier candidate than a river-name. Pinchbeck LIN probably refers to finches. Holbeach LIN was interpreted as 'hollow brook' in reference books earlier than Gelling 1984, but as pointed out there it is difficult to relate that etymology to the topography of this area bordering the Wash. 'Slightly raised ground in a hollow place' would give better sense.

barr PrW 'top', **barrǫg** PrW 'hilly'. *barr* occurs in Barr STF and (with the additon of ON **ey** 'island') in Barrow in Furness LNC. Berkshire is named from a wood called *Bearruc*, which is the OE form of a Welsh name derived from *barrǫg*, and this term occurs also in Barrock Fell CMB. Gaelic and Welsh names from this root occur in Scottish names such as Barra and Barrhead, and in Welsh names such as Barry. Padel 1985 lists some related Cornish names.

beorg, Anglian **berg**, 'rounded hill, tumulus'. Since 1984, when a brief account of this word appeared in *Place-Names in the Landscape*, field-work has led to the more precise definition given here. It cannot be claimed that every *beorg* name in the country has been matched to its visual setting, but this has been done in a sufficiently large number of instances for it to be asserted confidently that the defining characteristic of a *beorg* is a continuously rounded profile. This probably explains the use of the word for tumuli in the southern half of England, which led to the adoption of *barrow* as a technical term by archaeologists. Many tumuli would be *beorg*-shaped.

As so often with place-name elements, this specialised use is not to be explained by reference to ultimate etymology. Kitson (forthcoming) describes it as the inherited Germanic word for a mountain, but in settlement-names it usually refers to small hills, sometimes to glacial drumlins. Such hills are

28a

Granborough, BUC. — Rounded profile

Warborough, OXF. — Village beside _beorg_

Bromsberrow, GLO. — Church; Pronounced, rounded hill

Edlesborough, BUC. — Church occupies summit of _beorg_

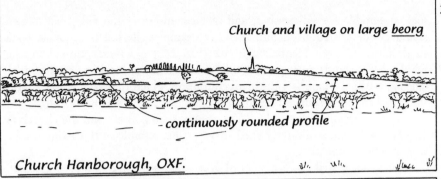

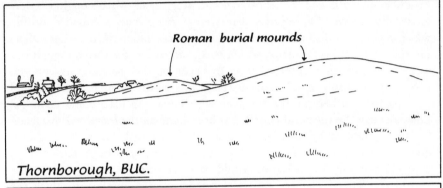

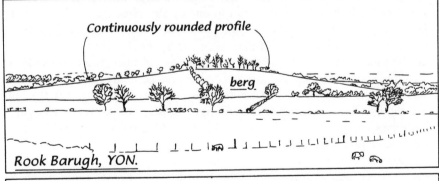

seldom suitable for village-sites, and in most instances the village so named stands beside the rounded hill. In other examples the hill is occupied by a single farm. Church Hanborough OXF (Fig. 28b) is an exception: here the whole village stands on a spacious hill which has the appropriate continously-curving outline. There is a distinctive group in which the *beorg* is not a free-standing hill, but a rounded knob on the end of a ridge. At Flawborough NTT, for instance, the church and Hall Fm occupy such a knob, and the church at Farnborough WAR is similarly sited. A very fine example of this usage can be seen at Bidborough KNT.

Particularly striking instances of *beorg* names include Warborough OXF, Bromsberrow GLO and Edlesborough BUC (Fig. 28a). Warborough is overlooked by Town Hill, a larger-than-average *beorg* with a rounded profile which makes it a distinctive feature from vantage points in the region. The summit of the hill is farmland, which makes it possible for the configurations to be seen, and when observed from this position the ground is never level. Bromsberrow stands beside a small round hill, relatively high, which would be a perfect marker for the traveller who knew the place-name code. At Edlesborough the medieval church has been built on the *beorg* and the shape has been modified by terracing to make the graveyard, but when a photograph of Bromsberrow is placed beside one of Edlesborough church it is apparent that the Anglo-Saxons would have seen closely similar hills at the two places. These instances are selected from a large number in which the sense 'perfectly rounded hill' is borne out by the evidence on the ground. Inkberrow WOR is OE *Intanbeorgas*, plural, and the village stands among several low, curved hills, one of them occupied by the church.

The use of *beorg* to define the rounded shapes of some hills called **crūg** is discussed under **crūg**. Crookbarrow WOR has been generally held to be hybrid Welsh/English and tautologous, with both elements indicating that the hill was thought to be man-made. It seems more satisfactory, however, to regard both the elements in Crookbarrow and in Creechbarrow SOM as references to the visual aspects of the hills.

beorg is used for tumuli in charter boundaries and in some settlement-names. Near Thornborough BUC (Fig. 28b) there are Roman burial mounds which are miniature versions of the hill at Bromsberrow GLO. In a WLT charter boundary *Wodnesbeorg*, 'tumulus sacred to the god Woden', is applied to a Neolithic long barrow near the Wansdyke now called Adam's Grave. Other instances in which *beorg* refers to tumuli were listed in Gelling 1978, 132–3, and to these should be added some others, such as Broughton HMP, overlooked by hill-top tumuli, and Barrowden RUT, where there are burial mounds on a ridge overlooking the river Welland. This use of *beorg* probably does not occur much farther north than Barrowden, however.

Woodnesborough KNT, etymologically identical with the WLT barrow-name, has been assumed to refer to a tumulus formerly to be seen near the church, but in fact the church occupies a low, rounded hill. Statenborough, a short distance south of Woodnesborough, is overlooked by a closely comparable feature, so it is probable that in these two names *beorg* has its non-archaeological sense.

It is not possible on formal grounds to distinguish between the Anglian form of *beorg* and the corresponding ON **berg**. A small number of names which have ON qualifiers have here been assigned to the latter, but the incidence of the ON word may have been higher than this. It seems probable, judging from the siting of some minor names in northern counties, that ON **berg** was used in the same sense as the English word. The topography of Roseberry Topping, discussed under **berg**, is particularly interesting in this connection.

There are about 100 names in Ekwall 1960 which contain *beorg*. It is probably much more frequent in minor names, and this means that only a rough estimate of the distribution pattern can be offered here. Somerset has the highest number, 14, of major names, Warwickshire and Buckinghamshire have eight, Devon has seven, Dorset five. The element is well-represented in the northwest, but most of the examples are minor names, so not listed here. It is probable that no special significance attaches to the distribution, the word being likely to occur anywhere if the topography rendered the appropriate type of hill a useful marker in a settlement-name. It is very frequent in minor names in WML, some of them, like Brownber and Bowber in Ravenstonedale, referring to hills with the classic *beorg* shape. A number of rounded hills which rise from the Somerset Levels are called Burrow or Barrow.

beorg has a variety of modern reflexes, with -borough being very much the commonest. Others are -berrow, -burgh, -barrow, -beare, -berry, -bury, -ber. The frequency of -borough and the occasional development to -bury and -burgh make it necessary to exercise care in distinguishing *beorg* names from those containing *burh*, dat. *byrig*, 'fort'.

A noteworthy modern development occurs in Yorkshire, where either Anglian *berg* or the ON word gives Barf, sometimes spelt Barugh. Northwest of Malton YON there are several settlements called Barugh, each associated with a rounded hill (Fig. 28b). Also in the North Riding, though some distance away, is Langbaurgh near Great Ayton, where the elongated *berg* is a notable landscape feature. *berg* does not always have this development in Yorkshire, however: the abbey of Rievaulx is overlooked by Ashberry Hill (*Escheberch* c.1140), a very large specimen of a rounded *berg*, and -ber is very common in minor names in the West Riding.

Analysis of qualifiers in compound *beorg* names reveals an uneven distribution among categories. Descriptive terms are commonest, with some, like 'green', 'rough' and 'stone', recurring. Next commonest are references to vegetation, and here also there are recurring items, 'fern' being the most

THE LANDSCAPE OF PLACE-NAMES

frequent. There is the usual uncertainty about the number of compounds with personal names, but there are probably 10 or 11, three of the names being dithematic. Wild creatures are an equally large group, but references to domestic creatures and to crops are rare. There are no compounds with qualifiers which are totally obscure, but a few are ambiguous.

Reference section for **beorg, berg**

As simplex names: Barrow RUT, North and South Barrow SOM, Barugh YON, YOW, Berrow SOM, WOR, Burgh SUR, Southburgh (*Berc* DB) NFK, Burrow SOM.

As first element: Barham CAM, HNT, SFK, Barholm LIN (*hām*); Barpham SSX (probably **hamm**); Barrowden RUT (*dūn*); Barway CAM (*ēg*); Bearsted KNT (**hāmstede**); Bergholt ESX, SFK (**holt**); Berkhamstead HRT ('homestead'); Broughton HMP, LIN (*tūn*); Burghfield BRK (**feld**).

As generic with descriptive terms: Blackberry Hill LEI, Blackborough DEV, NFK; Brokenborough WLT; Caldbergh YON ('cold'); Chiselborough SOM ('gravel'); Clannaborough DEV ('cloven'); Emborough SOM ('level', perhaps referring to the very slight curve); Flawburgh NTT (*flōh* 'stone chip'); Grandborough BUC, WAR ('green'); Hoborough KNT ('hollow'); Longborough GLO; Roborough DEV, Rowberrow SOM, Rowborough IOW(2) ('rough'); Singleborough BUC ('shingle'); Stoberry SOM, Stoborough DOR ('stone'); Thurlbear SOM (*thyrel* 'hole'); Wambarrow SOM, Wanborough SUR, WLT (*wenn* 'tumour'); Whitbarrow WML ('white').

With references to vegetation: Desborough BUC (*dwostle*, a plant-name); Farmborough SOM, Farnborough BRK, HMP, KNT, WAR ('fern'); Limber LIN ('lime-tree'); Malborough DEV, Marlborough WLT (*meargealla* 'gentian'); Mappleborough WAR ('maple'); Risborough BUC ('brushwood'); Thornborough BUC, Thornbrough YON(2); Treborough SOM ('tree'); Whinburg NFK; Woodbarrow SOM, Woodborough WLT.

With personal names: Attleborough WAR (*Ætla*); Aylesbeare DEV (**Ægel*); Baltonsborough SOM (*Bealdhūn*); Bidborough KNT (**Bitta*); Bromsberrow GLO (**Brēme*); Chedburgh SFK (*Cedda*); Edlesborough BUC (*Ēadwulf*); Hanborough OXF (*Hagena*); Inkberrow WOR (*Inta*); Luxborough SOM (*Lulluc*); Marshborough KNT (*Mæssa*); Statenborough KNT (*Stēapa*); Symondsbury DOR (*Sigemund*); Ugborough DEV (**Ugga*).

With references to wild creatures: Bagborough SOM (probably 'badger'); Caber CMB, Kaber WML (**cā* 'jackdaw'); Crowborough SSX; Durborough SOM ('deer'); Finborough SFK, Finburgh WAR ('woodpecker'); Oldberrow WAR ('owl'); Todber YOW ('fox'); Wigborough ESX ('beetles'); Willesborough KNT ('beetle'); Wolborough DEV ('wolves').

With references to domestic creatures: Ellesborough (earlier *Eselbergh*) BUC (*esol* 'ass'); Harborough LEI (*hæfer* 'he-goat'); Harborough WOR ('herds'); Hensborough WAR (*hengest* 'horse').

HILLS, SLOPES AND RIDGES

With references to topographical features: Backbarrow LNC (**bæc**); Chelborough DOR (probably *ceole* 'throat', i.e. 'deep stream-channel'); Pulborough SSX (**pōl**); Welbury YON (**well**); Wisborough Green SSX (**wisc**).

With references to man-made structures: Gawber YOW ('gallows'); Panborough SOM (**pæth**); Sedborough DEV (probably *geset* 'animal fold').

With numbers: Seaborough DOR ('seven'); Thrybergh YOW ('three').

With references to crops: Wadborough WOR ('woad'); Whatborough LEI ('wheat').

With references to supernatural beings: Shuckburgh WAR (*scucca* 'demon, goblin'); Woodnesborough KNT (Woden).

With references to people or activities: Lenborough BUC (an -*inga*- formation, perhaps based on **hlith**); Swanborough SSX ('peasants'); Warborough OXF (*weard* 'lookout').

With river-names: Charborough DOR; Smallburgh NFK.

Names with uncertain or ambiguous qualifiers:

Alburgh NFK. The first element could be 'old' or the personal name *Alda*.

Litchborough NTP. Formally the gen.pl. of *līc* 'corpse' would suit. An OE **lycce* 'enclosure' has been conjectured to explain some names with similar spellings. There may also have been an OE **lic(c)* 'stream'.

Pamber HMP (*Penberga* 1165); OE *penn* 'enclosure' or British **pen**.

Todber DOR. A different name from Todber YOW, as the early spellings have *Tote-* as well as *Tode-*. The qualifier could be the personal name *Tota* or *tōte* 'look-out place'.

Wellsborough LEI. The spellings suggest *hweowel* 'wheel', but the meaning of the compound is not clear. The low, curving hill is no more wheel-shaped than many other *beorg* features, so the reference is probably not to shape.

berg ON 'hill'. In the following names *berg* is compounded with ON elements: Barby NTP, Barrowby LIN, YOW(2), Borrowby YON(2) (*bý* 'settlement'); Berrier CMB (*erg* 'shieling'); Aigburth LNC, Aikber YON ('oak'); Brackenber WML, Brackenborough LIN; Gowbarrow CMB (*gol* 'gust of wind'); Kirkber WML ('church'); Roseberry YON (see below); Sadberge DRH, Sedberg YOW, Sedbury YON (*setberg* 'seat-hill', which also occurs as a place-name in Iceland and Norway); Scaleber LNC (*skáli* 'shieling'); Solberg YON, Sulber YOW (*sól* 'sun').

Roseberry, near Great Ayton YON (Fig. 28b) is *Othenesberg* 1119. It is the Norse equivalent of English Woodnesborough, meaning 'hill of Othin'. There are actually two hills close together. The one called Roseberry Topping is a spectacular peaked eminence, the other has the smooth, rounded shape normally associated with *berg* names in Yorkshire. The Topping part of the name has only been traced back to 1610, when Camden mentioned *Ounsberry or Rosebery Topping*, but it may well be older. It is a plausible ME, perhaps OE, term for a peaked hill. The *berg* element in Roseberry could be taken to refer to the rounded hill overlooked by the *topping*, which would bring the name into line with the general use of Anglian **berg** and ON *berg* in Yorkshire. This interpretation seems at least as likely as the suggestion in Fellows-Jensen 1981 (p. 135) that the Yorkshire name is a rare instance of a transfer from the Danish homeland, being a replica of Onsbjerg on Samsø.

blain PrW, PrCumb, **blaen** Welsh 'point, end, top' occurs in names of post-Roman origin in CMB: Blencarn, Blencogo, Blencow, Blencathra, Blennerhasset, Blindcrake. Ekwall 1960 considers that it also occurs in Plenmeller NTB.

breʒ PrW, PrCorn, PrCumb (British *brigā*) 'hill'. This word, which is cognate with OE **beorg**, is the commonest of several British words meaning 'hill' which survive in English place-names. In Mellor CHE, LNC, Mulfra CNW, the second part of Plenmeller NTB and the first part of Mallerstang WML *breʒ* is qualified by PrW, PrCorn, PrCumb *mēl* 'bald', and Mellor Knoll in Bowland Forest YOW may be another instance, though Smith (EPNS survey 6, p. 214) preferred to derive it from the surname Mellor. Moelfre is a common hill-name in Wales. Clumber NTT and Kinver STF have *breʒ* as final element. As a simplex name *breʒ* survives in High Bray DEV (but not in Bray BRK).

breʒ is the first element of the Welsh/English names Bredon WOR, Breedon LEI, Brewood STF and Brill BUC. Bredon, Breedon and Brill are tautological compounds which may have arisen through a mistaken belief by English speakers that Welsh speakers were using the name of the **dūn** or **hyll** when they were actually referring to it by a common noun. A similar process has led modern English speakers to add 'on the Hill' to Breedon LEI.

There is a discussion of the solitary recorded instance from Roman Britain, and of the use of this term in Continental names, in Rivet and Smith 1979 (pp. 277–8).

brekka ON 'slope' is fairly common in LNC and of very occasional occurrence in some other northern counties. No instances have been noted in CMB. Ekwall 1960 includes Haverback WML (ON *hafri* 'oats') and Norbreck, Scarisbrick, Warbreck LNC. Norbreck is 'north slope' and Warbreck contains ON *vǫrthr* 'look-out'. For Scarisbrick Fellows-Jensen (1985, 158) suggests ON *skǫr* 'depression'.

brig PrW (British *brīco-*) 'top'. Brickhill BUC and Brickhill near Bedford are believed to be tautological hybrids containing this word and **hyll**.

brinn PrW, **bryn** Welsh 'hill' occurs in Brynn LNC and in several instances of Bryn in SHR and of Brin, Bryn(n) in CHE. It is the final element of Malvern WOR ('bare hill') and probably of Yeavering NTB ('goat hill'). It is common in Welsh names.

camb 'comb'. The topographical use of this word is rare. There is little doubt that it occurs in Cambo NTB (hōh), and is the source of Combs SFK. Some hill-names in Yorkshire (e.g. Cam Fell by the Pennine Way, Cams Head and Cold Cam near Byland Abbey) are considered to have this origin, but it may be legitimate to suspect a British name from PrW, PrCumb *camm* (British *cambo-*) 'curved, crooked'.

Combs (south of Stowmarket) SFK may have been a district-name referring to the series of low ridges which run up to the river Gipping here. If so, the term was used for low, rounded ridges rather than (or as well as) cock's-comb-shaped ones. The word is difficult to identify, certainly rare, and probably not amenable to precise definition.

clif 'cliff' (Fig. 29). The main determining factor in the use of this word in place-names is probably the angle of the slope. It is regularly used of slopes which are 45° or steeper, **helde** being the term used for more gradual ones. A *clif* is frequently a riverside feature, but there are many examples which are not related to water. In Gelling 1984 much was said about the apparent slightness of some *clif* features as represented on Ordnance Survey maps. It has since become apparent, however, that all *clif* features need to be evaluated on the ground. This ideal is far from realisation, but enough field-work has been done to make it clear that features which look unimpressive on the map may be much more striking on the spot. A good instance of this is Clifton Hampden OXF, where the church occupies a rock outcrop, small in absolute terms, but most striking in its context of flat Thameside meadows. The feature referred to in Cliffe-at-Hoo KNT is not very high, but it too is very striking in its setting. Another *clif* wrongly defined from Ordnance Survey maps in Gelling 1984 is the one at Topcliffe YON, which is quite steep. Probably no *clif* referred to in a place-name should be described as a slight feature until field-work confirms that it is so. If there are some low river-banks to which the term is applied, allowance should perhaps be made for silting in post-Roman times.

There are many dramatically high river-banks (e.g. those at Cliveden BUC and Horncliffe NTB) to which the word is applied, and many inland escarpments and rock faces (such as those at Trottiscliffe KNT, Cliff Hill near Lewes SSX, Cleeve Cloud GLO), but less obvious examples need careful evaluation on the ground.

clif is common as a simplex name and as a first element, particularly with *tūn*. When it is a generic in compounds the qualifiers are not as varied as

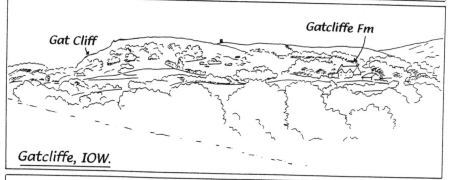

might be expected. Descriptive terms predominate heavily, 'red' being particularly common.

The list of names in the reference section includes a number gleaned from EPNS volumes which refer to places of insufficient status to be listed in Ekwall 1960. The selection is somewhat arbitrary, but it gives a fuller representation of qualifiers than could be obtained from names in Ekwall. *clif* is particularly common in DEV and YOW.

Reference section for clif

As simplex name: Cleeve GLO, SOM(2), WOR(2), Clevancy WLT (family name added in 13th century), Cleve HRE, Cliff CMB, WAR, YOE, Cliffe KNT(2), NTP (now King's Cliffe), WLT, YON, YOW, Clive CHE, SHR, Clyffe DOR. In minor names in DEV there are 22 instances of Cleave and three of Cle(e)ve.

As first element with tūn: Clifton BDF, BUC, CHE, CMB, DOR, DRB, GLO, LNC(3), NTB, NTT(2), OXF(2), STF, WAR, WML, WOR, YON, YOW(3).

With other generics: Cleadon DRH and Clevedon SOM (**dūn**); Cleveland YON (**land**); Cleveley LNC, OXF (**lēah**); Clevelode WOR (**gelād or lād**); Cliburn WML (**burna**); Clifford DEV, GLO, HRE, YOW (**ford**); Cliveden BUC (**denu**); Cliviger LNC (**æcer**). In Clewer BRK, SOM *clif* is combined with *-ware* 'dwellers' to make a group-name 'dwellers by a cliff'.

As generic with descriptive terms: Brancliffe YOW ('steep'); Brincliffe YOW ('burnt'); Buckleigh, Buckley (2) DEV ('bow'); Horncliffe NTB; Langcliffe YOW ('long'); Lightcliffe YOW; Radcliffe LNC, NTT, Radclive BUC, Ratcliff MDX, Ratcliffe LEI(2), NTT, Redcliff GLO, SOM, Great Rutleigh DEV (all 'red'); Rawcliffe LNC, YON(2), YOW, Rockcliff CMB, Roecliffe YOW ('red', OE *rēad* interchanging with ON *rauthr*); Scaitcliffe LNC(2) ('slate'); Scarcliff DRB ('gap'); Sharncliffe GLO ('dung'); Shirecliffe YOW ('bright'); Staincliffe YOW, Stancliffe DRB ('stone'); Turtley DEV ('dry', OE *thyrre*); Whitcliff(e) GLO, SHR, YOW, Whitecliff DOR (all 'white').

With personal names: Aldcliffe LNC (*Alda*); Austcliff WOR (*Ealhstān*); Baycliff WLT (*Bægloc*); Felliscliffe YOW (ON *Felágr*); Guy's Cliff WAR (*Kibbecliva* 12th century, *Cybba); Hatcliffe LIN (*Headda*); Trottiscliffe KNT (*Trott). Cliffe in Manfield YON was earlier *Ileclif, Ylclive*, probably from *Ylla*.

With references to wild creatures: Arncliff(e) YON, YOW, Arnecliff YON ('eagle'); Catcliffe YOW and Catsley DOR; Onecliffe (earlier *Wormeclif*) YOW (in Greetland, 'snake', OE *wyrm*); Raincliff(e) YOE, YON, Rams Cliff WLT, Ravenscliff DRB ('raven'); Swalcliff OXF, Swalecliffe KNT, Swallowcliffe WLT ('swallow').

With references to domestic creatures: Bowcliffe YOW ('bull'); Gatley CHE ('goat'); Swincliffe YOW and Swintley RUT ('swine').

THE LANDSCAPE OF PLACE-NAMES

With references to vegetation: Briercliffe LNC; Rushcliffe NTT ('brushwood', OE *hrīs*).

With references to ownership: Coniscliffe DRH (OE *cyning* 'king' replaced by ON *kunung*); Priestcliffe DRB.

With references to superstitions: Shincliffe DRH ('spectre', OE *scinna*); Winslow BRK (earlier *Wendelesclive*, Cleeve Cloud GLO was also *Wendlesclife* in OE boundaries: *Wendel* seems more likely to be a mythological than a human being).

With earlier place-names or topographical features: Aust Cliff GLO; Avoncliffe WLT; Wycliffe YON (probably 'river-bend').

With references to direction or to position of settlement: Sutcliff YOW ('south'); Topcliffe YON, Topley DRB.

With reference to a product: Wharncliffe YOW ('mill-stone', OE *cweorn*).

Names with obscure, uncertain or ambiguous qualifying elements:
Attercliffe YOW. There are no available spellings between *Ateclive* DB and *Atterclive* 1296, and this renders any etymology uncertain. The statement in Ekwall 1960 that this is 'at the cliff' with inorganic -r- is challenged in EPNS survey Part 1, pp. 208–9, on the cogent grounds that *æt* is not preserved in northern names. Also, where it is preserved, it is represented by T-, not At-. A. H. Smith in EPNS survey suggests a shortened form of a personal name such as *Æthelrēd* or *Eadrēd*.

Baycliff LNC (*Belleclive* 1212, *Beleclive* 1260). The first element could be *bēl* 'fire' or *belle* 'bell' used as a hill-name. Ekwall 1960, under Bealaugh, argues for a word meaning 'glade' in a number of names including this one.

Cookley WOR (*Culnan clif* 964). Possibly a personal name *Culna*.

Egglescliffe DRH. Generally accepted as a member of the corpus of names in which the first element is the OE derivative of British *eglēs* from Latin *ecclesia*. The early spellings are perhaps not wholly conclusive, but assuming the etymology to be sound it is still difficult to classify the name. OE *eclēs* occurs either as a simplex name (Eccles) or as the first part of compound name. It is a reference to a Celtic Christian church or community, but it is uncertain whether in this instance it would have been a simplex name before *clif* was added.

Hockcliffe BDF (*Hocgan clif* 1015). OE *hocg* 'hog' is a strong noun, so does not suit grammatically. There may have been a personal name *Hocga*.

Marlcliff WAR (*Marnan clive, Mearnan clyfe* in charter bounds). Possibly a personal name *Mearna*.

cnoll 'knoll'. Knowl(e) is a frequent alternative to Knoll. The northern form of the word is *know(e)*.

The most clearly evidenced sense in literary OE is 'summit of a hill', and (despite what was said in Gelling 1984) this may be the precise sense in some ancient place-names. Brent Knoll SOM (in which *cnoll* has been added to a British name meaning 'high place') is a truncated cone rising from a ridge. A miniature version of this can be seen south of Lichfield STF, where Knoll Fm stands on a spacious hill which has a small, flat-topped summit rising from it immediately behind the farm. There is another example near Gloucester: Robin Wood's Hill by Matson was *Mattesknoll(e)* in the 13th–17th centuries, and this also has a truncated cone rising from a ridge. The outline is not as sharp as that of Brent Knoll, but the similarity can be seen if both hills are observed by a traveller going north along the M5. The same shape can be discerned at Knowle WAR, where the village street climbs gradually up the hill to the church, and there is then a short stretch of dead-level road leading towards Warwick, followed by an abrupt descent. Here the truncated cone is set at one end of the ridge, as it is at Puncknowle DOR. At Knole SOM the small settlement is overlooked by a hill with a protuberance at one end.

Distinctive shapes of this sort may be present at some other instances not as yet inspected. The relative frequency of simplex examples in the small corpus of major names suggests that a *cnoll* was a distinctive type of hill, but the survival of the word into modern English means that it is liable to have been applied to any small hill in names of comparatively recent origin.

cnoll is commoner in minor names than in those included in Ekwall 1960. It is not represented in all counties. It is particularly common in DEV and SSX.

Knole occurs as a major name in KNT. In the form Knowle it occurs in DOR and WAR and five times in SOM. Knowlton DOR, south-west of Wimborne St Giles, is probably named from nearby Knowle Hill. Church Knowle DOR, on the Isle of Purbeck, was a simplex name until the 14th century. Close by are Bucknowle and Cocknowle, which may have personal names as first element. The word is used as an affix in Kirby Knowle YON (*Chirchebi sub Knol* 13th century).

There are only a few early recorded compound names with *cnoll* as generic. Chetnole DOR and Chipnall SHR probably have personal names as qualifiers. Bincknoll Castle WLT is probably 'knoll of the bees'. Whitnole DEV, a DB manor east of Rackenford, is 'white knoll'. For Puncknowle DOR, earlier *Pomecnolle*, Ekwall 1960 suggests *plūme* 'plum-tree', with loss of the first *-l-* for dissimilation from the *-ll-* of *cnoll*.

Knill RAD is considered to derive from a related term *cnyll(e)*. The settlement is on a hillock on a small bluff overlooking a stream.

cōc or cōce 'hill'. This word is not on record but may reasonably be inferred from Swedish and Norwegian parallels. In the EPNS survey of DOR (1, p. 229)

Professor Löfvenberg is quoted as suggesting that such a word may be the first element of Cookham BRK. Other names for which it would provide an excellent solution are Cook Hill and Cooksey, both in WOR.

OE *cocc* 'hillock', which survives in the word *haycock*, is fairly common in minor names (e.g. Weycock Hill in Waltham St Lawrence BRK) and field-names. It occurs occasionally in major names, being the generic in Withcote LEI (earlier *Withcoc*) and the probable first element of Cockhampstead HRT and Coughton HRE, WAR.

Cookham, Cook Hill and Cooksey cannot be derived from *cocc*, as only a word containing -ō- could have given the early spellings and the modern forms.

Other place-names beginning with Cock- are likely to contain *cocc* 'cock' (the bird) or a personal name *Cocca*.

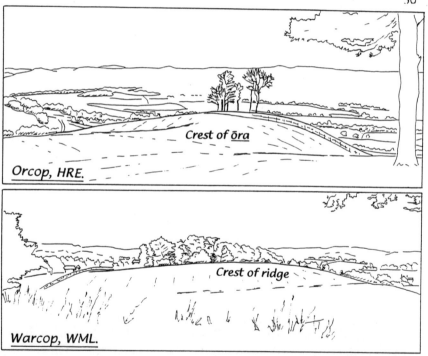

30

copp 'summit'. This rather rare place-name element is sometimes (perhaps always) used for a hill or ridge which has a narrow, crest-like summit.

Mow Copp CHE is a long, narrow hill. At Copp LNC, south-west of Great Eccleston, there is a low ridge made by the 50ft contour, and this may have a crest, as the village stands at 84ft. Orcop HRE (Fig. 30) is overlooked by a hill which in profile is an ōra, but which has steep sides, giving a crest-like appearance to the summit. Pickup Bank LNC, east of Darwen, may be a similar ridge. The large ridge of Warcop WML has a narrow crest running athwart its top (Fig. 30). The qualifier in this last name may be *weard* 'look-out', which would refer to the magnificent views along the valleys on either side of the ridge. Early spellings vary between *Warde-* and *Warthe-*, and ON *vǫrthr*, which has the same meaning, may be partly the origin.

copp occurs in a few minor names and field-names in some counties, but is not evidenced in all. In Copley Hill (in Babraham) and Coploe Hill (in Ickleton) CAM, *copp* qualifies **hlāw**, and this compound occurs also in minor names in DRB and YOW. Billesdon Coplow LEI is an impressive hill.

The meaning 'crest' is more clearly evidenced for *copp* than it is for **camb**.

Two names are unexplained. Coppull, south-west of Chorley, LNC, is said to mean 'peaked hill', but the village appears from maps to have a low, flat site. Sidcup GTL (*Cetecoppe* 1254, *Setecoppe* 1301) has parallels in ME field-names in CAM, ESX, HRT and NTT, and this suggests that there was a compound appellative, the sense of which has not been recovered: 'seat-shaped hill' has been suggested.

crūg PrW, PrCorn, PrCumb, **crūc, crȳc** OE 'hill, mound, tumulus'. The British ancestor of this word is found in a few of the place-names recorded from Roman Britain. The most notable example is *Pennocrucium*, modern Penkridge STF, which means 'headland tumulus' and refers to a burial mound at Rowley Hill Fm in the parish of Stretton, 1,200 yards north of Watling Street, at GR SJ 90251180. The tumulus has been much reduced by ploughing, but there is a description of it from 1797, and it was still 3ft high in 1907. The site is a typical one for a barrow, the brow of a shoulder of land from which the ground falls abruptly to the river Penk. The relevance of this archaeological monument to the Romano-British place-name was brought to the attention of scholars by Mr J. Gould.

The use of the British word in *Pennocrucium* indicates a meaning 'tumulus', and this may occur in some names of later origin. But the characteristic use in PrW and OE is for a natural hill with an abrupt outline which makes a specially striking visual impact. In some instances Anglo-Saxons were probably taking over the name by which they heard Welsh people call such eminences, but it is likely also that they adopted the word, since it occurs so frequently in place-names in England. Perhaps it filled a perceived gap in the OE range of hill-terms.

In Creechbarrow SOM and Crookbarrow WOR (Fig. 31) the word is combined with **beorg**. Both these hills are visible from the M5, which is the successor to the Roman road to Exeter. Creechbarrow is a particularly interesting example because it is mentioned in a charter of A.D. 682 which records a gift of land by a king of Wessex to the abbot of Glastonbury. The main area of the estate lay north of Taunton, but there was an extra piece of land described as lying 'on the south side of the River Tone at the island near the hill which in the British language is called *Cructan*, among us *Crycbeorh*'.

31

Crookbarrow, WOR.

Crutch, WOR.

Creechbarrow, SOM.

Creechbarrow, SOM.

This island is then defined as having the river Tone (earlier *Tan*) for its north boundary, and *Blacanbroc* on the south. A small tributary of the Tone is still called Black Brook, so the location of the island is absolutely certain. The description has in the past been taken to refer to Creech St Michael, a village two miles east of Taunton, an error which occurred because Creechbarrow was not shown on O.S. maps. The name is, however, well known locally, Creechbarrow Road being the route into Taunton from Creech Castle Hotel. The hotel was built in the late 19th century on the site of a house earlier called Creechbarrow. The hill is a more prominent landmark from the Taunton side than it is from the hotel. It is flanked by factories and supermarkets, and the Creech Castle roundabout (now a busy crossroads) encroached on its edge, but it still dominates the area as it did in 682.

The statement about the British (i.e. Welsh) name being *Cructan* and the English name *Crycbeorh* offers good evidence for several aspects of name-giving. The Welsh name is probably a post-Roman formation, since it has the generic first and the defining element, the river-name, second. English speakers would hear this name in the 7th century, by which time their own language had developed the rounded *u* sound which it had earlier lacked, so they were able to render the Welsh vowel by their own version of this sound which was spelt *y* in Old English. They discarded the Welsh qualifier *-tan* and glossed *crȳc* with the English word **beorg**. This is not tautology. It is a careful assessment of the shape of this particular *crȳc* which, as can be seen from the illustration in Fig. 30, conforms to the hill-shape which in OE was a **beorg**. Other *crūc, crȳc* hills, while having the same quality of abruptness, are not smoothly rounded so do not qualify for the term **beorg**.

Crookbarrow WOR has the same name as Creechbarrow SOM, but in this instance formed at an earlier date, when neither Welsh nor English had the *ü* sound. At this stage the Welsh word was likely, when adopted by English speakers, to develop into Crook, Crouch or Crutch. Crookbarrow is a remarkable landscape feature, situated by exit 7 on the M5. It has been suspected of being artificial, in which case it would be an ancient monument on the lines of Silbury Hill in WLT. It is very much a *crūc* in the impact it makes on the beholder, and from some aspects, one of which is shown in Fig. 31, it is very much a rounded **beorg**. It is now known locally as Whittington Tump.

Another striking example is Crutch WOR, immediately north of Droitwich. From the 12th to the 17th century the settlement by Crutch Hill was referred to simply as *Cruch, Crouch, Crowche*. The hill is shown on Fig. 31.

If the Anglo-Saxons had wanted to gloss Crutch WOR with an OE word they might have used **hyll**, which was their preferred term for hills which were neither smoothly rounded nor flat-topped. In the case of Crutch they were satisfied with the simplex name, but there are a number of names in which *crūc/crȳc* is compounded with **hyll**, and these are listed in the reference section.

The compound of *crūc/crȳc* with **hyll** has sometimes become modern Churchill. A documented instance is Church Hill south-east of Wells SOM, which is *crichhulle* in a ME version of an OE charter boundary. The development is due to association with the word *church*, and care must obviously be exercised in the separation of Churchills which have *crūc/crȳc* as first element from those which really do mean 'hill with a church'. If there are no early diagnostic spellings available, study of the site often supplies the answer. There are a number of instances of Church Hill applied to isolated hills with no settlement in the vicinity, and these are likely to contain *crūc/crȳc*. Examples are to be found north of Clows Top WOR at GR SO 710730, near Pontesbury SHR at SO 416041 and east of Leintwardine HRE at SO 410739.

Details are given in Gelling 1984 of three DEV Churchills (near Loxbeare, in East Down and in Broad Clyst) in which Church- has clearly replaced *crūc*. Other Churchills, in OXF and SOM, are satisfactorily explained as 'hill with or near a church'. There are two Churchills in WOR: at the one near Worcester the church is on a hill, but at the one near Kidderminster there is a small rounded hill not immediately adjacent to the church which could well have been called *crūc*.

Churchdown Hill near Gloucester is a large hill with an ancient church on its summit. The name is *Circesdune* in DB, *Chyrchesdon* in 12th- and 13th-century references, and the genitival -s- survives in the alternative name Chosen. This strong genitive would be appropriate if the first element were an earlier name of the hill, so 'dūn called *Crȳc*' has been accepted as the etymology. In a recent study of hill-names near Gloucester undertaken in connection with the Crickley Hill excavation report, however, it has been suggested that in Churchdown there may be an irregular treatment of the feminine noun *cirice* such as is apparently evidenced in a 10th-century estate boundary near Taunton SOM. The SOM estate is called *Cyricestune*, and it includes a property called *Cyriceswudu*. this is an area in which, although there is later confusion of the two words, the distinction bwetween *crȳc* and *cirice* is clearly preserved in pre-Conquest spellings, and it seems likely that *Cyricestune* and *Cyriceswudu* refer to a settlement and a wood associated with a church. If the three large eminences near Gloucester are considered as a group, Crickley Hill is the one most suited to the designation *Crȳc*. The beetling brow of the hill as seen from the southwest is very similar to the outline of the hill at Cricket SOM. The hill at Matson was designated **cnoll** (see above under that word). If the grammar does not absolutely prohibit it, Churchdown could satisfactorily be explained as 'dūn with a church'.

Other names derived from *crūc* or *crȳc* are listed in the reference section. The distribution is widespread, but SOM has more examples than any other county. Evercreech SOM is the only example in which the element appears as the second part of a compound, and here *crȳc* was probably a district name. The large settlement occupies a slight rise in an area enclosed by a horseshoe of hills, the most striking of which is called Creech Hill. Evercreech could be interpreted as 'boar farm in the district of Creech'. Cricket St Thomas SOM is *Cruche* in DB, and Cricket Malherbie is *Cruchet*, in which *-et* is a French diminutive ending applied to the smaller of the two settlements on this large hill. The *-et* form was later used for both.

The variety of forms in names derived from this element can be explained by the differential development of the vowel ü in the English and Welsh languages. It is not simply a matter of the date at which the Welsh word or a Welsh place-name derived from it was adopted by English speakers, however. That is an important factor, but the chronological sequence is not perfectly consistent. OE *y* did not correspond exactly to PrW *ü*, and the appearance of the form Crook in DEV and DOR may indicate that Anglo-Saxons sometimes continued to substitute *u* for the Welsh sound even after their language had developed the sound they represented by *y*.

It should be noted that in minor names and field-names Crouch most frequently derives from ME *crouche* 'cross', and Crook from ME *crōk* 'bend'. Crouch End GTL is from *crouche*. Several major names in northern counties derive from ON *krókr*, used of bends in rivers.

Welsh names containing *crŭg* include Criccieth CRN and Crickhowell and Crucadarn BRE. Crickheath SHR is the same name as Criccieth, but the nature of the qualifying element in these two names is uncertain. The Cornish word is discussed in Padel 1985 with lists of examples.

Reference section for crŭg, crūc, crȳc

As simplex names: Creech DOR, SOM, Crich DRB, Cricket SOM, Crook Hill DOR (southeast of Crewkerne, *Cruc* 1014, 1035), Crooke DEV, Crouch Hill OXF (*Cruche* 1268), Crutch WOR. Some hill-names for which there is no early documentation can confidently be placed in this category, e.g. Crouch Hill and Crook Hill WLT and Cold Crouch SSX, the last being a prominent knob on the South Downs escarpment overlooking Willingdon.

As first element: Christon SOM (*tūn*); Creechbarrow SOM and Crookbarrow WOR (**beorg**); Crichel DOR (**hyll**); Crickley GLO (**hyll + leah**); Crigglestone YOW (*crȳc-hyll*, with *tūn* added to genitive); Croichlow LNC (**hlāw**); Crooksbury SUR (**beorg** has been added to the genitive of a hill-name: *Cruc* was probably the hill, and **beorg** a reference to the tumulus which crowns it); Cruchfield BRK (*Cruchesfeld* 1251, **feld** has been added to the genitive of the hill-name).

As generic in a compound name: Evercreech SOM (*eofor* 'boar').

In British or Primitive Welsh names: Crickheath SHR (qualifier obscure); Cumcrook CMB (British *cumbo-* 'valley'); Gilcrux CMB (OWelsh *cil* 'retreat'); Penkridge STF (**penn**).

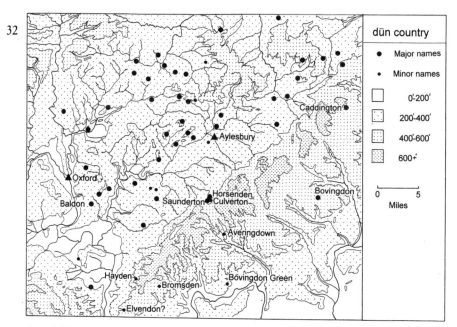

32

dūn 'hill, upland expanse'. This word is consistently used in settlement-names for a low hill with a fairly level and fairly extensive summit which provided a good settlement-site in open country. *dūn* was accorded extensive discussion in Gelling 1984, in which its special significance in ancient settlement-names was asserted for the first time. The following points are summarised from that discussion.

As regards ultimate origin and etymology, there is no easy association to be made with *dūno-*, the Romano-British form of a word which is one of the most important items in Celto-Latin toponymy throughout Europe, and which occurs in at least 16 of the names recorded from Roman Britain. The Celtic word meant 'fortified place', and its OE cognate is *tūn*, which originally meant 'enclosure'. The massive place-name evidence available for the usage of OE *dūn* precludes any association with fortification or enclosure such as might have been expected if it were a borrowing from British *dūno-*. It is likely that OE *dūn* is a different word whose origins in Germanic languages have not been ascertained.

dūn was probably in use for the formation of settlement-names from the beginning of the Anglo-Saxon period. It may not have been much employed in that capacity after A.D. 800, though it remained in use for field-name and minor name formation and for purely landscape features. The modern reflex is *down*.

The distribution of *dūn* names is governed by geography. It is most common as generic in village names in regions where there are clusters of level-topped hills suitable for settlement-sites. The most notable stretch of *dūn* country is a fairly compact one in the south midlands, comprising east OXF and BUC north of the Icknield Way (Fig. 32). Another important cluster, which includes Basildon, occurs in south central ESX. There is a smaller concentration in north-east WLT, and *dūn* is well represented in WAR and in NTB and DRH. Settlement-names in *dūn* are rare in SSX, HMP and BRK, perhaps because of the absence of water supplies on chalk hills.

There is a very high degree of consistency in the sites of villages with *dūn* names. Some examples in the south midland group are shown on Figs. 32 and 33. Most of the examples are on hills of 200ft–500ft O.D., though some are lower and a few are higher. At one end of the scale (where *dūn* overlaps with ēg, though generally applied to larger features) is Hedon YOE, at a height of about 25ft overlooking the river Humber. Some of the ESX examples are in the 50ft range. At the other end of the scale is Chelmorton DRB, at about 1,200ft in a recess in a hill which rises to 1,450ft. It is only here, in north STF and north DRB, that the term is well evidenced in really high country.

The use of *dūn* for some very low hills led to the suggestion in Stenton 1913 (p. 3 n.1, echoed in Smith 1956) that *dūn* sometimes meant an expanse of open land without reference to height, but there is no sound evidence for such a usage.

Where, as in the majority of instances, *dūn* is used for low hills in open country, it is obvious that the antecedents of most of the settlements must have been in these situations from prehistoric times. Settlement-sites like those depicted in Figs 32 and 33 cannot have been unoccupied when English speakers arrived. What we have here is not (as is frequently asserted) the coining of an OE hill-name which was subsequently transferred to a settlement. It is the application to ancient settlements of a new English name, the generic of which embraces both the habitation and the site. The exact nature of the site and the likelihood of the settlement being a relatively high status one would be conveyed to an Anglo-Saxon who knew the full meaning of *dūn* in place-names.

The suggested scenario is not applicable in all areas, however. The word is used differently in the Cotswolds, in a belt of country extending from Gloucester to the Oxfordshire border. This region does not have the settlement-friendly hills described above: it has larger massifs with settlements at the foot of them. The Rissingtons, Great, Little and Wyck, lie along the foot of a long ridge, and this ridge is the 'hill overgrown with brushwood' (*Risendune c.*1075) of the name. Oxenton and Dixton, which are neighbouring *dūn*s north of Cheltenham, lie on opposite sides of a large massif, and this use of *dūn* in names of settlements adjoining a large hill is also seen in Ilmington WAR, at the northern end of the Cotswolds. Near Gloucester the villages of Churchdown and Matson lie beside hills which would not be comfortable settlement-sites. Also near Gloucester, however, Swindon and Whaddon provide examples of the usage common elsewhere.

33

Village on summit of _dūn_
Toot Baldon, OXF.

Steeple prominent on large whale-backed hill
Steeple Claydon, BUC.

Village on whale-backed hill
Billington, BED.

Village on summit of _dūn_
Claydon, OXF.

Village dwarfed by hill
Stottesdon, SHR.

Other instances in which *dūn* refers to an uninhabited hill adjacent to the settlement include Bredon WOR, Breedon LEI, Blackdown DOR, Bleadon SOM, Raddon DEV, Brandon NFK, Puleston SHR, Quorndon and Bardon LEI, Baildon YOW, Billington LNC. But these are greatly outnumbered by examples in which the village and the hill form a single entity, and it is likely that for the most part settlement-names in *-dūn* were coined for pre-English settlements in recognition of this characteristic situation.

In later forms, as can be seen from the reference section, there is massive confusion of *-don* and *-ton*, and some weakening of *-don* to *-den*. Early spellings are vital for the correct identification of *dūn* names.

dūn is not common as a simplex name or as a first element. In the large corpus of compound names which have *dūn* as generic the largest category of qualifiers is that which comprises descriptive terms. Certain compounds recur, black, green, misshapen, long, sandy and stony hills being especially frequent.

Personal names form the next most frequent class of qualifiers, though with the usual caveat that some authorities might reject some of the proposed anthroponyms. A high proportion of them are starred, but there are grounds other than the place-names for considering them to have existed. Most of the certain or probable personal names are masculine and monothematic, which could be held to be consistent with the hypothesis of early origin. The only feminine names noted are *Bucge* in Bowden LEI, and *Ælfgȳth* in Aulden HRE, and the only dithematic names are **Eohhere* (Eggardon DOR), *Gǣrwald* (Garendon LEI), *Herebeald* (Harbledown KNT), *Cynemǣr* (Kilmersdon SOM) and *Sigebed* (Sibson LEI). There is a fairly large category consisting of references to vegetation, with 'broom', 'fern' and 'heath' making multiple appearances. The remaining qualifiers are divided fairly evenly among a number of categories. References to animals, both wild and domestic, are less common than might have been expected. It is noteworthy that *-inga-* compounds are very poorly represented. There are not many compounds in which the qualifier defies all attempts at interpretation, though a number are ambiguous.

Reference section for **dūn**

As a simplex name: Down(e) DEV(2), KNT; Downham LNC, NTB and Downholme YON (dat.pl.); Rousdon DEV (*Done* DB, Rous- from a later owner named Ralph).

As first element: Donhead WLT, Downhead SOM, *Dunheved* (later Launceston) CNW, Dunnett (in Compton Bishop) SOM (all **hēafod**); Downham CAM, ESX, NFK, SFK and Dunham CHE(2), NFK (*hām*); Downton HRE, SHR(2), WLT and Dunton ESX, LEI, NFK, WAR (*tūn*); Downwood HRE and Dunwood HMP (**wudu**); Dummer HMP (**mere**); Dundon SOM (**denu**); Dundry SOM (*dræg*, 'place where things are dragged'); Dunhamstead WOR ('homestead'); Dunhampton WOR

THE LANDSCAPE OF PLACE-NAMES

(hāmtūn 'settlement'); Dunstan NTB ('stone'); Duntish DOR (*etisc, probably 'pasture').

As generic with descriptive terms: Ballingdon SFK ('round'); Blackdown DOR, Blagdon DEV, DOR, SOM(2) (all 'black'); Bleadon SOM ('variegated'); Bowden DRB, Bowdon CHE ('bow-shaped', this is a frequent minor name in DEV); Butterton STF (a reference to good pasture); Caludon WAR ('bare'); Clandon SUR and Glendon NTP ('clean'); Claydon BUC, OXF, SFK; Edington WLT ('bare'); Fallowdon NTB ('fallow-coloured'); Fawdon NTB(2) ('variegated'); Garsden WLT, Garsington OXF ('grassy'); Gratton (in High Bray) DEV ('great'); Grendon BUC, NTP, WAR, Grindon DRH(2), NTB, STF (all 'green'); Hambledon DOR, HMP, LNC, RUT, SUR, YON(2), Hameldon LNC, Humbledon DRH, NTB (all 'misshapen'); Hendon GTL ('high'); Langdon DEV, DOR, ESX, KNT, WAR, Longden SHR, Longdon SHR, STF, WOR(2) (all 'long'); Leighton (*Lyhtedon* 1272) NTB (*lēoht* 'light'); Meldon DEV ('variegated'); Merrington SHR ('pleasant'); Neasden GTL ('nose-shaped'); Raddon DEV, Rawden, Rawdon YOW ('red'); Sandon BRK, ESX, HRT, STF, Sandown SUR ('sand'); Saunderton BUC ('sandy'); Sheldon WAR and Shildon DRH ('with shelves'); Shenington OXF ('beautiful'); Sholden KNT ('shovel', presumably referring to the shape of Sholden Down); Smithdown LNC ('smooth'); Standen IOW, Standon HRT, STF, Stondon BDF, ESX (all 'stone'); Wildon YON (probably 'uncultivated', OE *wilde*). Nackington KNT (*Natyngdun* 993) may contain a term **næting* 'wet place').

With personal names: Abingdon BRK (*Æbba*); Attington OXF (*Eatta*); Aulden HRE (fem. *Ælfgȳth*); Baldon OXF (**Bealda*); Basildon ESX (**Beorhtel*); Battlesden BDF (**Bæddel*); Billesdon LEI (**Bill*); Billington BDF (**Billa*); Blaisdon GLO and Bletchingdon OXF (**Blecci*); Blunsdon WLT (**Blunt*); Bosden CHE (*Bōsa*); Bowden LEI (fem. *Bucge*); Caddington BDF (**Cada*); Cartington NTB (**Certa*); Cheddington BUC (*Ceatta*); Cheldon DEV (**Ceadela*); Chessington GTL (*Cissa*); Crendon BUC (*Crēoda*); Criddon SHR (**Cridela*); Cuddesdon OXF (**Cūthen*); Duddon CHE (*Dudda*); Earsdon NTB(2) (**Eard*); Edgton SHR (*Ecga*); Eggardon DOR (**Eohhere*); Eggington BDF (*Ecca*); Eldon DRH (*Ella*); Evedon LIN (*Eafa*); Eydon NTP(**Æga*); Garendon LEI (*Gǣrwald*); Godington OXF (*Gōda*); Haddon HNT (*Headda*); Harbledown KNT (*Herebeald*); Hellesdon NFK (**Hægel*); Hillesden BUC (**Hild*); Hillingdon GTL (*Hilda*); Hixon STF (**Hyht*); Hoddesdon HRT (**Hod*); Hundon LIN (*Hūna*); Hunsdon HRT (*Hūn*); Kilmersdon SOM (*Cynemǣr*); Ockendon ESX (**Wocca*); Oddington OXF (**Ot(t)a*); Seckington WAR (*Secca*); Shebdon STF and Shobdon HRE (**Sceobba*); Sibson LEI (*Sigebed*); Sundon BDF (**Sunna*); Waddesdon BUC (**Wott*); Warndon WOR (**Wærma*); Wavendon BUC (**Wafa*); Wimbledon GTL (**Wynnmann*).

With references to vegetation: Ashdon ESX, Ashdown SSX, Ashendon BUC (all 'ash'); Baydon WLT ('berry', OE *bēg*); Brandon DHR, NFK, NTB, SFK, WAR, Brendon DEV, Brundon SFK (all 'broom'); Brereton STF ('briar'); Claverdon WAR ('clover'); Elmdon ESX, WAR; Eudon SHR ('yew'); Faringdon BRK, DOR, HMP,

HILLS, SLOPES AND RIDGES

Farndon CHE, NTP(2), NTT, Farringdon DEV (all 'fern'); Haddon DOR, DRB, NTP, Heddon NTB(2), Hedon YOE (all 'heath'); Horndon ESX ('thorn'); Ilmington WAR ('elm'); Lyndon RUT ('lime'); Marldon DEV ('gentian', OE *meargealla*); Meddon DEV ('meadow'); Meesden HRT ('moss'); Parndon ESX ('pear'); Rissington GLO ('brushwood'); Thorndon SFK.

With reference to wild creatures: Ambrosden OXF ('bunting'); Cronkston DRB ('crane'); Croydon SOM ('crow'); Dordon WAR ('deer'); Foulden NFK ('birds'); Harton DRH ('hart'); Hawkedon SFK; Pudlestone HRE ('mouse-hawk', OE *pyttel*); Rettendon ESX (possibly **rætten*, 'rat-infested'); Roxton BDF ('rook'); Wilden WOR ('beetle', OE *wifel*); Wooden NTB ('wolf'); Yagdon SHR ('cuckoo').

With references to domestic creatures: Ashingdon ESX ('ass'); Callerton NTB(3), Cauldon STF, Chaldon DOR, SUR (all 'calf'); Coundon DRH ('cow'); Endon STF ('lamb', OE **ēan*); Everdon NTP and Eversden CAM ('boar'); Horsenden BUC and Horsendon GTL ('horse'); Oxenton GLO; Staddon DEV ('herd of horses'); Swindon GLO, STF, WLT ('swine').

With references to crops: Bandon SUR ('beans'); Clayhidon DEV, Haydon DOR, SOM, WLT, Heydon NFK (all 'hay'); Raydon and Reydon SFK, Roydon ESX, NFK(2), SFK (all 'rye'); Waddington SUR, Waddon DOR, SUR, Whaddon BUC, CAM, GLO, WLT (all 'wheat').

With references to classes of people: Bishopton WAR; Hindon WLT ('monks'); Huntingdon HNT and Huntington STF, YON ('huntsmen'); Kingsdon SOM and Kingsdown KNT(3); Seisdon STF ('Saxons'); Stottesdon SHR ('stud-keeper'); Wembdon SOM (possibly **wǣtheman*, 'huntsman').

With -ingas group-names: Harlington BDF ('Herela's people'); Tillingdown SUR (**Tilla's people'); Toddington BDF ('Tuda's people').

With references to a product or an activity: Cowden YOE ('charcoal'); Findon SSX ('wood-pile'); Huntington CHE ('hunting'); Quarlton LNC, Quarndon DRB, Quarrington DRH, Quorndon LEI (all *cweorn*, 'millstone'); Spondon DRB (*spōn* 'wood-chip', perhaps 'wooden tile'); Warden BDF ('look-out').

With an earlier hill-name or river-name: Brandon LIN (river Brant); Bredon WOR, Breedon LEI (PrW **bre3**); Brendon SOM (*Brūne*, 'the brown one'); Charndon BUC (probably a British name from *carn* 'rock'); Horndon on the Hill ESX, Horrington SOM (a hill-name **Horning*); Laindon ESX (a pre-English river-name identical with Lea); Lansdown SOM (possibly *Langet* 'long strip' used as a hill-name); Longsdon STF, Longstone DRB (*Long*, used as a ridge-name).

With references to topographical features: Aldon SHR (**ǣwylle**); Cleadon DRH, Clevedon SOM (**clif**); Hemerdon DEV (probably *hennamere*, 'bird pond'); Morden CAM, DOR, SUR, Mordon DRH (**mōr**); Watton YOE ('wet place'); Weldon NTP (**well**).

With references to buildings or structures: Berrington NTB, Burdon YOW, Great Burden DRH, Burradon NTB(2) (the qualifier is *burh, byrig* 'fort': when this refers to a prehistoric earthwork it has been classified as a reference to topography, but when, as seems to be the case in these compounds with *dūn*, no ancient fortification is known, the reference may be to Anglo-Saxon structures); Churchdown GLO (here taken to contain a form of *cirice* 'church', see the discussion under **crūg**); Harrowden BDF, NTP (*hearg* 'heathen temple'); Malden SUR, Maldon ESX, Maulden BDF, Meldon NTB (*mǣl*, probably referring to a Christian cross); Weedon BUC, NTP(2) (*wēoh* 'heathen temple').

With references to the position of the hill, or to a settlement in relation to the hill: Bindon DOR ('inside the hill'); Bovingdon HRT and Bowden WLT ('above the hill'); Fordon YOE ('in front of the hill'); Marden SSX ('boundary'); Neadon DEV ('beneath the hill'); Wigton YOW (from the adjacent settlement of Wike).

Names with obscure or ambiguous qualifiers:

Artington SUR (a parish in Guildford). This name and Hartington DRB have similar spellings, and these suggest OE **Heortingdūn*. A noun **Heorting* 'place where there are harts' is possible, but the connective *-ing-* is frequently used with personal names and the base might be a personal name **Heort(a)*.

Baildon YOW. The meaning is 'circle hill', but there are several possibilities as regards the nature of the circle(s). OE *bēgel* is a derivative of *bēag* 'ring', and both words are sometimes used in place-names to denote river-bends. In Baildon, however, the reference may be to the shape of the hill, to the presence of prehistoric stone circles, or to cup-and-ring mark carvings on some of the stones.

Boldon DRH. Bol- could be from *bol* 'tree-stump' or **bole* 'smelting-place'. Ekwall 1960 suggests *bōthl-dūn* 'hill with a building'.

Bouldon SHR. Spellings like *Bolledon, Bulledon* alternate with others like *Bullardun*, and this makes the name impossible to interpret with any degree of certainty.

Brickendon HRT. A personal name **Brica* has been conjectured to suit this and a minor name, Brigden, in SSX.

Bullington HMP. Probably 'hill of the bull', but it seems likely that there was a personal name **Bul(l)a*.

Bursledon HMP. The spellings show an unusual degree of variation. They have been explained (Coates 1989) as deriving from the personal name *Beorhtsige*, alternating with a pet-form **Beorsa*, with complications due to association with the word *bristle*.

Buston NTB (*Buttesdune* 1166). Personal names **Buttel* and **Butta* have been conjectured from this and a few other place-names. There may, however, have

HILLS, SLOPES AND RIDGES

been an OE noun **butt*, perhaps meaning 'mound', of which OE *buttuc*, modern *buttock*, would be a diminutive.

Canewdon ESX (*Carenduna* DB, *Canuedon* 1181, *Canewaudon* 1201, *Kanewedun* 1224). This has been considered to be **Caningadūn*, 'hill of Cana's people', on the grounds that in some other names *-inga-* has developed into ME *-ew-*. Manuden ESX and Monewden SFK are cited as parallels, but the earliest spellings for these have *-ng-*, and there is no sign of this in the forms for Canewdon. Leaving aside the probably corrupt DB spelling, a base such as **can(e)we* seems to be indicated.

Clarendon WLT. The spelling *Clarendun* 1072 seems to rule out 'clover' as in Claverdon WAR. A Celtic river-name has been conjectured as the origin of Clare SFK and Clere HMP, and this seems a possibility for Clarendon. It might have been the earlier name of the river Bourne.

Crandon SOM. The two earliest spellings have *Gr-*, which would support derivation from *grēne* 'green', but other spellings have *Cr-*, which might indicate 'crane'. There is also, however, a puzzling alternation of *-e-* and *-a-*.

Dixton GLO. Earlier commentators favour the genitive of a compound **dīc-hyll*, 'hill with earthworks', but the spellings are more consistent with the hypothetical personal name **Diccel* which is believed to be the basis of Ditchling SSX.

Flordon NFK. The meaning is 'hill of the floor', but the nature of the referent is not known. OE *flōr(e)* is used in place-names for a Roman tesselated pavement (Fawler OXF, from *fāgan flōre* 'variegated floor') and also for a flagstone paving (Fawler BRK, earlier *Flagaflora*), but without a qualifying term *flōr(e)* cannot be precisely defined. Other possibilities such as 'threshing floor' may be considered.

Gaydon WAR. A personal name **Gǣge* has been conjectured to explain this name and a few others. Gayton is sometimes 'goat settlement' with ON *geit*, but this is not likely for Gayton STF and is not possible for the present name or for Gaywood NFK.

Gomeldon WLT. Possibly a personal name **Gumela*.

Hartington DRB. See Artington *supra*.

Hillerton DEV. There is an OE spelling *(on) healre dune*. The only available suggestion is Ekwall's (in EPNS survey p. 361) of an OE **hēalor* 'swelling' related to *hēala* 'rupture'. There is another instance in the bounds of Chaceley WOR, *healra mere*.

Latchingdon ESX. The spellings require an OE *læcen*, which is unexplained.

Leysdown KNT (*Legesdun* 11th century). The genitive of the masculine form of **lēah** would be formally acceptable, but there is no firm evidence for the use of lēah as a first element, and the word is usually feminine in KNT. The only suggestion available is that of Ekwall 1960, OE *lēg* 'fire', a word not otherwise noted in place-names.

Luddesdown KNT (*Hludesduna* 10th century). OE *hlūd* 'loud' may have been used as a personal name. A derivative, *Hlūde*, was certainly used as a river-name, becoming Loud LNC, Louth LIN. There is no river at Luddesdown, but a spring called *Hlūde* is perhaps worth considering.

Ludstone SHR. Said in Ekwall 1960 to be identical with Luddesdown KNT, but the early spellings make it clear that this name had original *-dd-*. Probably a strong form of the recorded personal name *Ludda*.

Matson GLO (earlier *Matteresdon*). OE *mǣthere* 'mower' is probably ruled out by the absence of *-th-* spellings. A late OE **mattere* 'mat-maker' has been suggested, but this is not convincing in a *dūn* compound.

Mundon ESX. As with Munden HRT the first element could be OE *mund* 'protection' or a personal name **Munda*. The first alternative could be considered to refer to protection from floods. Mundon is a very low *dūn* overlooking the marshes of the Blackwater estuary.

Peldon ESX. The suggestion in Ekwall 1960 of a term meaning 'thruster', from an OE **pyltan* 'to thrust' would suit the situation of Peldon, on a low *dūn* which (like Mundon) overlooks the Blackwater estuary. NED gives the hypothetical OE *pyltan* as the base of recorded ME *pilte, pulte*.

Pilsdon DOR, Puesdown (in Hazleton) GLO, Puleston SHR. A full range of forms for Puleston SHR and Puesdown GLO suggests an OE **Pēofelesdūn*. The smaller range of forms given in Ekwall 1960 for Pilsdon DOR includes *Pyulesdon* 1268, which could point to the same origin, but which requires support from more spellings of the same type. A personal name **Pēofel* may reasonably be conjectured as a diminutive of the recorded *Peuf(a)*, but a triple occurrence with *dūn* would suggest that there was a significant term, perhaps a hill-name, with *dūn* added to the genitive. The development to *Pil(l)es-* found in Pilsdon DOR is, however, not seen in Puesdown and Puleston, so Pilsdon should probably be dissociated and taken (in spite of the *Pyueles-* spelling) as having the gen. of *pīl* 'stake' as first element.

Polden SOM (*Pouldon* 1241, *Poweldun* c.1240). This probably incorporates the lost name of an estate called *Pouelt* 705, *Pouholt* 729, but the etymology of the estate-name and its relationship to Polden have not been securely established.

Poundon BUC. Only three ME spellings are available. These are consistent with derivation from OE *pāwa* 'peacock'.

Saredon STF. Ekwall 1960 suggests either a personal name **Searu* or a hill-name **Sēar* meaning 'dried, withered'. The use of the same first element in nearby Sarebrook (*Searesbroc* 996) appears to rule out an adjective. As the personal

HILLS, SLOPES AND RIDGES

name is of doubtful validity, and there is no parallel for the suggested hill-name, the nature of the qualifier is best left open.

Selsdon SUR. Three possibilities are set out in Ekwall 1960: a personal name *Sele, OE sele 'hall' and *sele 'sallow copse'. These are all plausible.

Shillington BDF (Scytlingedune 1060). V. Watts suggests a hill-name *Scytteling, from scyttel 'dart, bolt', modern shuttle, rather than an -ingas formation from an unrecorded personal name.

Slindon SSX, STF. A good case has been made for an OE *slinu 'slope', and Ekwall 1960 suggests that this word occurs also in Slinfold SSX and Slyne LNC. A word meaning 'slope' does not, however, seem meaningful in combination with dūn.

Stildon WOR (Stilladun c.957). Ekwall 1960 suggests the genitive plural of stiell, a form of stæll, which last is recorded in the sense 'place for catching fish'. He conjectures that it also meant 'trap for wild animals', or that it had the sense of northern dialect stell 'enclosure for giving shelter to sheep or cattle'. The Mercian form of this word is stæll, however, whereas stiell is the West Saxon form. The nature of the qualifier in Stildon has not been determined.

Winchendon BUC. The first element is OE wince, modern winch, and the reference could be to an installation for helping carts up a hill; but the word may have had other senses in place-names. Germanic cognates suggest 'bend, corner'. It is the second element in hleapewince 'lapwing', and it may have been used uncompounded as a bird-name.

Yeadon YOW. EPNS survey (Part 4, xi) suggests an OE *gǣh 'precipitous', a word not otherwise noted in place-names.

ecg 'edge'. There is little to add to the discussion of this word in Gelling 1984. It is a rather rare place-name element, better represented in minor than in major settlement-names, and fairly common only in a few counties such as DRB, GLO, YOW. It can be used of slight slopes, such as that at Drakenage ('dragon'), a well-recorded minor name in Kingsbury WAR, north-east of Coleshill, or rock scars in fairly low ground (e.g. Edge CHE), or long, low ridges (e.g. Edge in Pontesbury SHR). It is occasionally used of dramatic rocky escarpments, as in Hathersage DRB, probably 'ridge of the he-goat' referring to Millstone Edge. Harnage SHR (*hæren 'stony') refers to a long, fairly steep slope. Heage DRB ('high') is in very broken country north-east of Belper. Stanedge DRB and Standedge YOW are 'stone edge'.

Liversedge YOW is not a certain example. EPNS survey 3, p. 27, prefers 'sedge' on the grounds that nothing in the local topography could be called ecg. As noted above, however, the word does not always refer to dramatic landscape features.

THE LANDSCAPE OF PLACE-NAMES

Edge Hill WAR is an early-recorded example (from c.1250) of a hill-name which recurs in various counties. In some GLO names, Aston and Weston Subedge and Wootton under Edge, *ecg* is used as part of an affix, referring to the scarp slope of the Cotswolds. There are many instances, such as Alderley Edge CHE, Wenlock Edge SHR, of an escarpment or slope being named from a nearby settlement.

Ecg- is a common first theme in dithematic OE personal names, and this is sometimes the explanation of Edg- in modern place-names, as Edgbaston WAR ('Ecgbald's estate') and Edgerley SHR ('Ecghard's clearing'). Edgeworth GLO, LNC should probably be derived from the monothematic personal name *Ecgi*, which is evidenced in Edgware GTL (*Æcges wer* c.975). Edgefield NFK and Edgeley CHE, SHR contain *edisc* 'enclosed pasture'.

haugr ON 'tumulus, hill'. This word was not included in Gelling 1984 because that study (like the present one) was primarily concerned with settlement-names, and the great majority of names containing *haugr* do not fall into that category. A few do, however, so the element deserves a short article.

Smith 1956 says that in Scandinavia 'burial-mound' appears to be an older sense than 'hill'. There are many instances of the 'tumulus' sense in field and minor names, especially in north-eastern counties. Hoon DRB and probably Holme on the Wolds YOE, both from the dative plural, are likely to refer to groups of barrows, and some field-names in which *haugr* is compounded with a numeral probably have this significance. In Hallsenna CMB (earlier *Sevenhoghes*), however, the word is more likely to be used for the small hills which surround the settlement: the 1in map marks Stony How, Black How, Tarnhow and Bolt How.

In the Lake District *haugr* is used for mountains and large hills, as in Skiddaw and Ulpha CMB (Ulpha WML, though also a 'wolf' name, has a different generic). Some other CMB names, such as Whelpo, probably contain *haugr* in this sense, and LNC names which refer to hills include Clitheroe and Blacko in Colne. This use of *haugr* may be compared with the northern English and Scottish use of **hlāw** for large hills and mountains. In field-names, however, the likeliest meaning is 'tumulus'. Also like **hlāw**, *haugr* occurs in a number of meeting-place names.

ON *haugr* cannot always be distinguished from OE **hōh**, but spellings such as *-how(e)*, *-hou* are likely to be more frequent from *haugr*, and the Norse word is also to be preferred in compounds with ON qualifiers, such as *hvelpr* 'whelp' in Whelpo and the variant of *skúti* 'overhanging crag' probably found in Skiddaw. Cracoe YOW, for which Ekwall 1960 gives *hōh*, is more likely to contain *haugr*, as indicated in EPNS survey 6, p. 89. There are two other occurrences in YOW, and the name is probably an ON compound with *kráka* 'crow, raven'.

hēafod 'head'. This word was given a fairly detailed discussion in Gelling 1984, and the conclusions arrived at there still seem tenable. In the late Anglo-Saxon and the medieval periods *hēafod* was used in charter boundaries and in minor place-names in a variety of senses. These included 'end' (e.g. *cumbes heafod, mænan mores heafud* in BRK surveys) and 'source' (e.g. *holan broces heafod*, also in BRK). In field-names *hēafod* was used for the headland in a ploughed furlong. In ancient settlement-names, however, where it occurs between 30 and 40 times, the only sense observed is 'projecting piece of land'. It probably did not mean 'peak' or 'summit', as no instance has been noted in which it clearly refers to the highest part of a feature. There are a number of instances in which it clearly refers to a piece of land which juts out below the level of the rest of a massif, and this may be connected with the manner in which some animals, such as pigs and badgers, habitually carry their heads below the level of their shoulders. OE *dūn-hēafod* gives rise to four names: Donhead WLT, Dunnet Fm (in Compton Bishop) and Downhead SOM, and *Downhead* CNW, the old name of Launceston. This compound is wrongly translated in EPNS WLT survey and in Ekwall 1960 as 'top of the down', more correctly in Padel 1988 as 'hill-end'. The two SOM places are very precisely sited on flat ground at the lower points of narrow ridges.

hēafod is sometimes used of very low projections, as in Hatford BRK, where a very slight elevation raises the village above the marshy course of Frogmore Brook. Howden YOE (the only other name in which *hēafod* is a first element) poses special problems, and these are discussed under **denu**.

A distinguishing feature of the corpus of *hēafod* names is the numerical dominance of the type in which it is combined with the name of a living creature, usually in the genitive singular. This led to a suggestion by Henry Bradley in 1910, which was elaborated by Bruce Dickins in 1934, that there was a reference to pagan rituals in some, at least, of the names. Professor Dickins' presentation of this theory was published as an appendix to the EPNS survey of SUR. Such compounds are especially well represented in SUR minor names: examples include Eversheds Fm in Ockley (*eofor* 'boar'), Heronshead Fm in Leigh ('eagle', not 'heron') and Worms Heath (earlier *Wermeshevede*) in Chelsham. Minor names in other counties include Cats Head Lodge in Sudborough NTP and Ravenshead Wood northeast of Newstead NTT. Two instances of Manshead were also noted: Manshead Hundred BDF, and a lost *Mannesheved* in Hawton NTT. There are recurrences in minor names of Gateshead and Hartshead, but by far the commonest name in the category is the 'swine's head' type. Five instances are listed in the reference section, and at least another eight are known. One of the additional names is in Scotland: Swinside Hall in Oxnam ROX. There are occurrences in charter boundaries in WLT and IOW, and in field-names in BRK, CHE, HRT. There is also a hundred-name: Swinehead in GLO. It is probably no more than a coincidence that two names in the whole corpus refer to hundred meeting-places.

The suggestion of pagan sacrificial feasts, and, in Bradley's discussion, of the display of the sacrificed creature's head on a pole, postulates a manner of place-name formation which is at odds with present-day understanding of the Old English naming system. It was rejected by Ekwall (1960, p. 229) and Gelling (1961). A transferred topographical use of *hēafod* is to be preferred, and 'swine's head' was perhaps an appellative for projecting ridges in low ground, of a sort exemplified by Swinesherd (*Swinesheafod* in OE sources) southeast of Worcester. 'Swine's snout' might be a better translation.

Some coastal 'creature's head' names can be shown to refer to appropriately-shaped headlands. Padel 1988 says that Gurnard's Head CNW resembles the head of a gurnard fish, and it is generally agreed that Great Ormes Head near Llandudno, which is from ON *ǫrmr* 'snake', has a strikingly reptilian appearance. In Worms Head GLA, however, the name is less apt: the tongue of rocks has a wavy outline which might be thought snake-like, but there is no obvious reptile's head.

The 'animal's head' names do not necessarily all have the same significance. Hartshead and Gateshead could be headlands where such animals are frequently seen.

Other categories of first element are small, the largest of them consisting of references to vegetation. There are not many references to ownership or occupancy. Four instances have personal names as qualifier, and the headland referred to in the minor name Thiefside in Hesket CMB must have been associated with a thief. Conishead LNC is 'king's headland'.

Early forms are essential for the detection of -*hēafod* in a variety of modern disguises. It should be noted that while Hesket Newmarket CMB (*Eskhevid* c.1230) is 'ash headland' all other instances of Hesket are from ON *hestakeith* 'race-course'.

The corresponding ON *hǫfuth* occurs in Holleth and Preesall LNC and Whitehaven CMB, and sometimes interchanges with *hēafod* in names which are primarily of OE origin. Ormside WML and Conishead LNC could be ON. Bradda IOM, an ON name meaning 'steep headland', is named from Bradda Head near Port Erin, which is an excellent example of a headland lower than the ridge from which it projects.

Reference section for hēafod

As first element: Hatford BRK; Howden YOW (**denu**).

As generic with first elements referring to living creatures: Broxted ESX ('badger'); Farcet HNT ('bull'); Gateshead DRH ('goat'); Hartshead LNC, YOW, Hartside CMB, NTB; Ormside WML (ON *ǫrmr* 'snake'); Rampside LNC ('ram'); Read LNC (*rǣge* 'female roe-deer'); Shepshed LEI ('sheep'); Swinchurch STF, Swineshead BDF, Swinesherd WOR, Swinside (in Lorton) CMB (all 'swine's head'). Other instances are noted in the text, *supra*.

HILLS, SLOPES AND RIDGES

With references to vegetation: Birchitt (in Cold Aston) DRB, Birkenhead CHE ('birch'); Fearnhead LNC ('fern'); Hesket Newmarket CMB (ON *eski* 'ash'); Lindeth LNC, WML ('lime tree'); Sparket CMB (**spearca* 'brushwood'); Thicket YOE (*thicce* 'thicket').

With personal names: Arnside WML (*Earnwulf*); Burneside WML (perhaps Continental *Brunulf*); Fineshade NTP (*Finn*); Macknade (DB *Machehevet*) near Faversham KNT (*Maca*). The only other reference to ownership is ON *konungr* 'king' in Conishead LNC.

With topographical terms: Cams Head (in Kilburn) YON (**camb**); Donhead WLT, Downhead SOM, *Dunheved* (later Launceston) CNW, Dunnett SOM (all **dūn**); Portishead SOM ('harbour').

With an earlier hill-name: Mamhead DEV (*Mam* is probably a Celtic hill-name); Minehead SOM (possibly **mönith**); Quantoxhead SOM (*Cantuc*, believed to be the pre-English name of the hill-range).

With obscure first elements:

Consett DRH (*Conekesheved* 1183). This name and Conock WLT contain an obscure element which may be pre-English.

Lupsett YOW. A personal name **Hlupp(a)* has been conjectured to explain this name and Lupridge DEV, Lupton WML.

helde (Anglian, Kentish), **hielde** (WSaxon) 'slope'. This is a rather rare term in settlement-names, though fairly well evidenced in field and minor names in some counties, especially CHE, SHR, SUR, YOW.

helde may be the first element of Hilton DOR, Hylton DRH and Helton and Hilton WML. It is certainly the second element of Akeld and Learchild NTB (northwest of Wooler and southwest of Alnwick), Redhill in Reigate SUR (*Redehelde* 1301), Shooter's Hill GTL (*Sheteresheldo* 1292), Stockeld near Harrogate YOW, and Tylerhill near Canterbury KNT (*Tylerhelde* 1304).

Field-work undertaken since the discussion in Gelling 1984 suggests that *helde* was a specialised term for an inclined plane which was less steep than a **clif**. Plate 1 in Foxall 1980 is a photograph of a field called The Yelds in Berrington SHR, and this gives an excellent impression of the type of feature for which *helde* was used. Another relevant SHR name is Yell Bank, a ridge north of Cardington. A road runs along the 1,000ft summit, and on either side of the road are symmetrical slopes of less than 45°. This angle may be the deciding factor in the use of *helde* rather than **clif**. At Learchild NTB there is a long slope of similar character.

An alternative form *hilde*, seems likely to be the generic of Salkeld CMB, which is discussed further under **holt**. The first element is *salh* 'willow'.

hlāw, West Saxon **hlǣw** 'tumulus, hill'. This is primarily a term used for artificial mounds. It was generally preferred by the Anglo-Saxons for their own barrow burials, *beorg* being the commoner one for those of earlier cultures, but the distinction is not absolute. It has recently been established that *hlāw* was the term used for tumulus-like mounds which the Anglo-Saxons sometimes erected as markers for meeting places. This was demonstrated in Adkins and Petchey 1984, following the excavation of a mound called *Secklow* in Milton Keynes. The paper documents eleven meeting-place mounds which excavation has shown to be non-sepulchral, and six of these have *hlāw* names. The authors also point to the formation in 964 of a Worcestershire administrative district called *Oswaldslow* in honour of the contemporary bishop of Worcester. It was previously assumed that the frequent appearance of *hlāw* in hundred-names was due to the deliberate siting of meeting places at tumuli, but in the light of the Adkins and Petchey paper it seems probable that many of the mounds were constructed for the purpose.

There are, however, some *hlāw* names which certainly refer to late pagan Anglo-Saxon barrow burials, the most famous being Taplow BUC. Here there is a tumulus in the churchyard from which were recovered the rich grave-goods of an early 7th-century Saxon prince who may have been the eponymous Tæppa. At Lowbury Hill in Aston Upthorpe BRK there was another late Saxon burial. A number of the famous barrow burials of the Peak District were in tumuli with *hlāw* names.

In some instances where a *hlāw* name coincides with an Anglo-Saxon burial this is a 'secondary' in a prehistoric tumulus. There is an example in WLT, where a Bronze Age tumulus near Swallowcliffe, which is called *posses hlǣw* in a charter boundary, has revealed the splendid secondary burial of an Anglo-Saxon woman. Secondary burials may have been missed in some barrows subject to unskilled excavation, but it is not likely that this has happened in regard to all prehistoric tumuli with *hlāw* names.

At Cutteslowe near Oxford a barrow was ordered to be destroyed in 1261 because 'evil doers are wont to lurk in the hollow', and Dr Blair (1994, p. 39) comments that this sounds like a neolithic chambered tomb. Bartlow CAM refers to a cluster of Romano-British burial mounds. Ludlow SHR (Gelling 1988, p. 136) refers to a tumulus which was probably pre-English, and the recorded mound at Farlow SHR is believed to have contained a Bronze Age burial (Stanford 1980, p. 71). Compounds with numerals occur in minor names, as Twemlow CHE and Rumbelow STF ('two' and 'three barrows'), and folkloric associations are responsible for the recurrent names Drakelow ('dragon's mound') and Hordlow or Hurdlow ('treasure mound'). The WOR Drakelow, near Kidderminster, is, however, a rocky outcrop, not a tumulus.

The archaeological sense of *hlāw*, *hlǣw* is important, but the precise archaeological significance (or lack of it) has to be considered for each individual instance. Suggestions made in Gelling 1988 about the geographical parameters of the archaeological use still appear reasonably valid. In England south of a vague line from the Mersey to the Humber the term is more likely than not to refer to an artificial mound. In Harlow ESX, if the reference is to the hill on which the Roman temple stood, *hlāw* may have seemed appropriate because the hill had been scarped. North of this vague line *hlāw* is commonly used of natural hills, and sometimes of mountains. In Stanlow CHE the reference is considered to be to a rocky promontory. The 'mountain' sense is seen in such LNC names as Horelaw and Pike Law, and this is common in southern Scotland. In NTB *hlāw* is one of the commonest terms for a natural hill. In some DRH names, such as Kelloe and Moorsley, the nature of the *hlāw* is uncertain.

A major correction to the discussion of this element in Gelling 1988 and 1984 is the deletion of Lewes SSX, which does not derive from the plural of *hlǣw*. This derivation became established before the publication of the early 10th-century text known as the Burghal Hidage. The spelling *Lǣwe* in that text lacks the *H-* which would have been expected, and Coates 1991 musters additional phonological objections to derivation from *hlǣwas*.

Low occurs fairly frequently as a simplex minor name. Lew OXF, from *hlǣw*, is a parish named from a large tumulus said in Blair 1994 (p. 45) to be 'one of the most conspicuous landmarks in west Oxfordshire'. Use as a first element is rarer, but occurs in two CHE names, Buglawton and Churchlawton (both *Lautune* in DB).

In settlement-names *hlāw* is frequently combined with a personal name, as in Challow BRK, Bledlow, Cottesloe, Taplow and Winslow BUC, perhaps Thriplow CAM, Alcumlow, Bucklow, Chidlow, Hadlow, Hankelow and Wilmslow CHE, Boblow (in Helion Bumpstead) ESX, Wolferlow HRE, Betlow (in Tring) HRT, Cutslow OXF, Beslow, Longslow, Munslow, Pursloe, Walkerslow and Whittingslow SHR, Baldslow (in Hastings) SSX, probably Winterslow WLT. Such compounds are common also in hundred-names, and Kitson (forthcoming) notes that a personal name is the qualifier in more than half of the *hlāw/hlǣw* names in charter boundaries. It is obviously possible that many of the men named (there appear to be no women) are the Anglo-Saxon aristocrats buried in the tumuli.

Other qualifiers include descriptive terms: Henlow BDF was 'high', Copley and Coploe CAM contain **copp**, Shardlow DRB was 'notched' and Ploughley OXF was 'baggy'. There are some references to animals and birds, notably 'hawk' in Hawkesley (in King's Norton, Birmingham) and in a BRK charter boundary, and 'hound' in Hounslow GTL and in charter bounds in BRK and SOM. The BRK examples, *hafoceshlǣwe* and *hundeshlǣwe*, are surviving adjacent tumuli on the boundary of Uffington parish. Tadlow CAM may refer to toads. There are a few references to vegetation ('birch trees' in Bartlow CAM, 'cress' in Creslow BUC) and some to adjacent topographical features (Brinklow, Pathlow and Woodloes WAR).

THE LANDSCAPE OF PLACE-NAMES

Two hundred-names, Wenslow BDF and Thunderlow ESX, refer to pagan gods. References to dragons and treasure have been noted above. There are references, interesting though of uncertain significance, to classes of people in Cumberlow HRT ('Britons'), Knightlow Hundred WAR ('young warriors') and Harlow ESX ('army').

This impressionistic treatment of *hlāw/hlǣw* compounds has been included here instead of a thorough analysis in a reference section because only a small proportion of the material is contained in settlement-names. A reliable analysis of qualifiers would have to include occurrences in minor names, field-names and charter boundaries, and a full examination of these sources is outside the scope of the present study.

*hlenc 'extensive hill-slope'. This word is not recorded in literary OE. Its existence and meaning are inferred from its occurrence in the names of a line of villages stretching north from Evesham WOR. These are Lenchwick, Sheriff's Lench, Atch Lench, Church Lench, Abbots Lench and Rous Lench, which show that Lench must have been the name of a district extending 5 miles from north to south. The central feature of the area is a band of lower lias limestone, and the massif is referred to as *Lencdun* in the OE boundaries of Twyford. The word is related to **hlinc**.

There is a place called Lench in LNC, south of Rawtenstall, but this name is not recorded until 1526, and should perhaps be ascribed to the dialect derivative *lench* 'ledge of rock' rather than seen as evidence for the survival in the pre-Conquest place-name-forming vocabulary of the OE term as used in WOR. This consideration applies to a few other minor names in northern England.

hlinc 'bank, ledge' (Fig. 34). NED and Holthausen 1934 suggest a connection with *hlinian* 'to lie down, to lean'. Modern derivatives are dialect *linch* and its alternative form *link*, the latter being obsolete except for the use of the plural in golf-links. The word is fairly frequent in charter boundaries, where it can often be shown to refer to the cultivation terraces on hillsides which are known to modern students of landscape as strip lynchets. A major name in which *hlinc* refers to cultivation terraces is Stocklinch SOM, where the church of Stocklinch Ottersey has the dramatic background of a steep slope with five large strip lynchets.

There are fewer than 20 major settlement-names containing *hlinc*, but the word is better represented in minor names and field-names. Field work reveals that the reference is frequently to a terrace, natural or man-made, which carries a road.

Major names which refer to terrace roads include Linkenholt HMP and Moorlinch SOM. At Linkenholt there are several terrace roads on steep hillsides. At Moorlinch a terrace road winds from the village, on the edge of

King's Sedgemoor, up the Polden Hills. Sticklinch SOM, east of Glastonbury, is situated on a road which runs along one of the lower ledges of the Pennard Hill. This is probably the sense in Link WOR, at the north end of the Malvern Hills: the 19th-century O.S. map shows The Link as the name of the road running south-west from Malvern Link, climbing the lower slopes of the hills. Another instance can be seen at Lyng NFK, where a gently undulating road follows the curves of the river Wensum.

Man-made terraces can sometimes be seen at farms with names derived from *hlinc*. There is a particularly fine example at Lynch Fm west of Tenbury Wells HRE (GR 547676). Here there is a terrace alongside the modern sunken road but at a slightly higher level, which runs for half-a-mile from the farm to the river Teme. At Stonelink in Fairlight SSX the short approach road to the farm has been terraced into the slope. Margary 1955 (pp. 71-2) describes a Roman road running north-west from Chichester as 'a terrace way of the usual type' at Linch Fm between Treyford and Bepton SSX.

There are probably other terrace ways to be observed at places with *hlinc* names, but the O.S. map is usually too blunt an instrument for their detection. Major names and some well-recorded minor names are listed in the reference section, but a full study would have to take account of more minor names and of occurrences in charter boundaries.

As an element in minor names *hlinc* is particularly well represented in DEV and GLO. It does not occur in all counties: no examples have been noted in CMB or WML, in the east midland counties of DRB and NTT, or in Yorkshire.

Reference section for **hlinc**

As simplex name: Linch SSX, (Malvern) Link WOR, Lintz DRH, Lynch SOM, Lyng NFK.

As first element: Lingwood NFK, Linslade BUC (**gelād**), Linkenholt HMP (**holt**, the name probably contains an adjective **hlincen* which could be translated 'terraced').

As generic: Charlinch SOM (?personal name); Lydlinch DOR (river Lidden); Moorlinch SOM (*myrig* 'pleasant'); Sandlin (in Leigh) WOR, Sandling KNT(2) ('sand'); Shanklin IOW (*scenc* 'cup'); Stallenge DEV, Standlynch WLT and Stonelink SSX ('stone'); Sticklinch SOM ('steep'); Stocklinch SOM ('tree trunk'); Sydling DOR ('broad'); Swarling KNT ('sword'); Whitlinge WOR ('white').

hlith 'slope', specifically 'concave hillside' (Fig. 35). Field-work undertaken since the inconclusive discussion in Gelling 1984 has identified a specific sense for *hlith*. This meaning is very clearly attested in west midland counties, and is probably prevalent in southern areas. In northern counties it is not always possible to distinguish OE *hlith* from the ON equivalent **hlíth**, the place-name usage of which is discussed below.

hlith is not one of the commoner hill-terms used in name formation, and its distribution is patchy, a characteristic which often indicates dependence on regional landscape forms. The main concentration is in south SHR and north HRE. Here it occurs in settlement-names at Lyth (south of Shrewsbury),

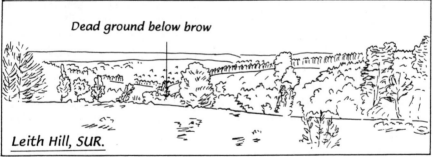

Evelith (in Shifnal) and Huglith (in Castle Pulverbatch) SHR, and in Gatley Park, Shirley, Underley and Wapley HRE. Wapley is now only the name of a hill, but the early documentation shows that it was once the name of a settlement also. These settlement-names are supported by hill-names such as Ragleth near Church Stretton: the Lawley, in the same range, was *Lallelyd* in 1221. The element is also found close to the Welsh border in Hoseley FLI, Pilleth RAD and Todlith MON.

The most striking SHR example is the great hill now called Pontesford Hill, which was *Ponteslith* in the 14th century and *Ponslythe* until about 1600. This massif has two peaks, with a great hollow between them on one side. Wapley Hill HRE, by the road from Shobdon to Presteigne, is a large massif with a deep hollow in the side which overlooks the road. On a smaller scale, but impressive, is Gatley Park HRE (earlier *Gattelyth*), where the manor house overlooks a natural amphitheatre in the hillside. At Huglith SHR and Shirley HRE the settlements are enclosed by horseshoe-shaped areas of high ground.

Such formations can sometimes be identified at relevant sites in southern counties. The Lyde in Bledlow BUC is a steep-sided hollow which has been landscaped into a park, and there is a narrow, wooded hollow in the hill at Warninglid, south-east of Horsham, SSX.

A related use of *hlith* is for hills and escarpments which have a hollow at the foot. Lyth Hill SHR is a striking instance, and at West Leith near Tring HRT there is a formation very similar to that on the east side of Lyth Hill. The farm at Lythe Brow near Lancaster occupies a hollow at the foot of a slope.

Another sense of *hlith* is found in Lytham St Annes LNC. Here the dative plural may be considered to have been used for the sand dunes which old maps show all along the coast. Sand dunes are often blown into concave-hill formations, and *sandhlith* is actually recorded in OE.

In other instances the sense of *hlith* is not so clear-cut. Leith Hill SUR, for instance, is a great massif with an unstable surface and hollows all over it. Sufficient instances of hillsides with a single large hollow in one side or a concavity at the foot have, however, been identified to establish that the word had this specialised meaning.

hlith is a landscape term which is used relatively frequently by the author of *Beowulf*, sometimes in the plural, *hleothu*. If it be accepted that the word had the specialised sense of 'concave hillside', the place-name evidence could contribute to the interpretation of some passages in the poem. In a number of instances the word has a menacing context. Grendel comes from the marsh 'under misthleothum' and when fatally wounded he returns to his lair 'under fenhleothu'. When Beowulf is told of the dangers he must traverse on his way to the monsters' lair these include 'wulfhleothu', and the description of the journey lists 'steap stanhlitho, stige nearwe, enge anpathas, uncuth gelad, neowle næssas' as well as 'nicorhusa fela' or 'many monster-lairs'. It has been suggested under **gelād** that the place-name use of that term is relevant to the understanding of this section of the poem, and the entries for **stīg** and **pæth** may also have something to offer. As regards *hlith*, the hollows in the sides of hills and those at the foot of slopes are dead ground in military terms, and could be lurking places for monsters if such were about. The implication of concavity could be relevant also to the use of *hlith* in the description of the lake, where monsters were seen lying 'on næshleothum'. A scooped-out hollow in a **næss** would provide a platform.

Because of its rarity *hlith* is frequently corrupted to -ley in modern forms, and it is clear that confusion with **lēah** occured in ME. Lyth SHR is *Lithe* or *Lythe* in most early spellings, but *La Lya* and *Leye* also occur. Hoseley FLI would not be suspected of containing *hlith* if a pre-Conquest form had not been available. Coreley SHR has been classified as a *hlith* name on the basis of *corna hlith* in a charter boundary, but in EPNS survey Part 1, p. 98, it is argued that Coreley is a *lēah* name, and that the bounds record other places in the vicinity – *cornahlith, cornabroc, cornawudu* – with the same qualifier but different generics.

Even when a number of minor names have been included in the reference section there are not enough compounds for analysis of first elements to be significant. The relatively high proportion of simplex names, however, supports the hypothesis of a specialised meaning. For a feature to be known as *hlith* it would be necessary for it to be different from other hills and slopes.

Reference section for **hlith**

As simplex name: Leith HRT, SUR, Lyde BUC, Lyth SHR, WML, Lytham LNC (dat.pl.), Lythe LNC, SUR, YON, YOW. Bowler's or Bowlhead Green near Thursley SUR is '(place) above the *hlith*' and Underley HRE is '(place) under the *hlith*'.

As first element: Leathley YOW (*lēah*); ?Litton DRB, YOW (*tūn*); Lytheside WML (*sīde*). Lenborough BUC, earlier *Lithingeberge*, is considered to contain an *-ingas* folk-name based on *hlith*.

As generic with personal names: Adgardley LNC (*Ēadgār*); Hoseley FLI (*Hod*); Huglith SHR (probably *Hugga*).

With other qualifiers: Evelith SHR and Ivelet YON ('ivy'); Gatley HRE ('goat'); Howler's Heath (in Bromsberrow) GLO ('owl'); Pilleth RAD (obscure); Risplith (in Sawley) YOW ('brushwood'); Shirley HRE ('bright'); Todlith (in Church Stoke) MON ('fox'); Wapley HRE (?'bubbling spring'); Warninglid SSX (obscure).

hlíth ON 'hillside'. Usage of the ON word differs from that of OE *hlith* in the frequency with which it designated districts. In Yorkshire there were districts called Grindalythe, *Hertfordlythe, Holdelythe* and *Pickeringlythe*, one of the Wards of Cumberland is called Leath, in Nottinghamshire the north part of Thurgarton Wapentake was called *Lythe*, and Lythe is the name of a district of WML east of Windermere. Grindalythe only survives as a affix in Kirby Grindalythe, but it is recorded in early references to a number of other settlements in the same valley, including Butterwick which is six miles away. *Holdelythe*, which contains ON *holdr* 'nobleman', is perhaps the only name in the list for which a certain ON origin can be asserted. In Grindalythe, Grindal- is earlier *Crandale*, which is probably a hybrid OE/ON name from *cran* 'crane' and **dalr**. In the other names the word could theoretically be the English one, but the district-name usage is so distinctive that a different origin is more probable. The identity of the Yorkshire districts is set out in EPNS YOE survey, pp. 12–13.

In settlement-names it is assumed that the ON word is the relevant one when it is combined with another ON word or an ON personal name, and such items are listed in the reference section. A systematic check of these sites might suggest a specific meaning, perhaps resembling the OE one, though nothing so specific can apply in the district-names. At Over Kellet LNC the spring (**kelda**) was in a hollow in the limestone below the church, with rising land all round.

Reference section for **hlíth** ON

As simplex name and first element: Kirk- and Upleatham YON (the forms indicate a dative plural formation, except for DB *Upelider*, which probably represents ON nominative plural *hlíthir*); Lythe Beck (in Whitby) YON (**bekkr**); Litherland(2) LNC (**land**); Litherskew (in Aysgarth) YON (**skógr**).

As generic: Ainstable CMB (ON *einstapi* 'bracken'); Bainley Bank (in Danby) YON (earlier *Bainwithelith* from ON *bein-vithr* 'holly'); Gatecliff (in Hetton) YOW (ON *geit* 'goat'); Hanlith YOW (ON pers.n. *Hagne*); Ireleth LNC ('of the Irish': the qualifier could be ON *Írar* or OE *Íras*, but the presence of Irishmen suggests a Viking-period name); Kellet LNC and Kelleth WML (*kelda* 'spring'); Stennerley LNC (ON pers.n. *Steinarr*). District-names are listed in the article.

hōh 'heel', used in place-names for a projecting piece of land. Since the discussion in Gelling 1984, fieldwork has identified a specialised use of *hōh* which is related to the anatomical sense.

Two main types of hill-spur are to be observed in the English countryside: level ridges with convex shoulders, for which **ofer** and **ōra** are used, and ridges which rise to a point and have a concave end. It is these last which are frequently referred to in *hōh* names, and the shape is in fact that of the foot of a person lying face down, with the highest point for the heel and the concavity for the instep. Classic instances occur in widely separated areas. Piddinghoe SSX, Ivinghoe BUC, Tysoe WAR, Rainow CHE, Weo SHR, Clougha LNC and Ingoe NTB are outstanding examples, and some of these are illustrated in Fig. 36.

This is, however, a land-form which does not occur in all regions, and where it is absent, as in East Anglia, *hōh* is used loosely of any spur. Also, the term is used occasionally of very low ridges which do not have diagnostic shapes, and this occurs even in areas where the classic *hōh* shape is found together with the precise use of the word. The exact application of *hōh* in any name must be determined by fieldwork. Minor names such as Hoo, The Hoo are found in areas where any ridge can only be a low one.

An unusual feature of the corpus of *hōh* names is the high proportion of compounds in which it is the first element, especially with *tūn*. The reference section lists 58 names derived from *hōh-tūn*, and there are a few other compounds, including seven with **land**. Together with 13 simplex names this

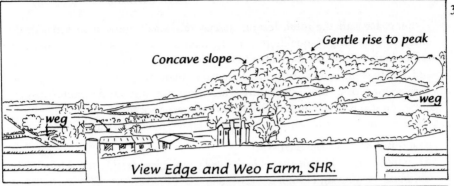

accounts for half the total, leaving a relatively small body of material to be analysed according to qualifiers. Among names in which *hōh* is the generic, a rather high proportion have qualifiers which present etymological difficulties, and *-inga-* compounds are more numerous than might have been expected with a word for a landscape feature which does not offer ideal settlement-sites. Some questions raised by the corpus have not been solved.

It is sometimes difficult to distinguish between *hōh* and **haugr**.

Reference section for hōh

As simplex name: Heugh DRH, Hoe HMP(2), NFK, Hoo KNT(2), SFK, Hooe DEV, SSX, Hose LEI (plural); Hough CHE(2), DRB.

As first element with tūn: Haughton NTT, Holton LIN(3), Hooton CHE, YOW(3), Houghton BDF(2), CMB, DRB, DRH(2), HMP, HNT, LEI, LIN, LNC, NFK(3), NTB(2), NTP(2), SSX, YOW, Hutton CMB(3), DRH, ESX, LNC(4), SOM, WML(2), YOE, YON(17), YOW.

With other generics: Hoby LEI (ON **bý**); Holland ESX, LIN, LNC, Hoyland YOW(3) and Hulland DRB **(land)**; Houghall DRH **(halh)**; Hubbridge ESX **(brycg)**; Huthwaite NTT, YON **(thweit)**; Huttoft LIN (ON *toft*).

As generic with personal names: Aynhoe NTP (*Æga); Bletso BDF (*Blecci); Cogenhoe NTP (*Cugga*); Duddo NTB (*Dudda*); Fitz SHR (*Fitt); Kersoe WOR (*Criddi); Lubenham LEI (*Luba*); Petsoe BUC (*Pēot*); Prudhoe NTB (*Prūda*); Silsoe BDF (*Sifel); Tudhoe DRH (*Tudda*); Wadenhoe NTP (*Wada*); Wixoe SFK (*Widuc*). Scottow NFK contains 'Scot', perhaps used as a pers.n.

With descriptive terms: Cambo NTB **(camb)**; Carrow NFK (*carr* 'rock'); Langenhoe ESX ('long'); Sandhoe NTB; Sharpenhoe BDF ('rugged'); Silpho YON **(scelf)**; Stanhoe NFK ('stone'); Trentishoe DEV (*trendel* 'circle').

With references to wild creatures: Cornsay DRH ('crane'); Cranoe LEI ('crows'); Rainow CHE ('raven'); Stagenhoe HRT ('stags').

With references to domestic creatures: Kyo DRH ('cows'); Moulsoe BUC ('mule'); Swinhoe NTB ('swine').

With references to vegetation: Ashow WAR; Pishobury HRT ('peas'); Salph BDF (*salh* 'willow').

With references to structures: Belsay (earlier *Belesho*) NTB (probably *bēl* 'fire, funeral pyre'); Millow BDF ('mill'); Staploe BDF ('post'); Totternhoe BDF (*tōtærn* 'look-out building').

With earlier hill-names: Cainhoe BDF, Cashio HRT, Kew SUR and Keysoe BDF (in Gelling 1984, p. 168, it is argued that the likeliest explanation of these 4 names is that *cǣg* 'key' was used as a hill-name); Ingoe NTB (*ing 'hill, peak' has been conjectured as the first element of several names, including this one; when the second part of the compound is a word for a hill or hill-spur it seems reasonable to conclude that these have been added to an earlier name *Ing*).

HILLS, SLOPES AND RIDGES

With references to topographical features: Clougha LNC (**clōh**); Furtho NTP (**ford**); Halse (earlier *Halsho*) NTP (*hals* 'col'); Weo Fm and View Edge SHR (**weg**).

With references to position: Midloe HNT; Sharow YOW ('boundary'); Southoe HNT.

With reference to an activity: Spelhoe Hundred NTP ('speech').

With a god's name: Tysoe WAR (*Tīw*).

With -ingas group-names: Bengeo HRT ('dwellers on the river Beane'); Dallinghoo SFK ('Dealla's people'); Fingringhoe ESX (? 'dwellers by a hill called Finger'); Ivinghoe BUC ('Ifa's people'); Martinhoe (earlier *Mattingeho*) DEV ('*Matta's people'); Piddinghoe SSX (*Pydda's people').

With obscure, uncertain or ambiguous qualifiers:

Croyde Hoe DEV (*Cridehoe* 1242). A noun **crȳde* meaning both 'headland' in this name, and 'weeds' in Creed (in Bosham) SSX, has been suggested. This would be a derivative of the verb *crūdan* 'to press on, drive', but the suggestion is not altogether convincing.

Culphoe SFK. The forms require a first element **culf*, which is not manifested anywhere else. Ekwall 1960 suggests a contracted form of personal name *Cūthwulf* without genitive inflection, and this is possible.

Farthinghoe NTP. 'Ferny hill-spur', with *fearnig* in an oblique case, *fearnigan*, is not impossible, but the early forms go better with an *-ing-* or *-inga-* formation. Other suggestions are 'hill-spur of the bracken dwellers', with a group-name from *fearn*, or a contraction of **Fearndūningahōh*, with a group name 'people of Farndon', from West Farndon, which is about 8 miles away.

Flecknoe WAR. The first element, which occurs also in Fleckney LEI, is probably best regarded as obscure, though a personal name **Fleca* has been conjectured from these names and Fletching SSX.

Limpenhoe NFK. The first element appears to be unique to this name. A personal name or nickname (from the verb *limp*) is possible.

Morthoe DEV. The element *mort*, which occurs also in Morcombelake DOR (*v.* **cumb**) and in Mortlake GTL (*v.* **lacu**) has not been sastisfactorily explained. It is difficult to envisage a topographical term which would be appropriate to a **cumb**, a **lacu**, and a *hōh*.

Pinhoe DEV. The Anglo-Saxon Chronicle, under the year 1001, gives this name as *Peonho*. Later forms, from 1086, give the first element as *Pin-* or *Pyn-*. Pace Ekwall 1960, neither *-eo-* nor the later *-i-*, *-y-* suits derivation from PrW **penn**. The qualifier might be *pinn* 'pin', as in Pinner, but in that case the Chronicle form is garbled.

Tattenhoe BUC. There is considerable variation in the early spellings (*Taddenho* 1167, *Tatenho* 1179, *Totenho* 1227), so this does not appear to be a safe instance of the personal name *Tāta*, which occurs in a number of names with consistent *Tat(t)en-* spellings. It is uncertain what the original form of Tattenhoe was.

Watnall NTT. Perhaps a personal name **Wata*, but the evidence for this is very slight.

Whessoe DRH. Perhaps a personal name **Hwessa*, derived from OE *hwæss* 'sharp'. The adjective would be a possible qualifier in its own right, but the word has not been clearly established as one used in place-name formation.

Wivenhoe ESX. OE *wīf*, 'woman, wife', was a neuter noun with *-es* gen.sg. and *-a* gen.pl. This is a likely first element in some names, such as Westow YOE, Westoe DRH, Winestead (earlier *Wifestede*) YOE, but it is grammatically unsuitable for Wiveton (earlier *Wiventone*) YOE and for Wivenhoe, which must have had *-n-* in the original form. It is possible that there was a personal name **Wifa*, which would have genitive **Wifan*.

Wysall NTT (earlier *Wisho*). Possibly the genitive of *wīg* 'heathen temple'.

hrycg 'ridge'. This occurs frequently in minor names, many of which, since the word remained in the language, may be of post-Conquest origin. No subtlety is required for interpretation: anything which qualifies for the modern term could have been described by the OE one. It is not one of the commoner topographical terms in ancient settlement-names, but the corpus is sufficiently large for analysis of qualifiers to be interesting. The references to wild creatures are evocative.

Distribution is obviously influenced by topography. There are some clusters, one of which is north-west of Chesham BUC, where the landscape is dominated by a series of long, parallel ridges. Here are Hawridge, Hundridge, Asheridge and Chartridge. There is another group near Luton BDF, comprising Cowridge, Ramridge and Putteridge, and in BRK Hartridge, Coleridge and Westridge are close together near Ashampstead.

ON **hryggr** is the first element of Ribby LNC and Rigsby LIN, and it has influenced the development of *hrycg* in Rigton YOW(2). In names ending in -rigg in the north of England it is often impossible to distinguish between ON and OE words, and as many of the names will be of ME origin the distinction is mostly irrelevant. Bigrigg CMB is likely to be wholly ON, since it has ON *bygg* 'barley' as qualifier. Crossrigg WML contains ME *cros* 'cross', probably a Norse borrowing from Irish. Grayrigg and Lambrigg WML may be hybrid OE/ON, or OE names modified by Norse speech.

Reference section for **hrycg** 'ridge'

As simplex name: Ridge HRT, STF, Rudge SHR, WLT.

HILLS, SLOPES AND RIDGES

As first element: Ridgewardine SHR (*worthign* 'enclosed settlement'); Rigton YOW(2); Rudgwick SSX; Rudloe WLT (**hlāw**); Rugeley STF (**lēah**).

As generic with personal names: Cotheridge WOR (**Coda*); Curdridge HMP (*Cūthrǣd*); Curridge BRK (*Cusa*); Cutteridge WLT (*Cūda*); Potheridge DEV (*Poda*); Putteridge HRT (*Puda*); Totteridge HRT (*Tāta*); Totteridge WLT (**Tetta*); Waldridge BUC (*Wealda*).

With references to vegetation: Asheridge BUC and Ashridge HRT; Elmbridge WOR; Iridge SSX ('yew'); Lindridge KNT(2), WOR ('lime-tree'); Timbridge WLT ('timber'); Witheridge DEV ('willow').

With references to wild creatures: Bageridge DOR, STF and Baggridge SOM ('badger'); Cheveridge WOR (*ceafor* 'beetle'); Eridge SSX ('eagle'); Hartridge BRK; Hawkridge BRK, SOM, WLT, Hawridge BUC and Ockeridge KNT ('hawk').

With references to domestic creatures: Cowridge BDF; Foulridge LNC ('foal'); Henstridge SOM (*hengest* 'horse'); Hundridge BUC ('hounds'); Ramridge BDF (probably 'ram').

With descriptive terms: Druridge NTB ('dry'); Longridge LNC, STF; Rowridge IOW ('rough'); Sandridge DEV, HRT, WLT; Smallridge DEV ('narrow').

With references to position, or to that of settlement in relation to feature: Boveridge DOR ('above'); Eastridge WLT; Middridge DRH; Norridge WLT ('north'); Westridge BRK.

With references to topographical features: Ditteridge WLT (*dīc*, i.e. the Fosse Way); Soldridge HMP (**sol**).

With references to structures: Lockeridge WLT ('enclosures'); Waldridge DRH (probably 'wall').

With reference to an activity: Coldridge BRK, DEV and Coleridge BRK ('charcoal').

With reference to superstition: Puckeridge HRT ('goblin').

With reference to ownership: Awbridge HMP ('abbot').

With obscure or uncertain first elements:

Chartridge BUC (earlier *Charderugge*). The regular *-d-* of the earliest spellings is against the suggestion (Ekwall 1960) of *ceart* 'rough ground'. A personal name **Card(a)* is inferred from some place-names, but this would not give *Ch-*. A personal name **Cearda* would suit, but the matter is best left open.

Kerridge CHE. The first element could be *cǣg* 'key', discussed in the reference section for **hōh**.

Lupridge DEV. A personal name **Hluppa* has been inferred from this name and Lupton WML, Lupset YOW.

Rawridge DEV, earlier *Roveruge*. The bounds of Ebbesbourne WLT include *of hrofan hrige*, which may be the same name. EPNS survey p. 650 suggests an adjective related to OE *hrēof* 'scabby'.

Tandridge SUR (earlier *Tenrigge, Tanrigge*). The first element appears to be unique to this name. Numerous *-a-* spellings rule out the numeral *tēn*, WSax *tȳn*.

hváll, hóll ON, ***hwæl** OE 'hill'. The OE word is inferred from Whalley LNC and Whalton NTB. The ON word occurs in Whale WML and probably in Falsgrave YON ('pit of the hill') and Staffield CMB (earlier *Stafhole*, 'post hill'). In the EPNS WML survey, 2 p. 183 s.n. Whale, Professor Smith advanced the specific sense 'an isolated rounded hill' for *hváll*. This would suit the supposed OE word as evidenced in Whalley LNC, where the church is overlooked by just such a hill.

hyll 'hill'. This word occupies a position among hill-terms resembling that of **halh** and **weg** among valley- and road-terms. It appears to be used for hills which do not have the clearly defined characteristics of those called **beorg** or **dūn**. There are no occurrences of *hyll* in names recorded by c.730, and it may, on the whole, belong to the later stages of Old English name-formation, perhaps coming into more frequent use as the precision of the earliest topographical vocabulary weakened.

hyll is commonest in minor names, but there is a substantial corpus of major names, and these are analysed in the reference section. It is well evidenced as a simplex name and as a first element, but in the latter capacity does not have anything like the frequency of **hōh**.

When *hyll* is the generic in a compound name, masculine personal names (certain or probable) are the most frequent qualifiers. There are three dithematic names: *Beornheard* in Barnsdale RUT, *Heardrēd* in Hartshill WAR and *Wulfhere* in Wolvershill WAR. Descriptive terms are only slightly less frequent than personal names.

A surprising factor to emerge from the analysis is the rather high proportion of names with obscure first elements, and this could be held to qualify what is said above about the possible late origin of *hyll* names.

Reference section for **hyll**

As simplex name: Hill CNW, GLO, HMP, SOM, WLT, WOR(2), Hillam YOW(2) (dative plural), Hull SOM. Hill Chorlton STF and Hillmorton WAR result from the amalgamation of simplex *hyll* with the name of a neighbouring settlement.

As first element: Hilfield DOR (**feld**); Hilton DRB, HNT, STF, YON, Hulton LNC, STF (*tūn*); Hillhampton WOR (*hāmtūn*, a habitative term of undetermined significance: Hillhampton is one of a remarkable group of *hāmtūn* names in north-west WOR).

HILLS, SLOPES AND RIDGES

As generic with personal names: Barnhill YOE (*Beorn*); Barnsdale RUT (*Beornheard*); Bubbenhall WAR and Bubnell DRB (*Bubba*); Buersill LNC (*Bridd*); Chaxhill GLO (*Cæc*); Credenhill HRE (*Creoda*); Hartshill WAR (*Heardrēd*); Hempshill NTT (**Hemede*); Hockenhull CHE (*Hocca*); Hodnell WAR (*Hoda*); Kelshall HRT (**Cylli*); Lilleshall SHR (*Lill*); Pamphill DOR (**Pempa*); Patshull STF (?**Pættel*); Pattishall NTP (?**Pætti*); Quixhill STF (probably **Cwic*); Ragnall NTT (ON *Ragni*); Rainhill LNC (**Regna*); Snarehill NFK (**Snear*); Stramshall STF (*Stronglic*); Tatenhill STF (*Tāta*); Tetsill SHR (*Tætel*); Tottenhill NFK and Tutnall WOR (*Totta*); Ughill YOW (?**Ucga*); Winshill STF (*Wine*); Wixhill SHR (**Wittuc*); Wolvershill WAR (*Wulfhere*); Wormhill DRB (**Wyrma*).

With descriptive terms: Bickenhill WAR (probably **bica* 'projection'); Burshill YOE (*byrst* 'landslip'); Calehill KNT ('bare'); Chishall ESX ('gravel'); Clennell NTB ('clean'); Coppull LNC (**copp**); Cropwell (*Crophille* DB) NTT ('hump'); Grindle SHR ('green'); Hernhill KNT ('grey'); Mansell HRE ('gravel'); Poughill DEV (*pohha* 'pouch' probably referring to shape); Shelfield (earlier *Shelfhull*) STF, WAR (**scelf**); Shottle DRB ('steep slope'); Smithills LNC ('smooth'); Snodhill HRE (possibly **snāwede* 'snowy'); Snowshill GLO; Solihull WAR (probably 'muddy'); Soothill YOW; Stapenhill STF ('steep'); Whitehill DRH and Whittle LNC(2), NTB(2) ('white'); Windhill YOW and Windle LNC ('wind').

With references to vegetation: Ashill SOM; Aspul LNC ('aspen'); Broomhill KNT; Buckenhill HRE ('beech'); Dodderhill WOR and Doddershall BUC (dialect *dodder*, perhaps 'bindweed'); Farnhill YOW ('fern'); Hethel NFK ('heath'); Lushill WLT (probably *lūs-thorn* 'spindle'); Pishill OXF ('pea'); Sedgehill WLT; Smerrill DRB (*smeoru* 'butter', implying good grass); Thornhill DOR(2), DRB, WLT, YOW; Withnell LNC ('willow').

With references to crops: Bearl NTB ('barley'); Benhilton (earlier *Benhull*) SUR ('bean'); Haverhill SFK ('oats'); Odell BDF ('woad'); Ryal and Ryle NTB, Ryhill YOE, YOW ('rye'); Wheathill SHR, SOM.

With references to domestic creatures: Bonehill (earlier *Bolenhull*) STF ('bullock'); Coole CHE, Cowhill GLO and Keele STF ('cow'); Goathill DOR; Henhull CHE; Hinxhill KNT ('horse'); Tickhill YOW ('kid').

With references to wild creatures: Ampthill BDF ('ant'); Beal NTB ('bee'); Catshall SUR and Catshill WOR; Harthill CHE, DRB, YOW; Hawkhill NTB; Roxhill BDF ('buzzard').

With earlier hill-names: Brickhill BDF, BUC (**brig**); Brill BUC (**breʒ**); Churchill DEV(3), WOR (near Kidderminster), Crichel DOR, Crickley GLO and Crigglestone YOW (*crūg*); Cook Hill WOR (*cōc*); Hockerill HRT (Ekwall 1960 makes a convincing case for an OE **hocer* 'hump'); Pendle LNC and Penhill YON (**penn**); Pexall CHE (**pēac**).

With earlier place- or river-names: Coleshill WAR (river Cole); Dosthill WAR (*Dercet*, ME reflex of British name meaning 'oak wood'); Earnshill SOM (river

193

Earn); Penkhull STF (PrW *penn-gēd* 'wood's end'); Redmarshall DRH (*hrēod-mere* 'reed pond'); Wrinehill CHE/STF (*Wrine*, a district-name perhaps meaning 'river bend').

With references to structures: Burghill HRE, Burrill YON and Burlton HRE, SHR (*burh* 'fort'); Churchill OXF, WOR (near Worcester); Earle NTB ('yard'); Throphill NTB (*throp* 'hamlet').

With references to religion or superstition: Gadshill KNT and Godshill HMP, IOW ('God'); Greenhill (earlier *Grimanhyll*) WOR ('spectre').

With topographical terms: Brindle LNC (**burna**); Whitehill OXF (probably *wiht* 'river-bend').

With 'look-out' terms: Tothill LIN, MDX (**tōt*); Wardle CHE, LNC and Warthill YON (*weard*).

With an -ingas name: Sunninghill BRK.

Miscellaneous: Hordle HMP ('hoard'); Queenhill WOR (probably 'royal'); Thinghill HRE ('meeting'); Wreighill NTB (literally 'felon', possibly 'wolf').

Names with obscure or ambiguous qualifiers:

Blymhill STF. Ekwall 1960 suggests *plȳme* 'plum tree' with change of *p-* to *b-*, but no such development is evidenced in other names containing *plȳme*.

Bucknell OXF, SHR. Could be the personal name *Bucca* or the noun which means 'he-goat'.

Caughall CHE. Spellings vary between *Cochull* and *Coghull*: *cocc* 'heap' and *cogg* 'cog of a wheel' are both possible.

Coleshill BRK. The first element could be a British stream-name, a personal name or an OE **coll* 'hill'. The case differs from that of Coleshill WAR, on the river Cole, in that there is no record of a stream here called Coll.

Coleshill BUC. For this name and for the apparently identical Coleshill FLI no spellings have been noted with *-ll-*, so it may be a different name from Coleshill BRK and WAR. A personal name **Col*, related to the recorded *Cola*, is a reasonable conjecture.

Deuxhill SHR. The only suggestion available is a personal name **Deowuc*, but this is no more than an inference from Deuxhill and a charter-boundary name *Diowuces pæth*, also in SHR.

Foleshill WAR, earlier *Folkeshull*. This is ambiguous: *folc* 'folk' could refer to the use of the hill for some communal activity, or **Folc* could be a personal name.

Gopshall LEI. The only word on record which would suit is *gōp*, recorded once in a Riddle, apparently meaning 'captive, slave'.

HILLS, SLOPES AND RIDGES

Grinshill SHR. The spellings require *gryneles-. There may have been an OE *grynel*, a derivative of *gryn* 'noose, snare'.

Harnhill GLO. Could be 'grey' or 'of the hares'.

Howsell WOR. Professor Zachrisson's suggestion in EPNS survey that the first element is the ME personal name *Huwe, Howe* seems preferable to that in Ekwall 1960 of *hōges*, gen. of *hōh*. The name is first noted *c*.1230, so is not necessarily of pre-Conquest origin.

Ightenhill LNC. Ekwall's suggestion (1922 and 1960) of the British ancestor of Welsh *eithin* 'furze' is rejected by Jackson (1953, p. 310). The spellings suggest an OE *ihta*.

Inkersall DRB. Professor Dickins, quoted in EPNS survey, suggested a nickname *Hynkere meaning 'limper', which he inferred from dialect *hink* 'to limp'. There is no other evidence for this.

Marnhull DOR. A careful discussion in EPNS survey 3, pp. 166–7, concludes that the original form may have been *Marenhyll. No etymology has been found for this.

Prawle DEV. The spellings suggest *prā(w)*, apparently unique to this name. Professor Ekwall's suggestion, quoted in EPNS survey, of 'look-out hill', from a word related to OE *beprīwan* 'to wink', modern *pry*, is not semantically convincing.

Pulloxhill BDF. The same first element occurs in some DEV names, Poulston (*Polochestona* DB), Pulsford in Woodland (*Pollokysford* 1330) and a surname, *de Pollokesfenne*, 1333 in Newton Ferrers. Since Pulloxhill and Poulston are both in DB the surname Pollock, which is in any case Scots (from a place in Strathclyde) or Jewish ('from Poland'), is not relevant. There is no obvious source for an OE personal name *Pulloc, so the matter is best left open.

Sugnall STF. Ambiguous: *Sucga* is a personal name and a bird name.

Tintinhull SOM. The spellings suggest *tinta*, which appears to be unique to this name.

Wincle CHE: see the discussion of *wince* under Winchendon in the reference section for **dūn**.

Wuerdle LNC. The spellings suggest *weorod*. There is a noun of this form meaning 'army', and an adjective meaning 'sweet': neither seems appropriate.

mönith PrW, PrCorn, PrCumb, **mynydd** Welsh, **meneth** Corn 'mountain, hill'. These Celtic words are cognate with Latin *mons*. The Welsh word is used in Mynde HRE and Long Mynd SHR for large areas of high ground. In CHE, parts of a ridge of high moorland on the boundaries of Sutton, Wincle, Gawsworth and Bosley townships are called Bosley Minn and Wincle Minn.

Mindrum NTB is explained in Ekwall 1960 as a compound of *mönith* and the ancestor of Welsh *trum* 'ridge', with which he compares Mynydd Drymmau GLA.

In GLO, in the Forest of Dean, there is evidence that Welsh *mynydd* was borrowed into ME as *munede* and used in a new sense which Smith (EPNS survey 3, p. 218) renders as 'a piece of waste or open ground in the forest, a forest-glade'. This has become Meend in a number of names in this area.

Ekwall 1960 derives East and West Myne and Minehead SOM from *mönith*, although the early spellings for Myne do not have the *-d* which should be present, and which is well represented in forms for Minn CHE. He suggests that the hill was called *Mynydd*, and English speakers adopted this, adding **dūn** to give *Myned-dūn, which became *Mynedun*, and was then misunderstood, yielding a hill-name *Myne*. The form *Menedun*, recorded in 1255, could fairly be held to support this conjecture. It is possible, however, that the hill had a pre-English name similar to that of Meon Hill GLO, and that *Mene*, the DB form for Myne, is to be taken as a genuine form, not as arising from a misunderstanding by English speakers of a compound with **dūn**.

Ekwall 1960 also ascribes Mendip Hills SOM to *mönith*, suggesting that the second part of the name is OE **hop**. The valley containing Winscombe could be a **hop**, but it is difficult to envisage the transference of such a village-name to the massif. The main objection, however, is that early spellings for Mendip mostly have *-ep*, which is not evidenced in other **-hop** names. Mendip is probably best regarded as unsolved.

As a first element *mynydd* is found in three SHR names. Mynd Town and Minton are by the Long Mynd. Menutton is just below the 1,000ft contour by a spur of the massif which includes Clun Hill.

Examples of Cornish names containing *meneth* are given in Padel 1985.

næss, ness 'projecting piece of land'. This word is used rather sparingly in place-names: it does not occur in all counties, and in some others (e.g. CAM, NTT) there is only one instance. The commonest use is for low-lying land jutting into water or marsh, and the county with the largest number of examples is ESX.

Flat, marshy, coastal promontories are referred to in, for example, Foulness ESX, Bowness on Solway CMB, Sheerness KNT and Widnes LNC. Promontories with slightly raised ground include The Naze and Wrabness ESX and Ness and Sharpness GLO. Low projections into marsh are referred to in East and West Ness YON, Neswick YOE and Claines WOR, and excellent examples in minor names include Ness Fm in Burwell CAM and Outerness Wd in Conington HNT. this last use of *næss* resembles one of the uses of **hamm** which is evidenced particularly in SOM and in GLO by the river Severn.

HILLS, SLOPES AND RIDGES

In some names the promontory is a tongue of land between streams, as in Haynes BDF and Noss Mayo DEV. Phynis (in Slaidburn) YOW is on a high promontory between the river Hodder and Phynis Beck.

The occasional use of *næss, ness* for striking inland hills or ridges seems arbitrary. Instances are Ness SHR, Ashness CMB and Totnes DEV. It is not obvious from the map why the hill at Neston WLT should be the only *ness* in that county.

Related OE words are occasionally used in place-name formation. Hackness northwest of Scarborough YON is *Hacanos* in Bede's *Ecclesiastical History*, so is considered to contain *nōs*, which occurs in *Beowulf* in the sense 'headland, promontory'. This had been replaced by the commoner *næss* before 1086. The 'hook-promontory' is the distinctively-shaped ridge which runs northwest from the village. In Dunnose IOW this word is combined with **dūn**. Another related word, *nēs*, is found in Neasden GTL (with **dūn**) and in Neasham DRH, where there is a nose-shaped bend of the river Tees.

nes ON 'projecting piece of land'. In Hornsea YOE (ON *horn-nes-sǽ*), which refers to the horn-like peninsula projecting into the nearby lake, and Skegness LIN, which probably refers to the 'beard-shaped' promontory south of the town, the ON word is used in a similar way to its OE cognate.

There are, however, some names in which *nes* appears to be used for much larger areas than would be denoted by the OE word. It is possible that this results from the extension to a wide area of a name originally referring to a point of land at the end of it, but this has not happened with names containing **næss**, and there is a possibility of a significant difference in usage.

The names requiring discussion are Holderness YOE and Amounderness and Furness LNC. It is possible that only Furness is relevant. Dr Fellows-Jensen (1989) has made the interesting suggestion that the generic in Amounderness and Holderness is OE *hērness* 'obedience, district in subjection to a secular or ecclesiastical authority', instanced in district-names such as Berkeley Harness GLO and Lugharness HRE. This would mean that Amounderness and Holderness were hybrids, as the first elements are the Norse personal name *Agmundr* and the ON word *holdr* 'nobleman'. Furness, however, cannot easily be accounted for in this manner. The earliest spelling (*c.*1150) is *Fuththernesse*, and the first element appears to be shared with Fouldray, earlier *Fotherey*, which was the old name of Peel Island off the southern tip of the peninsula. Ekwall (1922, p. 201) put forward a highly ingenious solution for the two names. He noted that Peel island has a cleft running south to north, and he suggested that this led to its being named *Futh*, from an ON word for the human posterior. Fouldray would be 'island of *Futh*' (with ON *ey* added to the genitive), originally referring just to the headland at Rampside. A possible objection to this is that Peel Island is an insignificant feature when viewed from Rampside, and it is difficult to envisage the extension of a name with such a small referrent to the whole area of the Furness peninsula. Interpretation of these three *nes* names is best left open.

In IOM *nes* is used of a rocky coastal promontory in Langness, and of an inland promontory in Agneash. Cregneash, now a village name, may have been applied originally to the great promontory opposite the Calf which the village overlooks. In ORK *nes* is very common in the names of coastal promontories.

Reference section for **næss, ness**

As simplex name: Nass GLO, Ness CAM, CHE, SHR, YON(3), The Ness (in West Muskham) NTT, Noss Mayo DEV. Nassaborough Hundred NTP is 'ness of Peterborough', so classifiable as simplex.

As first element: Naze Wick ESX (2, in Fobbing and Foulness); Neston CHE, WLT; Neswick YOE. Nassington NTP and Navestock ESX require consideration here. Early spellings for both names suggest formations in which *-ing-* links *næss* to a habitative term (Navestock is *Næsingstoc c*.970). The *-ing-* formation is unusual with topographical words, however, and it is possible that a derivative term *næsing should be considered for these two names. Nazeing ESX has early spellings with plural endings, so is an *-ingas* formation, 'people of the ness'.

As generic with personal names: Allfleet's Fm (in Wallasea) ESX (*Alfledenesse* 1285, feminine *Ælflæd*); Levens WML (*Lēofa*); Russellhead (in Canvey) ESX (*Rikeresnesse* 1200, *Rīchere*); Sharpness GLO (*Scobba). The Naze ESX was originally a compound name with *Ēadwulf* or *Ealdwulf*.

With descriptive terms: Bowness CMB ('bow-shaped'); Claines WOR ('clay'); Sheerness KNT ('bright'); Widnes LNC ('wide').

With references to structures: Cotness YOE ('cottage'); Crossens (in North Meols) LNC ('cross'); Haynes BDF ('enclosure').

With references to position: Eastness (in Great Clacton) ESX; Outerness Wood (in Conington) HNT.

With reference to a wild creature: Foulness ESX ('bird').

With reference to a domestic creature: Bowness WML ('bull').

With reference to vegetation: Reedness YOE.

With an earlier place-name: Skinburness CMB (*Schineburgh* 1185 'demon fort', *Skynburneyse* 1298).

Names with obscure or uncertain first elements:
Phynis YOW. Names which require a first element *fin(a)* are often ambiguous, as *fin* 'heap of wood' cannot always be distinguished from *fina* 'woodpecker'. EPNS survey (6, p. 204) prefers *finn OE/ON 'coarse grass' for the present name, but this is a doubtful place-name element.

HILLS, SLOPES AND RIDGES

Totnes DEV. EPNS survey (1, p. 334) points out correctly that 10th- and 11th-century spellings *Totanæs* and *Tottaness* accord better with the personal name *Totta* than with **tōte* 'lookout'. This last element, which is well represented in place-names, would make excellent sense in relation to Totnes, where the town commands a magnificent view down the estuary of the river Dart. As the first element of compounds, however, *tōte* is usually uninflected and represented in early spellings by *Tot-*. It is probably not admissible here, but a lingering suspicion that we may be dealing with a side-form or a derivative leads to the placing of Totnes in this category.

Wrabness ESX. An element *wrabba* is required for this name and for Rapton SFK. Ekwall 1960 suggests a personal name related to *wrabbed*, a word recorded in the 16th-century meaning 'perverse'. This is obviously a long shot: it would be just as reasonable to postulate an adjective with some such sense as 'crooked'.

ofer, ufer (Figs 37-9) 'flat-topped ridge with a convex shoulder'. Fieldwork since 1984 has amply confirmed the definition put forward in *Place-Names in the Landscape*. This topographical term is employed with remarkable consistency from near the south coast (e.g. the hill in SSX called High and Over) to NTB (where Wooler is a northern outlier). It describes ridges which contrast in shape with those for which *hōh* is used. Usually the ridge is level, but sometimes it has a slight downward slope, and a good instance of this is to be seen overlooking the town of Wooler. In southern counties *ofer* overlaps with *ōra*, an alternative term for the same landform. As will be seen from the names listed in the reference section *ofer* is frequently shortened to *-or* in modern forms, but early spellings usually enable it to be distinguished from *ōra*.

In spite of the fact that it is not on independent record, the existence of this term is universally accepted among place-name scholars. The recorded word *ōfer*, 'border, margin, edge, river-bank, sea-shore', while phonologically possible for most *-over* names, will not suit those which have early spellings with *u*. Such are: Over CHE, *Ufre* c.1000, *Huuere* c.1232: Southowram and Northowram YOW, *Huuerum* c.1210, *Suthuuerum* 13th century, and *Ufrun* DB: Tidover YOW, a 'lost' settlement in Kirkby Overblow, *(on) tiddanufri* c.715: Mickleover DRB, *Ufre* 1011.

In later forms for these name *o* spellings predominate heavily. It seems necessary to conclude that *ufer* and *ofer* were variant forms of a word which had a short vowel, as opposed to the long vowel of *ōfer*.

While *ōfer* in the sense 'river-bank' is theoretically possible for names which do not have *u* spellings, the geographical situation of most of the names discussed here renders it inappropriate, and the consistency with which the names refer to a well-defined landform strongly supports the existence of a term for this particular type of ridge.

It is not certain that *ōfer* 'edge, shore' occurs in place-names, though it has naturally been assumed to be present in its 'river-bank' sense in a number of -over names which are situated by rivers or streams. Over and Elmore, west of Gloucester, for instance, are generally considered to refer to positions by the river Severn. There is, however, nothing distinctive in a position by the Severn, as many other settlements lie similarly by the river. At both places there are low ridges with convex shoulders, which may have been noteworthy landmarks for river traffic. The suggestion of a reference to characteristically-shaped ridges on river banks is here offered tentatively for the various examples of Northover, Southover and Westover in SSX, DOR, IOW and SOM. At some, perhaps all, of those which lie on river banks there are ridges which would be more conspicuous from the river than they are from the land. This is certainly true of Southover in Frampton DOR and of Westover in Muchelney SOM. Stour DOR and Langport SOM have districts called Estover and Westover, and in both cases ridges overlook the rivers in the relevant areas. It would be unwise to make a dogmatic assertion about the generic in all these names, but it is by no means certain that *ōfer* rather than *ofer/ufer* is involved. Most of the Northover-type names are at riverine sites, but Westover by Calbourne IOW is not by a river.

Moving back onto safer ground, some instances may be cited of names in non-riverine, non-coastal situations for which derivation from *ofer* in the sense defined here is particularly convincing.

A striking *ofer* landform can be seen at Haselor WAR, where the church is some distance back from the ridge-end, which is clear of buildings. This is shown on Fig. 37. Wentnor SHR (also on Fig. 37) is another splendid example, with the village on the tip of a level ridge jutting out into the low-lying ground of Prolley Moor. The ridge at Shotover OXF is a well-defined specimen. A substantial proportion of *ofer* names has been investigated on the ground, and a collection of photographs is available to support the contention that a particular shape of ridge is consistently found at these sites.

The map of *ofer* names (Fig. 38) shows what may fairly be considered a more than random association with ancient roads, and the perceptions of travellers may well have played a part in the emergence of these names. Hadzor WOR is seen very clearly from the M5 (which follows the line of the Roman road to the southwest), and this and Haselor WAR are among the *ofer* names which bear a close relationship to the saltways which converge on Droitwich (Fig. 39). Wychnor STF is a landmark from the A38, and Hunsingore YOW from the A1. In compounds referring to vegetation all but one of the qualifiers is a tree-name, and this also may reflect the perception of the ridges from nearby roads.

The common name Overton has been omitted, as there is no safe way of distinguishing between derivatives of a hypothetical **ofer-tūn* 'ridge settlement' and those from OE *ufera tūn* 'higher settlement'. Ekwall 1960 adduces

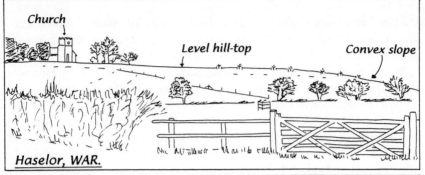
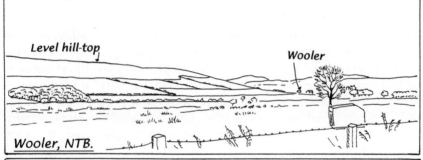
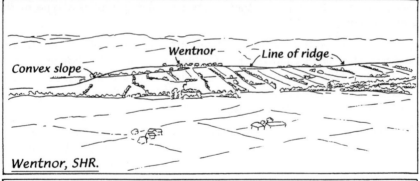

ōfer-tūn 'river-bank settlement' as a third possibility. West Overton WLT is *Uferan tune, Oferan tunes* 959, and Overton HMP is *Uferantun* 909, so certainly 'higher settlement', but there are no OE forms for any of the other Overtons. **ofer, ufer* is related to the preposition *over*, but the preposition cannot, of course, be involved in simplex names, or in those with *ofer* as generic.

A few settlement-names derive from a word **yfre*, which is a derivative of **ofer, ufer*, and appears to be used in exactly the same way. Examples are Iver BUC, Rivar WLT, River SSX and Hever KNT. Rivar and River have R- from the phrase *æt thære yfre*, Hever (*Heanyfre* 814) is 'high ridge'. Bignor SSX ('Bica's ridge' or 'pointed ridge') has many spellings such as *-evere* which indicate *yfre*, but there are some with *-ov(e)re* which indicate interchange with *ofer*.

Reference section for ofer

As simplex name: Over CAM, CHE, DRB, GLO, Overs SHR. Mickleover and Littleover DRB are two settlements named from the same ridge. Mickleover was *Ufre* 1011, *Ufre, Overe* in DB, and the simplex form was still in use in the mid-13th century. Littleover is *Parva Ufre* in DB, but 'little' is for distinction from the (presumably) earlier settlement, so the name is not a compound one. In Northowram and Southowram YOW, *-owram* is from the dative plural of *ofer*. There are early spellings *Huverum* for both the places, which are two miles apart, on ridges east and west of Shibden Dale. *Uferum* was perhaps a district-name, indicating that there were several ridges of this type in the broken country east of Halifax.

As generic with a personal name: Adzor HRE (*Æddi*); Badger SHR (**Bæcg*); Bolsover DRB (**Bull*); Chadnor HRE (*Ceabba*); Codnor DRB (**Codda*); Edensor DRB (**Ēden*); Hadzor WOR (**Headd*); Hunsingore YOW (*Hūnsige*); Seanor DRB (?**Sǣga*); Tidnor HRE (*Tudda*); Tittensor STF (**Titten*); Wentnor SHR (**Wenta* or **Wonta*); Yazor HRE (Welsh *Iago*).

With references to vegetation: Ashover DRB; Bircher HRE and Birchover DRB; Buckover GLO ('beech'); Elmore GLO; Haselor WAR, WOR, Haselour STF; Lineover GLO ('lime'); Okeover STF ('oak'); Ramshorn (earlier *Romesovere*) STF (probably 'wild garlic'); Thorner YOW.

With descriptive terms: Bicknor GLO/HRE (**bica* 'point'); Heanor DRB, Hennor HRE, Highover (in Hitchin) HRT, High and Over (in Alfriston) SSX ('high'); Longnor STF; Radnor CHE, RAD ('red').

With references to domestic creatures: Calver DRB; Tixover RUT (*ticcen* 'young goat').

With references to a wild creature: Yarner (in Bovey Tracey) DEV (*earn* 'eagle').

With references to groups of people: Wellingore LIN (possibly *Wellingas* 'dwellers at a spring'); Wychnor STF (*Hwiccenofre* 11th/12th century, tribal name *Hwicce*).

HILLS, SLOPES AND RIDGES

With references to topographical features: Scottsquar (earlier *Schottesovera*) GLO and Shotover OXF (*scēot 'steep place'); Wooler NTB (**welle**).

With a river-name: Condover SHR (river Cound).

With references to direction: Northover SOM(2), Southover DOR(2), SSX, Westover DOR(2), IOW, SOM. Eastnor HRE is '(place) east of the ridge'.

With obscure qualifiers:

Spernall (earlier *Sperenoura*) WAR. Ekwall 1960 suggests *spæren* 'of plaster or mortar', but this has not been noted in any other place-name.

Tansor NTP. Other names with *Tanes-* are Tansley DRB and (possibly) Tansterne YOW. OE *tān* means 'twig, sprout, shoot, branch'. Ekwall 1960 suggests that the word might have had a transferred topographical sense 'branch of a valley or a river', but it seems more likely that if *Tanes-* is in fact from *tān* these names exhibit a collective use of the genitive singular.

ōra 'bank'. The precise sense which emerges from the study of place-name usage is the one also given here for *ofer*, 'flat-topped ridge with a convex shoulder'. There appear to be only two literary occurrences (leaving aside the *Cymenesora, Cerdicesora* of the Anglo-Saxon Chronicle). One of these is poetic: a cuckoo is heard calling from a grove *on hlithes oran*. This can fairly be translated 'on the edge of the slope', though a more specific sense for **hlith** can be deduced from place-names. The second is a translation of a Latin phrase, *in oram vestimenti eius*, by *on oran his hrægles*: this may indicate that OE *ōra* meant something like 'hem', but the choice of word may be an echoing of the Latin rather than a record of common usage. As with other words treated in this book, the place-name corpus is the main source for evaluation of meaning.

Study of the place-name corpus has shown that *ōra* is used in place-names in the south of England in the same sense as **ofer**. Both terms are evidenced in southern counties, but *ōra* predominates there and is especially common in the Chilterns and in SSX, KNT and DEV. The distribution is shown on Fig. 38.

The case for identification of *ōra* with this landform was first presented by Ann Cole in 1990, and illustrations can be seen on Fig. 40. This superseded the discussion in Gelling 1984, which failed to reach a satisfactory conclusion.

Ridges referred to in *ōra* names in the south are often bigger than those characteristically described by **ofer** further north. This difference of scale means that they sometimes form the background to a fairly distant view of the settlements named from them, and a less myopic approach is needed in order to appreciate the relationship than is needed to see the connection between a ridge and a settlement with an **ofer** name. Also, it is uncommon for the settlement to be on top of the *ōra*, as is frequently the case with **ofer** names. Instances of large *ōra* forms include the great Chiltern ridge which has

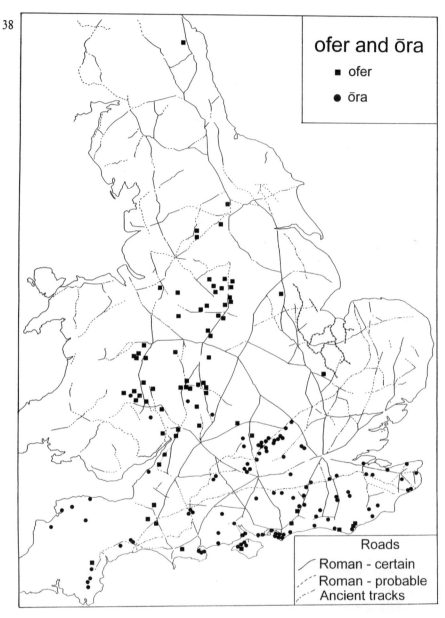

38

Chinnor at its foot, and the ridges which stand behind the shoreline on the south coast, were the *ōra* names cluster in w.ssx, HMP and DOR. The difference from **ofer** is, however, only in scale: the two terms consistently refer to ridges with the same configuration.

OE *ōra* is related, perhaps directly by borrowing, to Latin *ōra* 'rim, bank, shore'. Ann Cole makes the bold suggestion that the ridges behind the southern coastline would be seen rising above the horizon by Latin-speaking

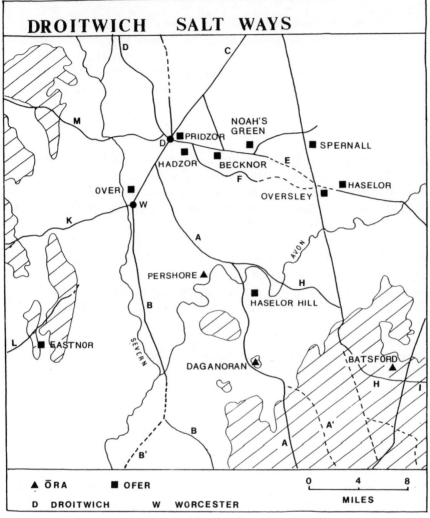

sailors, and their use of this word for the first land they saw in Britain might have led ultimately to its adoption with the sense 'ridge' by Germanic settlers. In the 1990 article she observes:

> A sailor coming up Spithead would see the long level crest of Portsdown rising behind such settlements as *Rowner* and *Copnor*. Near Chichester Harbour he would see the line of the South Downs in the distance beyond *Bognor, Itchenor, Keynor* and *Eleanor Farm*. Portsdown and the South Downs are shown on the coastal profiles drawn for sailors in the Channel Pilot and would have been as relevant for Anglo-Saxon navigators as for modern sailors wishing to identify their position and make a landfall. From the Solent the three coastal examples of *ora* in the Isle of Wight, namely *Bouldnor, Elmes-*

worth and *Gurnard*, would also be visible as flat-topped hills ... Across Poole harbour looking south the Purbeck Hills rise in a level line, with shoulders descending to the sea at the eastern end and to the Corfe Castle gap at the western end. Between the observer and the Purbecks lie *Goathorn, Fitzworth* and *Ower*. It can be shown that the great majority of coastal examples of *ora* refer to flat-topped hills ... The only exception is Stonar, Kent, which refers to a long, shingle ridge.

A similar exception to Stonar KNT may be the feature called The Owers, now under the sea off Selsey Bill, which is generally held to be the *Cymenes ora* of the Anglo-Saxon Chronicle, where the ships of the first South Saxons were reputed to have touched shore. It is possible, however, that the South Downs were visible overlooking this landing-place.

At Oare near Faversham and Oar Fm near Reculver there are identically-shaped low ridges which would have been landmarks for Jutish settlers approaching the north coast of Kent. Ekwall 1960 recognised 'ridge' as an inescapable meaning of *ōra* in inland, non-riverine, situations, and his discussion hints at a sense-progression from 'border, margin, bank'. He says (p. 350) "Senses such as 'edge of a hill, steep slope' or even 'hill, ridge' are found in Oare BRK, W, Ore SX, Bicknor K, Orcop". The possibility of a 'foot of the slope' meaning was explored in Gelling 1984, but the material did not fit easily with it, and the topographical evidence now assembled gives overwhelming support to the Cole interpretation, whatever the semantic development which led to it.

In the discussion of **ofer** consideration was given to whether, in names referring to settlements on river banks, the simple sense 'river-bank' might be considered wholly satisfactory. Similar considerations arise in relation to *ōra* names which are on the coast or by rivers. It was recognised by Ekwall that for some names on the south coast a more specialised sense than 'shore' was required in order to account for the clustering, and he suggested 'firm foreshore or gravelly landing-place'. As regards riverine situations, it was pointed out in Gelling 1984 that Pershore WOR is one of many settlements on the banks of the river Avon, so a more specialised sense than 'river-bank' would be appropriate. In fact, at Pershore there is a fine example of an *ōra*, which shows on early prints as a striking background to the town as viewed from the river. The qualifier is believed to be an OE precursor of dialect *persh* 'osier', and while 'osier bank' is a sensible etymology, 'osier bed overlooked by an *ōra*-shaped ridge' is much more specific.

Simplex names from *ōra* frequently have N- prefixed from the phrase *æt thǣm ōran*, ME *atten ore*. In southern counties there are many instances of Nore, Nower in minor names and field-names which can be ascribed to this origin. The hill which gave name to Pinner GTL is called Nower Hill. As a

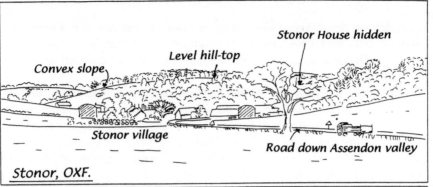

Stonor, OXF.

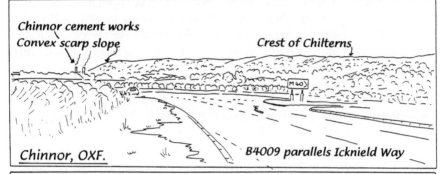

Chinnor, OXF.

Oare, WLT.

Bradnor, HRE.

first element *ōra* is very rare, and some possible examples (e.g. Orford SFK, Worton OXF) have been omitted as insufficiently certain. Some names with Or-, e.g. Orgrave LNC, Orsett ESX, contain the formally identical word *ōra*, 'metal ore'.

An exceptional number of minor names has been included in the analysis of qualifiers in the reference section: this is on account of a comprehensive corpus having been assembled by Ann Cole. When *ōra* is used as a generic, a significant proportion of qualifiers consists of descriptive terms and tree-names, and this may reflect the function as travellers' guides which is also suggested for **ofer**. The personal-name category has rather a lot of starred items, but none of these is conjectured solely from place-names.

The recurrent name Windsor was discussed carefully in Gelling 1984, and there is little to add to that discussion. There must have been a compound term *windelsōra* in which the word did not have a primarily visual sense. Ekwall's suggestion that the first element is an OE **windels* 'windlass' has met with general acceptance, but when the five instances of the name – Windsor BRK, WAR, Winsor DEV, HMP, Broadwindsor and Little Windsor DOR – are considered carefully it becomes impossible to accept Ekwall's interpretation of a 'windlass bank' in terms of an apparatus for helping boats along stretches of river. The DOR and HMP places are by watercourses which could never have carried any sort of traffic. It is possible that the reference is to a windlass which helped carts up steep banks. Such banks might be riverine: at Broadwindsor carts crossing the stream would meet with a stiff climb up the banks. Or they might be on stretches of road not involving water negotiation, as seems likely to have been the case at Winsor near Eling HMP, where there is a steep, winding, approach road from the east. As regards ascent from a river, a modern parallel is recorded in a CHE name, Twitch Hooks Lane in Farndon, where there was a windlass at the top of a steep lane to help waggons up from the river Dee (EPNS survey 4, p. 75).

Owermoigne DOR is no longer to be included in the lists of **ofer* or *ōra* names. Coates 1995a points out that frequent spellings *Ogre(s)* are not consistent with derivation from either of the English words. He suggests a British name which would be an ancient form of modern Welsh *oerddrws* 'wind-gap', with some analogical reformation due to association with **ofer* and *ōra*.

Reference section for *ōra*

As simplex name (sometimes with N- or R- from *atten, atter*): Nore Fm (in Bramley) SUR, Nower Fm (in Kilmington) DEV, Nowhurst Fm (in Slinfold) SSX, Oare BRK, KNT, WLT, Ore SSX, Ower HMP(2), Rora (in Ilsington) DEV. In Bure Homage HMP, Bure is '(place) by the *ōra*'.

As first element: Orcop HRE (**copp**); Oreham SSX (*hām*).

As generic with personal names: Ballinger BUC (*Beald, -ing-*); Batsford GLO (*Bæcci*); Bognor SSX (fem. *Bucge*); Chinnor OXF (**Ceonna*); Chivenor DEV

HILLS, SLOPES AND RIDGES

(*Cifa*); Copnor HMP (**Coppa*); Cumnor BRK (**Cuma*); Eleanor Fm (in Wittering) SSX (*Ealda*); Galsworthy DEV (possibly British *Gall*); Hazard DEV (*Hereweald*); Hedsor BUC (*Hedde* or **Heddel*); Itchenor SSX (*Ycca*); Launder's Fm (in Watlington) OXF (*Lāfa*); Lewknor OXF (*Lēofeca*); Loxhore DEV (*Locc*); Pednor (in Chesham) BUC (*Peada*).

With descriptive terms: Bicknor DEV (in Bishop's Nympton), KNT (**bica* 'point', cf. Bicknor GLO/HRE, under **ofer**); Bradnor HRE ('broad'); Clare OXF ('clay'); Gurnard IOW (probably 'mud'); Honer SSX ('hollow'); Honor End BUC ('stone'); Pinner GTL ('pin'); Radnor SUR, *Radnor* (later Pyrton) OXF ('red'); Rowner HMP, SSX ('rough'); Stanners Hill (in Chobham) SUR, Stonar KNT, Stonor OXF ('stone').

With references to vegetation: Ashour (in Leigh) KNT and Axford (in Nutley) HMP ('ash'); Barker's Hill (in Semley) WLT (*Beorcora* 956, 'birch'); Boxford BRK (*Boxora* 821); Elmsworth (near Porchfield) IOW; Golder OXF (probably referring to yellow flowers); Hodore (in Hartfield) SSX ('heath'); Pershore WOR ('osier'); Vexour (in Penshurst) KNT (*feax* 'hair' assumed to refer to coarse grass when used in place-names).

With references to wild creatures: Bagnor BRK ('badger'); *Kitnor* (later Culbone) SOM ('kite'); Woolvers Barn (in East Ilsley) BRK ('wolf').

With references to domestic creatures: Bouldnor IOW ('bull'); Chalder (in Sidlesham) SSX ('calf'); Goathorn DOR ('goat'); Keynor SSX ('cows').

With references to topographical features: Denner BUC (**denu**); Horner (in Diptford) DEV ('horn'); Lynsore (in Upper Hardres) KNT (**hlinc**).

With reference to position: Marker (in West Thorney) SSX (*mearc* 'boundary', i.e. between SSX and HMP).

With reference to an activity: Fitzworth DOR (?*fitt* 'contest'); Wardour WLT ('watch').

With reference to a structure: Werrar (near Northwood) IOW ('weir').

Names with obscure or ambiguous qualifiers:

Brickworth (in Whiteparish) WLT (earlier *Bricore*). Possibly a hill-name from British *-brīco* (**brīg**), but this is only otherwise evidenced in two compounds with **hyll**, and a compound with **ōra** does not seem altogether convincing.

Drellingore (in Alkham) KNT (earlier *Dyllyngore*). Assuming that the generic is **ōra** rather than **gāra** 'gore', this name is comparable to Dillington HNT, SOM, for which a personal name **Dylla* has been conjectured. The spellings for the two Dillingtons are compatible with this and **-ingtūn**, but a triple occurrence of the supposed personal name with a connective **-ing-** strains credulity. An element **dylling* seems to be required.

Gobsore (in Honiton) DEV (earlier *Coppeshore*). The first element could be either the genitive of **copp** or a personal name *Copp*.

Icknor (in Stocksbury) KNT. Only one spelling – *Ictenore* 1300 – is available.

Kitchenour SSX. Perhaps the same qualifier as in Kitchingham, Kitchenham SSX, discussed under **hamm**.

Poulner (in Ringwood) HMP. The spellings indicate *pola*, which appears to be unique to this name.

Upnor (in Frindsbury) KNT was *atte Nore* 1292, *Upnore* 1374. It was probably a simplex name with *N-* from *atten*, and *upp* prefixed later. The significance of *upp* 'higher' is uncertain. The settlement, which straggles along the north bank of the Medway estuary, has two parts called Upper and Lower Upnor. The likeliest candidate for the *ōra* is the long bank which runs behind the shore to the north of Lower Upnor. It is possible that the whole line of settlement was called *Nore* from this bank, and that the southern part became Upnor because it was higher up the Medway, then when the name Upnor had spread to the other half they were distinguished by Lower and Upper. Further documentation is desirable.

Warningore SSX (earlier *Waningore*). A personal name *Wænna has been conjectured. There is, however, a river-name Wenning (LNC/YOW), which is a derivative of *wann* 'dark', and this would suit Warningore, as *e* became *æ* in SSX dialect. There is a small stream by the farm.

pēac 'peak', used of pointed hills, is occasionally found, simplex or as first element, in settlement-names. Peek in Ugborough DEV is well recorded from DB on, and compounds include Peakdean SSX, Peckham KNT, SUR, Petton SHR and Peckforton CHE. The genitive occurs in Pegsdon BDF, earlier *Pekesdene*, in Pexall CHE (**hyll**), in *Pekesbru*, the ME name of Ganton Brow YOE, and in Peak's Arse, the earlier name of Peak Cavern in Castleton DRB. This last name is *Pechesers* in DB, and it is, of course, in the district called The Peak. The first reference to the Peak District is in the late 7th-century text known as the Tribal Hidage, where a province of Mercia is called *Pecsætna lond*, 'Peak-dwellers' land'. The Anglo-Saxon Chronicle calls it *Peaclond*, but from DB onwards it is *Pec, Pek*, later *Peke, Peek*.

penn PrW, PrCorn, PrCumb, **pen** Welsh, Cornish. The literal meaning is 'head'. The transferred place-name senses certainly include 'end', which is probably the most frequent, 'headland', and (as adjective) 'chief'. 'End' can be 'place at the end of something', as in Penge, or 'coastal promontory', as in Penzance. It is a moot point whether *penn* could also mean 'hill'. Place-name evidence which can be adduced in support of this is set out below, but Dr Oliver Padel regards this as inconclusive and would prefer to fit all the *penn* names into other categories.

HILLS, SLOPES AND RIDGES

penn is one of the commonest Celtic words in English place-names, but its true frequency is difficult to assess because it is almost impossible to distinguish it from OE *penn* 'pen for animals'. This is always a possible alternative in simplex names and in compounds in -pen, but it is somewhat less likely when *penn* is the first element in a compound.

British or PrW compounds with *penn*- may be considered first. There is a notable group of names which derive from PrW **penn-cēd* 'wood's end'. Here belong Penge GTL, Pencoyd HRE, Penquit DEV, Penquite CNW, Penketh LNC. Clarendon Forest WLT was called *Penchet* or *Paunchet* until the 15th century, and this compound is the first part of Penkhull STF. Another recurrent compound has become Pennard SOM and Penyard HRE. The qualifier in this name is either **arth* or **garth*, both of which mean 'height'; and 'place at the end of the high ground' is the likely meaning. This compound occurs eleven times in CNW, is also found in Brittany, and has become Peniarth in Wales. Other PrW *penn*- compounds are Penrith CMB, discussed under **ritu**- in Chapter 3, Pencraig HRE with the ancester of W *craig* 'rock', and Pengethly HRE which is a synonym of the **penn-cēd* names using another word for 'wood'.

The 'end' meaning of *penn* shades into 'coastal promontory', i.e. 'land's end'. Penryn CNW is from a compound appellative which means 'promontory'. The Cornish name of Land's End is Penwith: Padel 1988 thinks this may mean 'end district', but other Celtic scholars, cited in Ekwall 1960, suggest 'promontory seen from afar'. Penzance CNW is 'holy promontory. Pembroke, Wales, is British *Pennbroga*, Welsh *Penfro* 'end land'.

There is an exhaustive discussion of the large category of Cornish Pen-names in Padel 1985, pp. 177-83, where many recurrent names derived from compound appellatives and phrases are listed in addition to the two discussed above. These include, for instance, **pen-fenten* 'stream-source', **pen hal* 'moor's head', **pen nans* 'valley's head'. Dr Padel recognises 'headland' as well as 'promontory, point', and some of the 'headland' names cited are inland. He is, however, unwilling to allow that *pen* ever meant simply 'hill' in Cornish names, and in correspondence and discussion he has made it clear that he does not accept 'hill' as a proven sense of the British and PrW word in place-names in England, though this has been generally advocated by English place-name scholars.

A compromise position might be obtained by considering that *penn* in some names in England refers to a high promontory-type ridge. The simplex name Penn occurs in BUC and STF, and Pendomer SOM is *Penne* in the earliest records. The BUC and STF names can be interpreted as referring to ridges, and so probably can Pendomer. The great hill called Penyghent YOW is also ridge-shaped. There are two hybrid names in LNC, Pendle and Pendlebury, in which *penn* has been glossed by **hyll**. Pendle, if it had been named wholly in English, might well have been a **hōh**. Pendlebury, in Greater Manchester, is difficult to assess. There are other instances of the compound with **hyll**, including Penhill YON. Pensnett STF is a non-tautological compound with *snǣd* 'detached piece of ground': this place, like Pendlebury, is not amenable to topographical investigation.

Penistone YOW may have as its first element a hill- or ridge-name *Penning*, formed by the addition of OE *-ing* to *penn*. This possible indication that *penn* was adopted by English speakers as a term for a hill or ridge is supported by the two occurrences of Hackpen Hill, in DEV and WLT, in which *penn* is qualified by OE *haca* 'hook'.

The only recorded instance of the British ancestor of PrW *penn* occurs in *Pennocrucium*, now Penkridge, which is discussed under **crūg**. This may fairly be taken to mean 'tumulus on a headland'.

Another group of *penn* names which remains to be discussed consists of compounds with animal names. In Pentridge DOR, Pentrich DRB and Pentyrch GLA the qualifier is a fossilised genitive of Welsh *twrch* 'boar'. Dr Padel associates these with the Swineshead names discussed under **hēafod**, and considers that they refer to the promontories shaped like a boar's head. In Padel 1985, p. 179, he lists a number of Cornish names – Pencarrow, Pinnik, Penkevil, Penvith, Penbough and Pemboa, Penfranc, Pedn Kei, Pedn Tenjack – which refer to deer, horses, cows, goats, ravens, dogs and (in Pedn Tenjack) to a fish. These also he associates with English **hēafod** names which refer to animals and birds, and he takes a less dismissive view than the present writer of suggestions that some of them have a ritual significance.

There is a difference of opinion between Dr Padel and myself about the significance of the WOR name Pensax. In Gelling 1988, p. 99, I adopted Ekwall's suggestion that Pensax meant 'hill of the Saxons', taking it to be a reference by Welsh speakers to the presence of a minority of English speakers in the vicinity of the massif of Clows Top. Dr Padel, influenced partly by his reluctance to admit 'hill' as a meaning of *penn*, prefers to take the name as a reference to a rock or protuberance which reminded Welsh speakers of an Englishman's head. In Pensax, which has a pre-Conquest spelling *Pensaxan*, 'Saxon' may be plural, whereas in the comparable name Pennersax DMF, it is certainly singular. The DMF name means '*penn* of the Saxon', with a definite article. This could be a collective singular, as could 'goat' in Gateshead and the animal names in Dr Padel's Pencarrow, Pinnick etc. where these are definitely singular. The precise meaning of 'animal's' or 'animals' head' names, whether English or Welsh, remains debatable, and the category is not necessarily a unitary one.

As noted above, it is not possible to compile a definitive list of names containing PrW *penn*. Inkpen BRK has been omitted, as the generic may be OE *penn* 'fold', used figuratively of the hill-fort on the parish boundary. Pilsdon Pen in DOR is also crowned by a great fort. Kilpin YOE is considered to be 'calf pen', and some names with OE personal names as qualifiers – Hampen and Owlpen GLO and Ipplepen DEV – are generally ascribed to the OE word, and

HILLS, SLOPES AND RIDGES

this seems appropriate to Pencombe HRE. Pamber HMP has been treated as ambiguous in the reference section for **beorg**, perhaps illogically since Pensnett STF is included here.

Penselwood SOM could be assigned to PrW *penn* if an OE form **peon* were accepted as a possible derivative. The forest-name Selwood has been added to an earlier *Penne*, which is generally identified with the place called *Peonnum* and *Peonnan* in the Anglo-Saxon Chronicle. Ekwall 1960 considers this name and Pinhoe DEV (*Peonho* in the Chronicle) to belong here, but Pinhoe (discussed under **hōh**) is more problematical. Pendock HRE (*Pendoc* and *Peonedoc* in OE sources) is explained in Ekwall 1960 as 'head of the barley field' from *penn* and PrW **heiddioc*.

pīc 'point'. This is a commoner term than *pēac* for a pointed hill, but some of its other senses also occur in place-names. It means 'prickle' in Pickthorne SHR and 'pike' (the fish) in Pickmere CHE and Pickburn YOW. The 'pointed hill' sense occurs in Pickhill ESX, Pick Hill KNT, Pickhurst KNT, Picton PEM, Pickup LNC (**copp**), Pickwell LEI, Pickwick WLT, Pyecombe SSX.

Individual hill-names in -Pike are common in northern England.

*ræc. The meaning 'raised straight strip' was suggested in Gelling 1984. The term is a derivative of the verb *ræcan*, modern *reach*. It is found in eastern England, where there are two major names, Reach BDF and CAM, and two minor names, The Reaches in Eye NTP and Reach in Whittlesey CAM.

Reach CAM (Fig. 15) offers the clearest evidence for the meaning. It is at the fenland end of the Devil's Dyke, a great post-Roman earthwork of dramatic straightness which runs for seven miles to the south-east, barring the route from the midlands into East Anglia. The Dyke forms the boundary betweeen the ancient parishes of Burwell and Swaffham Prior, and the village of Reach was split between them. The settlement stands at the exact junction of fen and firm ground, and there are well-preserved installations of the hythe which controlled traffic on the canal called Reach Lode which runs out into The Fens. The place could have been called Hythe, but its relationship to the earthwork must have seemed more important. It would be perverse not to connect the name with the Devil's Dyke.

Reach BDF is north of Leighton Buzzard, where there is a straggling village called Heath and Reach. The parish is bounded by Watling Street, which has a very high agger in this stretch, and it is possible that **ræc* was here applied to the raised Roman road.

The two minor names are both in The Fens. The NTP one belongs to a road near Eye. This runs alongside a stream called Cat's Water which is the old course of the river Nene, and there could have been levées on the river banks. At Reach near Whittlesey CAM the Fen Causeway could have had an agger and the King's Ditch could have had raised banks.

**ric* 'strip'. It has long been recognised that this is a place-name element which is related to **rǣc* and to other Germanic words such as ON *reik* 'hair-parting', Swedish dialect *reik, rek* 'stripe, furrow'. Place-names constitute the only evidence for its precise meaning in OE.

Ekwall (1928, pp. 369–71) considered the meaning to be 'stream, ditch'. The small corpus of names which has the word as generic does not, however, support this interpretation if we exclude the recurrent stream-name Skitterick, on which Ekwall's suggestion was mainly based. Ekwall collected eight examples of Skitterick, and more could probably be found. The name is more likely to be an extended version of the recorded stream-name *Shiter*, Skitter, which is considered to refer to sewage.

ric* is not a common place-name element, and it is liable to be replaced by **hrycg in modern forms. There should be a preponderance of early spellings in *-ric, -rich(e)* (as *Cameric* 1086, *Kimerich* 1212 for Kimmeridge DOR) before a name is classified in this small category. The corpus examined in the reference section is not definitive. Puckeridge HRT, for which Ekwall 1960 gives **ric*, has been omitted as being just as likely to contain **hrycg**, but Marrick YON, assigned by previous reference books to ON **hryggr**, has been included. Lostrigg Beck CMB has been omitted on the grounds that **ric* does not otherwise occur in the northwest. The distribution suggests that it was in use only during the earlier part of the Anglo-Saxon period.

It is sometimes possible to associate **ric* with a straight strip of raised ground. The clearest instances are probably Escrick and Wheldrake, south-east of York (Fig. 41). These stand on the stretch of the Escrick Moraine which runs between the rivers Ouse and Derwent. At Marrick YON there is a moraine with an earthwork on top. In other instances a reference can be postulated to straight, narrow ridges.

At Reighton, south of Filey YOE, Speeton Hills and Speeton Cliffs make a straight line running to the east of the settlement. Rastrick, south of Brighouse YOW, is on the side of a long narrow ridge, with straight lengths on either side of the settlement. At Chatteris CAM there is a low ridge which shows clearly on the hachured 19th-century 1in. map.

In some instances it is possible that **ric* refers to the agger of a Roman road, in the manner postulated for **rǣc*. This is possible for Cookridge YOW, where the agger is 16ft wide, and Margary 1957 (which gives details of these roads) says that there is a visible agger near Mouldridge Grange DRB.

As can be seen from the reference section, in an unusually high proportion of names with **ric* as generic the nature of the qualifying element presents difficulties. This could be held to support the hypothesis of an early date. Among compounds with identifiable first elements all but one refer to trees. Askrigg YON (for which no landscape feature has been identified) and Escrick YOE contain ON *askr* or a Norse-influenced form of OE *æsc*. The occurrence of four compounds with OE *lind* 'lime-tree' is unexplained. If the

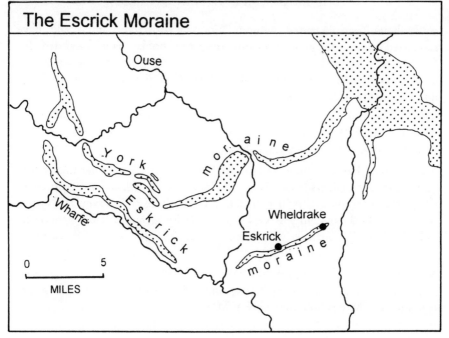

The Escrick Moraine 41

meaning were simply 'row of trees' (a possibility glanced at in Smith 1956) there should be more compounds with the names of other species.

There are two simplex names, *Riche*, a lost village southwest of Boston LIN, and Glynde Reach SSX, which is referred to in the surname of Thomas *atte Riche*, recorded 1332. Glynde Reach is *the sewer called Ritche* in 1544, and this has naturally been held to support Ekwall's explanation of Skitterick. It is possible, however, that in both these simplex names **ric* refers to the upcast banks of drainage channels, or, in the LIN name, to a pre-Conquest sea-bank.

Reference section for **ric*

As simplex name: Riche LIN (a lost village southwest of Boston), Glynde Reach SSX (referred to in the surname of Thomas *atte Riche* 1332).

As first element: Reighton YOE.

As generic with reference to trees: Askrigg YON and Escrick YOE ('ash'); Lendrick Hills YOW, Lindrick NTT, YOW, Lindridge (in Stanton by Dale) DRB ('lime').

With a personal name: Chatteris CAM (*Ceatta*).

With obscure or ambiguous qualifiers:
Cookridge YOW. The first element is probably 'cuckoo'. A strong case has been made by Professor Richard Coates (1995b) for the existence of an OE **cucu*, and its occurrence in a number of place-names, including this one. Conventional wisdom has it that *cuckoo*, not noted in NED until the 13th century, is of French origin, but Coates argues that this onomatopoeic name

could have been invented without Norman assistance, and could have co-existed with OE *gēac*. Previous reference books have suggested for Cookridge a personal name *Cuca* or a stream-name *Cuce*, both envisaged as derivatives of OE *cwicu* 'living', or a derivative of that word used for a quick-set hedge or for couch grass.

Kimmeridge DOR. The first element could be *cȳme* 'beautiful, splendid', but a personal name derived from the adjective, which has been conjectured for Kimpton HMP, HRT, would be equally satisfactory.

Marrick YON. Early spellings and modern form suggest **ric* rather than **hrycg** or **hryggr**. Dr Fellows-Jensen (1972, p. 257) supports Ekwall's suggestion of OE *(ge)mǣre* 'boundary' as first element, rather than EPNS survey's ON *marr* 'horse'. A third possibility is ON *marr* 'marshy ground'. If the generic be indeed **ric* an OE qualifier will be more likely: but *(ge)mǣre* usually becomes Mear-, Meer- in Yorkshire, and Marrick is not closely related to a boundary.

Mouldridge DRB. Two suggestions have been made: *molda* 'top of the head' (used of a hill in Mouldsworth CHE) and *molde* 'earth', but neither seems convincing. There is no notable hill at Mouldridge.

Rastrick YOW. The only suggestion available (excluding a very unlikely one in Ekwall 1960) is that the first element is ON *rost* or a Scandinavianised form of OE *rest*, meaning 'resting-place'. The element appears to be unique to this name.

Wheldrake YOE. OE *cweld* 'destruction, death' is favoured by reference books, but the word has not been noted in any other place-name and the compound is not convincing.

scelf (Anglian, Kentish), **sci(e)lf**, **scylf** (West Saxon), **scylfe** (Anglian, West Saxon), **scelfe** (Kentish) 'shelf', Fig. 42. The different forms of this word are discussed in Smith 1956 (2, pp. 104–6).

This element was generally translated 'slope' or 'shelving terrain' until that interpretation was challenged in Gelling 1984. The suggestion made there, that in place-names it refers to exceptionally level ground, has been strongly supported by subsequent fieldwork.

Level ground which is designated *scelf* may be at high or low altitudes. At Bramshall STF there is a large hill which has a startlingly flat wide summit. The church stands on the rim of this, and on passing through the churchyard the observer is confronted by a great level field, not built on, probably because it is badly drained. At Shelton-under-Harley STF, by contrast, the farm overlooks a fan-shaped broadening of a valley. At Shelland SFK, Shilton OXF and WAR, and Skelton YON, there are wide expanses of level ground. A smaller 'shelf' is the remarkably level field across the road from Shovelstrode Fm in East Grinstead SSX.

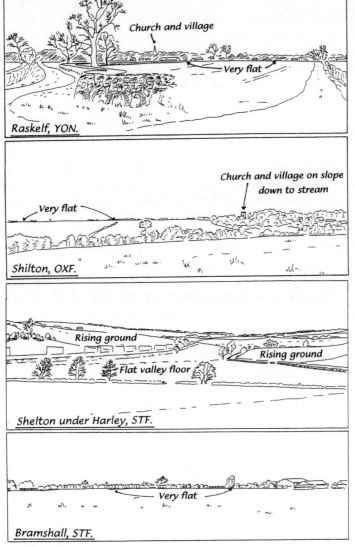

Sometimes the flat ground is noteworthy because it is surrounded by hills. Shelf YOW, occupying a rare level space in the mountainous country southwest of Bradford, is a striking instance of this use. Sometimes the level ground among hills is quite small, as at Shelve SHR. Sometimes it is a small platform on a hillside which offers a convenient site for a farm: an example is Shelf Fm in Luppitt DEV.

At Bothamsall NTT a long level bank overlooks the **botm** formed by the parallel courses of the rivers Meden and Maun. This rises to a point at one end, which probably accounts for the alternation of **hyll** with *scelf* shown by early spellings. In Bashall YOW (discussed under **bæc**) and Silpho YON the 'shelf' is the flat top of a ridge.

Appreciation of this element requires fieldwork. By no means all of the sites have been visited, but sufficient material has been amassed to suggest that level ground may be expected rather than slopes or shelving terrain.

In DEV, the recurrent minor name Shilstone refers in one instance (in Drewsteignton) to a surviving cromlech. 'Stone monument with a shelf' is probably the meaning, *scylf* referring to the flat stone lying on top of the uprights.

scelf, scylf is not a common place-name element, and it is rare in minor names and field-names. As a generic it is frequently confused with commoner elements, particularly **halh** and **hyll**. In Bothamsall NTT, however, there was probably genuine alternation with **hyll**.

As the analysis in the reference section shows, *scelf, scylf* are commonest as first elements, especially with **tūn**. When they are used as generics they are mostly qualified by personal names.

The corresponding ON word **skjalf** may be found in the Yorkshire names Hinderskelf, Raskelf, Skutterskelf and Ulleskelf, but Sk- in Skelton is probably due to the influence of Norse speech on the OE word.

Reference section for **scelf**, ON **skjalf**

As simplex name: Shelf DRB, YOW, Shell WOR, Shelve SHR.

As first element with tūn: Choulton SHR, Shelton BDF(2), NFK, NTT, SHR, STF, Shilton LEI, OXF, WAR, Silton YON, Skelton CMB, YOE, YON(3), YOW(2). Shilvinghampton DOR is '*tūn* of the shelf-dwellers'.

With **lēah**: Selly (Oak) WOR, Shelley ESX, SFK, SSX, YOW.

With **dūn**: Sheldon DEV, WAR, Shildon DRH.

With **hyll**: Shelfield STF, WAR (both earlier *Shelfhull*).

With other generics: Shelfanger NFK (**hangra**); Shelland SFK (**land**); Shovelstrode SSX (**strōd**); Silpho YON (**hōh**). Shulbrede SSX was mistakenly indexed as a 'shelf' name in Gelling 1984: it is probably an early-recorded example of a recurrent minor name meaning 'shovel broad', i.e. a narrow strip of land.

As generic with personal names: Gomshall SUR (*Guma*); Hinderskelfe YON (ON fem. *Hildr*); Hunshelf YOW (*Hūn*); Minshull CHE (*Mann*); Oxhill WAR (*Ohta*); Paxhill SSX (OE *Bacanscylf, Baca*); Tanshelf YOW (**Tædden*); Tibshelf DRB (*Tibba*); Ulleskelf YOW (ON Ulf); Wadshelf DRB (*Wada*); Waldershelf YOW (*Waldhere*).

With references to vegetation: Bramshall STF ('broom'); Hathershelf YOW ('heather').

With topographical terms: Bashall YOW (**bæc**); Bothamsall NTT (**botm**: in this name *scelf* alternates with **hyll**).

With an earlier place-name: Skutterskelfe YON (probably an ON stream-name, 'Chatterer').

With an animal name: Raskelf YON ('roe').

Names with obscure or ambiguous qualifiers:
Litchfield HMP. The name is much corrupted. Early spellings vary between those with *-v-* as in *Livesulve* 1212, and those with *-d-*, as in *Lidesulve* 1238. A careful discussion in Coates 1989 concludes that the first element was probably **(h)lyf*, with the vowel-length indeterminable. He tentatively suggests a meaning 'sheltered place'.

Moxhull WAR. A word *mox* or a personal name **Mox* occurs in a boundary-mark, *moxes dun*, in Alton HMP.

Ranskill NTT. The first element is 'raven', which could be a personal name.

Shareshill STF (*Sharesweshulf* 1252). Ekwall's suggestion of a metathesised form of *scræf* is probably the best available. *scræf*, which has recorded meanings 'cave' and 'hovel', occurs in a few place-names but has seldom been identified with a surviving feature. In Salford Bridge (earlier *Shrafford*) in Aston WAR it refers to caves in a sandstone cliff.

sīde 'side'. This word is sometimes used for a hillside. It is well-evidenced in that sense in northern counties but rare in the south of England. Some names containing *sīde* refer to land beside a river or a wood, so only part of the corpus is relevant here.

There is one simplex name, Syde GLO. In compounds *sīde* is most frequently qualified by descriptive terms, as in Barnside YOW (*Barnedeside* 13th century, 'burnt'), Brownside DRB and Langsett YOW. There is a recurrent compound with *fāg* 'variegated'. This gives Facit LNC, Fawcett Forest WML, Fawcettlees (in Askerton) CMB, Fawside (in Lanchester) DRH, and Phoside in Hayfield DRB, which was 13th-century *Fauside*. EPNS CMB survey (p. 56) notes that there are two examples in Berwick. One reference has been noted to vegetation, Birkenside NTB, and one to animals, Lambside (in Holbeton) DEV. In Whernside YOW(2) the first element is *cweorn* 'millstone'. ON *garthr* 'enclosure' occurs in Gartside LNC. Personal names occur in Padside YOW (*Pada*) and Tosside YOW (ON *Thórir*). In two other YOW names, Bishopside and Swanside, the reference is to late Anglo-Saxon or medieval ownership, i.e. the Archbishop of York, and an owner called *Suanus filius Suani* who is mentioned c.1147.

The evidence suggests that names containing *sīde* are likely to be of comparatively late origin. Some north-country names with modern -side contain ON *sætr* 'shieling'.

CHAPTER 6

Woods and Clearings

Modern landscape historians have rejected the views of their predecessors about both the extent of woodland in post-Roman Britain and the role which it played in the lives of Anglo-Saxon settlers. The extent of woodland in the 5th century is now considered to have been roughly comparable to that of the present day, and it seems probable that farmers, both British and Anglo-Saxon, regarded the woods more as vital resources than as obstacles to travel and settlement. The main exception is in south-east England, where the Weald was a formidable barrier, isolating the kingdom of Sussex. The most fervent advocate of continuity in settlement-patterns would probably not dispute that permanent settlements in the Weald were scarce in the Roman and post-Roman periods, though mineral, pastoral and timber resources would always be exploited there.

By far the most important word discussed in this chapter is lēah, incomparably the commonest topographical term in English place-names. The main use of lēah is here taken to be for naming settlements which flourished in a woodland environment. The resources offered by this environment would influence the pattern of farming and would elevate the standard of living above that available to farmers in treeless countryside. Later, as is revealed by Anglo-Saxon charters and the Domesday Survey, the burden of taxation would be lighter for them than for communities with great expanses of ploughland.

It is not at all likely that all, or even most, of the vast number of settlements with lēah names were new creations of the Anglo-Saxon period. The term should probably be seen mainly as a recognition by Old English speakers that these places had a different economy from those in areas where woodland was sparse. But there would, of course, be continuous encroachment on forest as demand for arable increased, and lēah would also be used for new settlements on the fringes of ancient woods, such as Kinver and Wyre Forests in the west midlands. Readers may care to study the noughts and crosses map which is reproduced as Fig. 44, and to consider which of these alternative explanations best suits each cluster of noughts.

Although technically 'topographical', lēah is for the most part a quasi-habitative term in settlement-names. It could, however, be used in its ancient sense of 'wood' if the wood in question was an isolated one in mainly treeless countryside. This is akin to the occasional use of *tūn* in its archaic sense of 'enclosure'.

Since woodland is no longer believed to have been virtually unlimited in the early Anglo-Saxon period it is reasonable to interpret some woodland terms as referring to the management of a greatly valued resource. Coppicing is a management technique which is shown by archaeology to have been of immemorial antiquity in Britain, and the 'grove' names discussed in this chapter may reflect the Anglo-Saxon appreciation and continuance of this practice. Deliberate plantation may have created some of the single-species woods referred to in the numerous compounds with tree-names listed under **holt**. Other terms in this chapter, particularly **fyrhth** and **hyrst**, gave information about the quality of the woodland. A special function in the exploitation of good woodland is probably indicated by the recurrent compound *wudu-tūn*.

Of the 17 words discussed here four – **lundr**, **skógr**, **thveit** and **vithr** – are Norse, and one, **cęd**, is British. This last is one of the commonest British place-name terms which survived the language change, but the evidence does not suggest that the Anglo-Saxons adopted it as a word. Most of the compounds in which it appears are wholly British, and tautological compounds with Old English **lēah**, **wudu** and **hyrst** suggest that **cęd** was understood to be the name of the wood in question.

bearu 'small wood'. The gen. sg. is *bearwes*, dat. sg. *bearwe*, nom. pl. *bearwas*. There is evidence (for which see the EPNS survey 107–8) that in DEV *bearu* developed an irregular dat. sg. *beara*, and this is likely to be the source of -beer, -bear and Bere, Beere, Beare, Beera in names in that county. When only ME spellings are available, *bearu* may be difficult to distinguish from **beorg** 'hill' and *bǣr* 'swine pasture'. It is probable that most names containing *bearu* have been identified, but some in DOR – Bere Regis and Beer Hackett – and in SOM – Beer Crocombe and Beer near Cannington – could be from *bearu* or *bǣr*. In WLT, where Barrow usually comes from **beorg** and refers to tumuli, two instances, in Langley Burrell and Bishopstrow, are ascribed to *bearu* because they have early spellings like *Barwe*, *Barewe*, *Baruwe* instead of spellings like *Berewe* which are appropriate to **beorg**.

Except in DEV and SOM *bearu* is a rare element, absent from many counties and represented by only one or two examples in a number of others. In DEV it is predominantly a minor-name element. The EPNS survey (p. 657) lists 15 examples of Beara, 10 of Beer, eight of Beera, and instances of Beare, Bere, Berry, many of these being recorded in ME surnames of the *atte Bere* type. Elsewhere, however, a high proportion of the sparse names containing *bearu* are parishes and Domesday manors. See Fig. 43.

It seems likely that, apart from **lēah** and **wudu**, woodland terms had specific meanings, but the material available for *bearu* does not suggest a

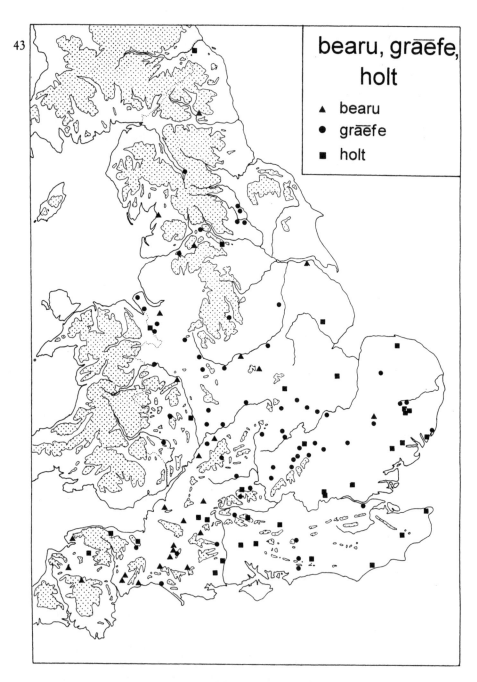

definition comparable to those offered here for the 'grove' words and for **holt** and **hyrst**. It was probably used for woods of limited extent, and literary references show that it was considered to be the equivalent of Latin *nemus*.

WOODS AND CLEARINGS

Reference section for **bearu**

As simplex name: Bare LNC, Barrow CHE, DRB, GLO, LEI, LIN, SFK, SHR, SOM(2), Beer DEV(2). The frequent use of *bearu* as a simplex minor name in DEV is discussed in the article *supra*. There are occasional minor names in other counties, e.g. Beare Green in Abinger SUR, Barrow Fm and Ho in Langley Burrell and Bishopstrow WLT, Barrow Hill in Hampstead GTL, Barrow in Wentworth YOW.

As first element: the only example noted is the well recorded minor name Bar Bridge in Uckington GLO.

As generic with personal names: Adber DOR (*Eata*); Aylesbeare DEV (*Ægel*); Eggbear and Eggbeer DEV (*Ecga*); Kentisbeare DEV (**Centel*); Loxbeare DEV (*Locc*); Muxbere (in Halberton) DEV (*Mucel*); Sedgeberrow WOR (**Secg*).

With references to wild creatures: Bagbear(e) DEV(3) (probably 'badger', the minor name Bagburrow Wd in Mathon WOR is another example); Kigbeare DEV ('jackdaw'); Larkbeare DEV(2); Rockbeare DEV ('rook', also in DEV three minor names Rookabear, Rockbear(e).

With references to trees: Haselbury DOR, SOM (a 3rd instance is the minor name Hazlebarrow DRB, near Sheffield); Ogbear (in Tavistock) DEV ('oak').

With references to products: Shebbear DEV ('shaft'); Timsbury SOM ('timber').

With references to topography: Mooresbarrow CHE ('of the marsh'); Woodbeer DEV.

cẹ̄d, **coid** PrW, PrCorn, PrCumb, **coed** W 'wood, forest', cognate with English *heath*. The word became *cos* in Cornish.

This is one of the commonest Celtic words in English place-names, though, as with **penn**, its British ancestor only occurs once in names recorded from Roman Britain. This single instance is *Letocetum*, preserved in the first part of Lichfield STF, *Liccidfeld* in Bede's *Ecclesiastical History*. In an early Welsh poem the Romano-British name appears as *Luitcoed*. This compound, which means 'grey wood', was probably an appellative in British speech. There are two examples in Wales, Llwytgoed in Aberdare GLA and in Llandrillo-yn-Rhos DEN. There are several in CNW, where it has become Lidcott, Lidcutt, Lydcott, Ludcott, and another example in an English-speaking county is Lytchett DOR. Litchett HMP, which was mistakenly cited in Gelling 1988, 1984 and from there in EPNS DOR survey 2, p. 29, is a ghost name.

In the forest-name Chute on the WLT/HMP border *cẹ̄d* is used alone, and another simplex example may have been incorporated into Chidden HMP, discussed under **denu**. In Melchet, another forest-name on the WLT/HMP border, it is qualified by PrW *mẹ̄l* 'bare', perhaps used substantively to mean 'bare hill'. Culcheth LNC and Culgaith CMB, identical names with *cẹ̄d* as generic, are variously explained as 'nook wood' or 'narrow wood'. Morchard DEV means 'great wood': there are two settlements, Morchard Bishop and Cruwys Morchard, 10 miles apart, and if these refer to a single wooded area the appropriate translation is 'forest'. Dassett, earlier *Dercet*, in Avon Dassett and Burton Dassett in south-east WAR is probably 'oak forest', referring to the western fringe of an area of 'wold' in NTP, defined in Fox 1989, p. 79. Orchard DOR is 'place beside the wood'.

Names derived from PrW *penn-cēd*, 'wood's end', are listed under **penn**. This compound, like the 'grey wood' one, occurs so frequently that it may fairly be regarded as an appellative, rather than as a separate coinage in each instance. Pendoggett CNW is '(place at) the end of two woods'.

cos is a frequent element in Cornish names, and a generous selection is given in Padel 1985, pp. 67–8.

cēd occurs in a number of hybrid British/OE names. Some of these are tautological: in Cheadle CHE, STF, *cēd* is glossed by **lēah**, and in Chetwode BUC by **wudu**. Chithurst SSX may be a comparable formation with **hyrst**. In Chatham ESX, KNT and Cheetham LNC *hām* 'village' is combined with *cēd*. Kesteven, the name of one of the three divisions of LIN, is explained in Ekwall 1960 as formed by the addition of ON *stefna*, used in the sense 'administrative district', to an ancient district-name formed from *cēd*.

There is as yet no comprehensive study of Scottish place-names which would reveal the frequency of the Cumbric form of this word in southern Scotland. Nicolaisen 1976, p. 172, cites Bathgate WLO (earlier *Batket*, 'boar wood') and Dalkeith MLO (earlier *Dolchet*, 'meadow' or 'valley wood') in addition to Pencaitland ELO, one of the 'wood's end' names discussed under **penn**.

An adjective *cēdiōg* 'wooded' occurs in Chideock DOR, Dunchideock DEV and Quethiock CNW.

There are a number of English place-names beginning with Chat- or Chet- which are not usually derived from *cēd* because the ME spellings have *Chate*- or *Chete*-, and the *-e-* between the first and second elements is likely to indicate a genitive inflection, suitable to derivation from the OE personal name *Ceatta*.

Datchet BUC and Watchet SOM are sometimes cited as names in *-cēd*, but neither is a certain example.

fyrhth(e) 'land overgrown with brushwood, scrubland on the edge of forest'. A form *gefyrhth* is sometimes found in charter boundaries. The accepted translation before Gelling 1984 was 'wood', but the definition given there and repeated here suits the use of the term in settlement-names and the sense-development traced in NED under *frith*. It would also explain the adoption of the word into Welsh, where in the 14th century *ffridd* meant 'barren land'. In modern Welsh *fridd*, pl. *friddoedd*, is used of land at a certain altitude, above the belts of good arable. The mid-11th-century document known as Gospatric's Writ, which deals with land tenure in CMB, refers to rights *on weald, on freyth, on heyninga*: this could be translated 'in forest, in heathland, in enclosed arable'.

fyrhth is much more common in minor names than it is in major settlement-names. The most frequent modern forms are Frith (found in CAM, CHE, CMB, DOR, DRB, ESX, GLO, MDX, RUT, SSX, SUR, WML, WOR) and Thrift (BDF, BRK, CAM, ESX, HRT, NTP, OXF, SUR, WOR). Other forms include Frieth (in Hambledon) BUC, Firths (in Cottingham) YOE, The Frythe (in Welwyn) HRT. In Kentish dialect *fyrhth* became *fright*, and there is also a Fright Fm in Rotherfield SSX. It is nowhere common, and it is not possible to ascertain its presence or absence in counties for which there is as yet no detailed survey, but the above lists suggest that the term had a country-wide distribution.

fyrhth occurs as generic in Pirbright SUR (first element *pirige* 'pear-tree'), which lies in a vast expanse of heathland. As first element it is found in Firbank WML, in an area of rough grassland northwest of Sedbergh, probably in Firbeck YOW, on the edge of a forest area southwest of Tickhill, and, in the genitive, in Frithsden HRT, at the edge of Berkhamsted Common. Another possible instance, though with only a few available spellings, is Fritham HMP, in which the generic may be **hamm** in the sense 'enclosure in marginal ground'. Coates 1989 says of Fritham 'Possibly coincidentally, the site of the present village is heath surrounded pretty well on three sides by woodland'. If the first element be in fact *fyrhth*, or a variant **fryhth(e)*, this situation is certainly not coincidence.

Frithville LIN was earlier a simplex name *Le Frith*. Here the word may have been used for an area of fen where brushwood grew: the EPNS survey of CAM, pp. 179-80, notes that *frithfen* was an appellative with that meaning.

In two DRB names the ME reflex of *fyrhth* refers to medieval hunting forests. Duffield Frith, having been *foresta de Duffeld* and *Duffel Chace*, appears as *Duffeldfrith* in 1332, and the affix of Chapel en le Frith refers to the Peak Forest. Neither of these, however, consisted of uninterrupted woodland.

Other relatively well-recorded names containing *fyrhth* or its ME reflex include Bigfrith in Bisham BRK, with the personal name *Bicga*, and Fredley in Mickleham SUR, a compound with -lēah. A lost *Akefrith* in Melling LNC, with a Scandinavianised form of *āc* 'oak', is the only compound with a tree-name noted in addition to Pirbright SUR.

Flyford WOR should probably be deleted from the small corpus of major names containing *fyrhth*. It is recorded in the pre-Conquest spellings *Fleferth* 930, *Flæferth* 972, *Fleferht* 1002. It was the name of a wood – the charter of 930 grants *locis silvaticis ad Fleferth* – and this has encouraged derivation from *fyrhth*. Ekwall 1960 suggests that *-ferth* is 'a weakened form of OE *fyrhth*', and says 'the OE forms are in transcripts and *Fleferth* etc. may well be ME developments of *Flæde-fyrhth*'. This will not do, however, as the other numerous place-name spellings in the relevant charters show no sign of weakening or of ME development.

Fleferth, Flæferth, Fleferht must be accepted as faithful copyings of spellings from 930, 972, 1002, and it is (as Ekwall perceived) difficult to see them as renderings of a name in *-fyrhth*. Both the derivation from *fyrhth* and the association of the first element with a lady called *Ælflæd*, named in a lost *Ælflædetun*, which were accepted in Gelling 1984, must, with regret, be abandoned, leaving Flyford unexplained.

græfe, grāf, grāfa, grāfe 'coppiced wood'. In Gelling 1984 it was suggested that the various forms were synonymous, and that the likely significance of all of them was 'coppice' in the technical sense, i.e. managed woodland, where trees were cut to form stools, and the resulting pole growth was harvested at regular intervals. Both suggestions still seem convincing, and in accordance with the first one no distinction is made in the reference section between *græfe* and the forms with *-ā-*. Modern forms Greave, -greave derive from *græfe*. In some areas, particularly YOW, there is a great deal of interchange between *-greve* and *-grove* or *-grave*. Kitson (forthcoming) concludes that in charter boundaries 'the rationale of the variant *græfe* does not seem geographical': in place-names, however, it is particularly common in the west midlands and in the West Riding of Yorkshire. See Fig. 43.

As regards derivation, Kitson strongly endorses the association with *grafan* 'to dig', which was tentatively advanced in Gelling 1984. He defines a 'grove' as wood of limited extent, in private ownership, digged about with a boundary ditch. The obvious reason for treating a wood in that way would be that it was coppiced.

Other place-name elements which certainly derive from *grafan* 'to dig' are *grafa* and *græf*, both meaning 'pit, trench'. Neither of these should give *-grove* in ME or modern forms, and this constitutes a useful formal distinction from *grāf(a)* and *grāfe* in some names. Grammatical formation helps in the case of Gravesend KNT, as the *-s* genitive is only appropriate to the strong 'grove' word, *grāf*. In many names with *-grave*, *-greve* spellings, however, derivation from either the 'grove' or the 'pit' words is theoretically possible, and because of this the items listed in the reference section cannot be taken as a totally reliable corpus of 'grove' names. The principle adopted is to consider 'grove' the likelier unless the first element clearly demands 'pit' (as it does in Orgrave LNC and Orgreave YOW which refer to iron ore), or unless there is a pit or trench in the vicinity. Wargrave BRK, which was included in the 1984 account, has been omitted due to information that there is a great trench by the weir at that place. Cole 1987 (pp. 50–51) demonstrates that 'chalk pits' is the likely etymology for Chalgrave BDF and Chalgrove OXF, though confusion with *-grove* has occurred in the latter. Bygrave HRT could be '(place) by the pit' or '(place) by the grove'.

It is not likely that *græf* and *grafa* 'pit' were common in settlement-names, and it can be confidently asserted that most of the names analysed in the reference section refer to a particular type of wood. The qualifiers provide interesting support for the suggestion that woods for which this was the appropriate term were small and clearly defined. The descriptive terms which are much the commonest qualifiers imply that the wood could be perceived as a single entity. This is especially true of the colour adjectives black, red, white, yellow and hoar. Kitson 1994, pp. 37-8, should be consulted on the last of these. 'Groves' could also be defined by reference to a topographical feature, such as a hill-spur, a hollow, a gully or a bridle-path, and they could be 'north' or 'south' or on parish boundaries, which also implies something fairly small. They were seldom associated with animals, either wild or domestic. There are not many references to tree-species, in which respect there is a contrast with **holt**.

In order to obtain sufficient material for analysis this study includes well-recorded minor names from counties which have detailed place-name surveys, and the locations of these have been given in the reference section. The next characteristic which requires discussion is the proportion of major to minor names.

The 'grove' words are well-evidenced in major names. They are commoner in minor names, but the proportions are not so unequal as they are for **hyrst**. The most striking instance of a major name is Bromsgrove WOR, which in 1086 was the central place in an estate stretching 12 miles from Timberhanger, south of Bromsgrove, to Moseley in Birmingham. Northfield and King's Norton in the southern suburbs of Birmingham are 'north' because of their position in this estate. One of the WOR Graftons is adjacent to Bromsgrove. The Domesday account details the quantity of wood, 300 cartloads a year, which was sent from this manor to the saltworks at Droitwich, and some 500 years later John Leland remarked that the wood used for salt production was 'young pole wood, easy to be cloven'. It seems reasonable to conclude that the presence of managed woodland in the Bromsgrove estate was felt to be an important characteristic, worthy of note in the name of a very high-status settlement.

In Bromsgrove the coppiced woodland may, exceptionally, have been extensive. Elsewhere, however, the tenor of the material suggests that groves were relatively small, and this makes it specially significant that a considerable number of these names denote Domesday manors and parishes. Some groves must have been of noteworthy economic importance. Among the *c.*130 names listed in the reference section, about 30 refer to ancient parishes, while Domesday manors (an overlapping category) number 38. The Domesday manors, apart from Bromsgrove, have moderate taxation assessments, Wingrave BUC being probably the largest at 13 hides. But achievement of DB and/or parish status suggests that in many instances a grove was a significant feature in the economy of an estate, and the same inference may be drawn from the use of the 'grove' words as qualifiers for *hām* and *tūn*. Apart from **wudu**, other woodland terms do not qualify habitative elements, with the single exception of Haltham LIN.

Reference section for **græfe, grāf, grāfa, grǣfe**

As simplex name: Grove BUC, BRK, NTT, *passim* as minor name. Grovebury in Leighton Buzzard BDF was *Grava* 1195, *La Grave* 1245. The three major names have forms with the definite article.

As first element with hām *and* tūn: Grafham HNT, Graffham SSX; Grafton CHE, GLO, HRE, NTP(2), OXF, SHR, WAR, WLT, WOR(2), YOW.

With other generics: Graston DOR ('stone'); Graveley CAM, Grovely Wood WLT (lēah); Gravenhill in Bicester OXF; Gravenhunger SHR (**hangra**); Gravenhurst BDF, SSX; Gravesend KNT; Grayshott HMP ('projection').

As generic with colour adjectives: Blackgrave in King's Norton WOR, Blackgreaves in Lea Marston WAR, Blagrave in Lambourn BRK, Blagrove in Wootton BRK, in East Worlington DEV, in Wroughton WLT (all 'black'); Hargrave CHE, NTP, SFK, WAR, Hargrove in Stalbridge DOR and in Sapperton GLO, Harragrove in Petertavy DEV, Hartgrove in Musbury DEV and in East Orchard DOR (probably all 'hoar'); Redgrave SFK; Whitgreave STF ('white'); Youlgreave DRB ('yellow').

With other descriptive terms: Faws Grove in Swinbrook OXF (*fæst* 'thick'); Horgrove in Tythegston Higher GLA (probably 'dirty'); Michel Grove in Clapham SSX ('big'); Notgrove GLO (*næt* 'wet'); Shortengrove in Grovely Wood WLT, Shortgrove in Whipsnade BDF, in Newport ESX and in Syresham NTP; Sidgreaves in Preston LNC ('wide'); Tingreave in Eccleston LNC ('fenced'). In Gargrave YOW and Orgreave STF, with first elements 'gore' and 'point', the references may be to the shape of the woods or to adjacent topographical features.

With personal names: Addingrove in Oakley BUC (*Æddi*); Bromsgrove WOR (*Brēme*); Chedgrove NFK (**Ceatta*); Chicksgrove in Tisbury WLT (**Cic(c)a*); Copgrove YOW (**Coppa*); Cosgrove NTP (**Cōf*); Cotgrave NTT (**Cotta*); Filgrave BUC (**Fygla*); Katesgrove in Reading BRK (**Cadel*); Kingrove in Old Sodbury GLO (*Cēna*); Palgrave (earlier *Pagrave*) NFK (*Paga*); Paygrove in Wotton St Mary GLO (*Pǣga*): Pinsgrove in Sunningwell BRK (*Pinn*); Weathergrove in Sandford Orcas DOR (**Wedera*); Woolsgrove in Sandford DEV (*Wulfmǣr*); Wools Grove in West Overton WLT (*Wulfsige*).

With references to topographical features: Congreave STF (**cumb**); Dungrove in Tarrant Gunville DOR (**dūn**); Egrove in South Hinksey BRK (**ēa**); Howgrave

WOODS AND CLEARINGS

YON and Hulgrave in Tiverton CHE ('hollow'); Sulgrave NTP ('gully'); Warpsgrove OXF (***werpels**).

With references to owners or users: Maidensgrove in Pishill OXF (*Menygrove* 1432, possibly an *-ing* derivative of *gemǣne* 'common'); Mangrove in Offley HRT ('common'); Piper's Grove in Higham GLO; Priest Grove in Ascot under Wychwood OXF, Swangrove in Hawkesbury GLO ('herdsman', cf. the synonymous, lost, *Hurgrove* in Steventon BRK); Tungrove in Horton GLO (*tūn*, presumably implying 'common').

With references to position: Hen's Grove in Swinbrook OXF (*Hemegrove* 1300, 'boundary'); Norgrove in Feckenham WOR ('north'); Morgrove in Spernall WAR ('boundary'); Southgrove in Burbage WLT, Sudgrove in Miserden GLO, Sutgrove in Gloucester (all 'south').

With references to trees: Ashgrove in Tollard Royal WLT; Birchengrove in Westbury on Severn GLO; Boxgrove SSX; Hezzlegreave in Saddleworth YOW ('hazel'); Thorngrove in Grimley WOR and in Gillingham DOR, – Hundred WLT.

With references to wild creatures: Algrave in Shipley DRB ('owl'); *Merdegrave* LEI, later Belgrave ('marten'); Musgrave WML ('mouse'); Woolgreaves in Sandal Magna YOW ('wolf').

With references to domestic creatures: Cowgrove in Pamphill DOR; Gedgrave SFK ('goats'); Ramsgreave in Blackburn LNC.

With references to products: Palgrave SFK ('poles'); Staplegrove SOM ('post'); Yardgrove in Marnhull DOR ('rod').

With references to structures: Bangrave in Weldon NTP ('barn'); Hexgreave in Farnsfield NTT (ME *hekke* 'wooden framework').

With older place-names: Walgrave NTP (Old); Wingrave BUC (Wing).

With other 'wood' terms: Copsegrove in Bisley GLO; Lyegrove in Old Sodbury GLO (**lēah**).

Miscellaneous: Bengrove in Teddington GLO (**begen* 'growing with berries'); Mortgrove in Hexton HRT (first element *morth* 'murder').

With obscure or ambiguous qualifiers:

Bedgrove in Weston Turville BUC. Documentation, starting in DB, is reasonably full, but the spellings show such variation (*Begrave* DB, *Bebegrave* 1247, *Belegrave* 1284, *Bedgrave* 1461, *Belgrove* 1627) that no etymology is likely to be wholly satisfactory.

Bengrove in Sandhurst GLO (*Bendegrave* 1271). First element 'bend', which could refer to the course of a road, or could be a plant-name, as in *wudu-bend* 'woodbine'. The first alternative suits the situation of Bengrove Fm.

Birchgrove in West Hoathly SSX, earlier *Buntsgrove*. As in Bonsall (under **halh**) the first element could be a bird-name related to ME *bunting*.

Kesgrave SFK. Available spellings suggest 'cress', but as Kesgrave is in heathland east of Ipswich this seems unlikely.

Leagrave BDF. The modern form is due to assocation with the river Lea. Early spellings include *Littegrave* 1224, *Littlegrave, Lichte*grave 1227, Lightgrave 14th century. The solution proposed in EPNS BDF survey, a personal name *Lihtla*, is probably more convincing than that in Ekwall 1960. This last is 'light-coloured grove', with variation due to *leoht* being associated with *lȳtel* 'little'. 'Light grove' would, however, fit well with the class of colour adjectives listed above.

Potsgrove BDF. ME *potte* 'deep hole' occurs in a number of names but only in the north country, so it is not likely in BDF. The late OE word *pot*, referring to pottery vessels, occurs in some south-country names, but the genitive of this word seems unsatisfactory with *grāf(a)*.

Seagrave LEI. Early spellings require *set-* or *sat-*. *(ge)set* 'animal fold' is possible, but seems otherwise only to be noted as a generic.

Slow Grove in Cookham BRK. The first element could be **slōh** 'mud' or *slāh* 'sloe'.

hangra 'sloping wood'. This derivative of the verb 'to hang' is well evidenced in the boundary surveys of charters but is not otherwise on record in OE. It is usually translated 'wood on a steep slope', which is the sense in which *hanger* is recorded in the 18th century, and this may suit most of the boundary marks, many of which are on appreciable slopes. In settlement-names, however, the reference is more frequently to very gentle slopes, and this may be an earlier usage. The word 'steep' should probably be omitted from etymologies of settlement-names containing *hangra*. There may be a special feature of the gentle slope, perhaps a slight concavity, which gave woods in this position a hanging appearance.

Examples were given in Gelling 1984 of many sites where the slopes referred to by *hangra* can only have been gentle. There are some high sites in the corpus: Woolhanger in Lynton DEV is at 1,000ft, Binegar SOM, northeast of Wells, is between the 700ft and 750ft contours, and Clayhanger DEV, east of Bampton, lies athwart a steepish slope between 500ft and 550ft. Sometimes a visit to a site reveals a slope not clearly apparent on maps: at Gravenhunger in Woore SHR, for instance, there is an appreciable bank to the west of the settlement. Sites like these are, however, the exceptions rather than the rule, and most of the settlements with *hangra* names are in areas with low relief. Two GLO instances by the Severn estuary, south of Sharpness, illustrate the inadvisability of automatically using the word 'steep' in this context. Sanigar and Oakhunger are farm-names, so the positions of the woods can be located with some precision. The farms are on slightly raised ground just below the 50ft contour. In the fields across the road from Oakhunger Fm, directly overlooking the estuary, there are a few ancient oak trees, presumably the remains of the wood.

hangra is only common in DEV and GLO, but WLT has five examples and BDF four. The BDF statistic is an exception to the generally south-western bias of the distribution. Occasional occurrences are widespread, and although no instances have been noted further north than CHE and YOE it seems probable that the word was current in most of England. Both NFK and SFK have one example in major names. In charter boundaries Kitson (forthcoming) notes that two-thirds of occurrences are in BRK and HMP.

The analyses in the reference section include minor names in counties for which there are detailed surveys. This enables it to be seen that a number of compound names recur. There are multiple instances of Oakhanger and three references to cats, and several other names occur twice. Much the most frequently recurring name is Clayhanger, and this perhaps reflects the geology most likely to give rise to land formations on which woods appeared to be 'hanging'. Recurrence of compound names is not unusual in itself, but it is noteworthy when it occurs to this extent in a relatively small corpus of material.

Eight of the places whose names are analysed in the reference section were DB manors, but none of them had a high tax assessment. A *hangra* may have been a highly visible feature and so a suitable marker with which to identify a settlement, but it is not likely that such woods had a great deal of economic value.

Reference section for **hangra**

As simplex name: Hang YON, Hanger, minor name, in BDF, GTL, SUR etc.

As first element: Hangerberry in West Dean GLO (**beorg**).

As generic with references to trees: Birchanger ESX, in Bratton WLT; Hazel Hanger in Brimpsfield GLO; Oakhanger HMP, in Welford BRK, Oakhunger in Hamfallow GLO (other instances occur in minor names and field-names); Solinger in Kimble BUC ('sallow willow').

With references to other vegetation: Bramingham (earlier *Bramblehanger*) in Luton BDF; Clattinger in Oaksey WLT ('burdock'); Rishangles SFK ('brushwood').

With references to wild creatures: Cathanger SOM, in Woodend NTP and in Chute WLT; Foxhanger's Fms in Rowde WLT; Hartanger KNT; Hinnegar in Didmarton GLO ('birds'); Pitshanger in Ealing GTL (*pyttel* 'hawk'); Saniger in Hamfallow GLO ('swans').

With reference to a domestic creature: Mullingar in Instow DEV ('mule').

THE LANDSCAPE OF PLACE-NAMES

With references to soil: Cheesehunger in Hinton GLO ('gravel'); Clamhunger in Mere CHE, recurring as a CHE field-name (*clām* 'mud'); Clayhanger CHE, DEV, STF, Clehonger HRE, Clinger in Buckland Newton DOR, Clingre GLO (all 'clay'). Clayhill Fm in Tottenham GTL is believed to be the *Clæighangran* 'clayey hanger' mentioned in the Anglo-Saxon Chronicle under the year 1016.

With other descriptive terms: Chalhanger in Buckland Filleigh DEV and Challenger in Membury DEV ('cold'); Goldhanger ESX; Henegar in Culmstock DEV ('high'); Smallhanger in Plympton St Mary DEV ('narrow', this also occurs as a field-name in Hurley BRK).

With references to topographical features: Cleaveanger in Nymet Rowland DEV (**clif**); Halshanger in Asburton DEV and Halsinger in Braunton DEV ('neck of land'); Polehanger in Meppershall BDF (**pōl**); Wishanger in Miserden GLO (**wisc**); Woolhanger in Lynton DEV (**well**).

With references to products: Shutlanger NTP ('shuttle'); Sparhanger in Lynton DEV ('spear'); Timberhanger in Bromsgrove WOR.

With reference to a structure: Barnacle WAR (*Bernhangre* DB, 'barn').

With a reference to position: Songar in Langley WAR ('south'). Anger's Fm in Alveston GLO was earlier *Middlehanger*.

With another woodland term: Gravenhunger SHR (**grāfa**).

With personal names: Chadlanger in Warminster WLT, Chaddlehanger in Lamerton DEV and Channy Grove in Hurley BRK (all *Ceadela*); Deanshanger NTP (*Dynne*); Shoppenhangers in Maidenhead BRK (*Sc(e)obba*); Tyttenhanger in Ridge HRT (*Tida*).

With obscure or ambiguous first elements:

Betteshanger KNT (*Betleshangre* 1176). This may have the same, unexplained, first element as Biddlesden BUC and Biddlestone NTB, discussed under **denu**.

Binegar SOM (*Begenhangra* 1065). Either fem. personal name *Bēage* or **begen* 'growing with berries'.

Moggerhanger BDF. *Mokerhanger* 1276 may be the most reliable form, but the first element is unexplained. It may be related to Modern English *muck*, with reference to soft ground.

Panshanger HRT. The early spellings vary between *Pales-*, *Penes-* and *Panes-*. *penn* 'enclosure' in its ESaxon form *pænn* is possible but far from certain.

Shuthanger in Halberton DEV. Other instances are found in charter boundaries and field-names. A wood in West Overton WLT was *scyt hangran* in 939, and in BRK there are ME field-names *Shitehangercroftes* in Shottesbrooke and *Shutehangre* in Kintbury. There is also a Shuthonger in Twyning GLO for which there are no early spellings. The material is just sufficient to suggest that OE *scythangra* was an appellative. The name is placed in this section because the precise significance of *scyte* is uncertain. It could indicate that the wood was on a steep slope or that it had a jutting-out shape. The latter is perhaps the more likely.

Tring HRT. This name is not certain to contain *hangra*, though association with that word is indicated by spellings like *Trehanger* which occur c.1200. Other early forms include *Trunga* 1086, *Trawinge* 1176, *Tresange* 1208. Commentators who take this to be a *hangra* name consider the first element to be *trēow* 'tree', which does not give a very meaningful compound.

holt 'single-species wood'. This specialised sense does not appear in literary sources, where 'small wood' is the meaning, but it predominates in place-names and charter boundaries. Kitson (forthcoming) notes that tree-species form the largest category of qualifiers for *holt* in charter boundaries, and it can be seen from the analysis in the reference section that this is so in place-names. In the GLO EPNS survey, Part 1, p. 145, Smith notes that Buckle Wood in Brimpsfield is described as 'a beech wood called *Bocholte*' in 1338. References to wild creatures constitute the only substantial category of qualifiers apart from tree-names. A considerable number of minor names has been included in the analysis since, as was the case for **bearu** and **grāf**, the bias of the material cannot be clearly demonstrated without them.

As regards distribution, *holt* is most frequent in CHE, ESX, GLO, SSX. As with **hangra** it has not been noted in counties north of CHE and YOW. See Fig. 43.

A variant *hylte* occurs in a few charter boundary-marks, and has been considered to occur in Emlett in Woolfardisworthy DEV (*Emhylte* 1249) and Navant in West Lavington SSX. Forms for the latter are far from conclusive, however. The derivation of Salkeld CMB from *hylte*, proposed in Ekwall 1960 and accepted in EPNS CMB survey and in Gelling 1984, seems doubtful in view of the failure of *holt* to appear further north than CHE. Early spellings for Salkeld have -*hild*, and a variant of **helde** is more likely than *hylte*.

Reference section for **holt**

As simplex name: Holt DEN, DOR, HMP, LEI, NFK, SOM, STF, WAR, WLT, WOR. There are many instances in minor names and field-names.

As first element: Haltham LIN (*hām*); Howtel NTB (**halh**).

As generic with references to tree-species: Acol KNT, Knockhall and Knockholt KNT, Oakhall in Grimley WOR and in Almondsbury and Hawkesbury GLO, Occold SFK, Ockold in Upton St Leonard GLO, Ockwells in Bray BRK (all 'oak'); Aisholt SOM, Ashold in Aston WAR, Esholt YOW (all 'ash'); Alderholt DOR; Apes Hall in Littleport CAM ('aspen'); Birchall in Hertingfordbury HRT and in Grimley WOR, Bircholt KNT; Buckhold in Bradfield BRK, Buckholt HMP, SSX and in Cranham, West Dean and Frocester GLO (all 'beech'); Hallshot in Cadbury DEV, Hazel Hall in Shere with Gomshall SUR, Hazeland in Bremhill

WLT, Hazelholt in Old Shoreham SSX, Hazelwood in Preston Bagot WAR and in Avening GLO (all 'hazel'); Lineholt in Ombersley WOR ('lime'); Wiggenholt SSX ('wych-elm').

With references to other vegetation: Ramsholt SFK (probably *hramsa* 'wild garlic'); Unhill Wd in Cholsey BRK (*hūne* 'hoarhound').

With references to wild creatures: Eversholt BDF ('boar'); Gledholt YOW(2) ('kite'); Ravens' Hall in Borough Green CAM; *Rawerholt* between Whittlesey Mere and King's Delph HNT (*hragra* 'heron'); Rockwell in Hambleden BUC, Ruckholt in Leyton ESX, Rucklers Green in King's Langley HRT (all 'rook'); Wormwood (earlier *Wormholt*) Scrubs GTL ('snakes'). There are several 'crow holts' in CHE.

With references to products: Sparcell's Fm in Lydiard Millicent WLT, Sparsholt BRK, HMP ('spear'); Tanholt in Eye NTP (*tænel* 'basket'); Throckenholt CAM ('beams'). Stockholt in Akeley BUC, with *stocc* 'stump', may belong here.

With references to ownership: Chittlehamholt DEV ('wood of the people of Chittlehampton'); Hainault Forest ESX ('monastic community'); Kensal Green GTL ('king').

With references to topography: Bergholt ESX, SFK (**beorg**); Linkenholt HMP (**hlinc**).

With a personal name: Poulshot WLT (*Paul*).

With a direction: Southolt SFK.

With obscure or ambiguous first elements:
Conholt in Chute WLT. The name was earlier *Covenholt*, one of several names with ME *Coven-*. This could be the ME reflex of OE *cofan*, oblique case of *cofa*, recorded in the sense 'inner chamber' but used in place-names in senses which have not been satisfactorily explained. Some of these names (e.g. Coventry) probably contain a personal name *Cofa*, but personal names are so rare with *holt* that it seems safer to consider that in Conholt we have *cofa* in one of its topographical uses. Conholt House is in hilly country.

Singlesole in Eye NTP. Spellings of 12th- and 13th-century dates have *Sengles-*, which appears to be unique to this name.

hyrst 'wooded hill', cognate with Welsh *prys* 'brushwood'. Occurrences in OE texts are ambiguous as between 'hillock' and 'wood'. Both senses are evidenced in Middle and Modern English, sometimes combined, so 'wooded hill' seems the best rendering.

hyrst occurs as a simplex name or generic in about 65 major settlement-names in England. The major names stand out from a much greater number of minor names and field-names, especially in KNT, SSX and SUR, where *hyrst* names are concentrated in The Weald. In SSX there are 20 major names out of

WOODS AND CLEARINGS

a total of about 125, and in SUR three out of 72 (excluding names not recorded before A.D. 1500 and field-names). A similar imbalance is found in some counties where *hyrst* is less common. GLO has three major and three minor names, but in WAR the figures are one and 11, and in YOW two and 18. It seems likely that settlements with names in *hyrst* are of relatively late origin and may be in areas not recognised by the earliest Anglo-Saxons as appropriate to arable farming. Hurstpierpoint SSX, atypically, was a large estate in 1086, but this may have become the administrative centre for a number of settlements which are not named in Domesday Book.

In HNT there was probably a district called *Hyrst*. Old Hurst and Woodhurst, with **wald** and **wudu** prefixed, are adjacent parishes, and on the boundary between them there is a rough stone seat, now the Abbot's Chair, earlier the *Hurstingeston*, 'stone of the *hyrst* people'. This was the meeting-place of Hurstingstone Hundred.

No attempt has been made to categorise the qualifiers of minor names. Analysis of those names which are treated in the reference section shows an unusually even spread among the categories of personal names, vegetation, descriptive terms, and wild and domestic creatures. In the last category most of the references are to goats. Where soil is noted it is gravelly, sandy or stony, which suggests that the wood was not likely to be dense. The four -*inga*- compounds may be due to *hyrst* settlements having a relatively late origin on the borders of tribal areas.

Old Hurst and Woodhurst HNT stand on a broad, low ridge made by the 100ft contour. Deerhurst and Sandhurst GLO also give a good idea of the sort of low hill to which the term *hyrst* was felt appropriate.

In ME *hurst* developed a meaning 'sandbank in the sea or a river', and this is the sense in Hurst Castle in Milford-on-Sea HMP.

Reference section for **hyrst**

As simplex name: Herstmonceux SSX, Hirst NTB, YOW, Hurst BRK, KNT, SOM, WAR, Hurstpierpoint SSX. Old Hurst and Woodhurst HNT, with **wald** and **wudu** prefixed, were probably originally simplex.

As generic with personal names: Bathurst SSX (*Bada*); Baughurst HMP (**Beagga*); Billingshurst SSX (*Billing*); Chippinghurst OXF (**Cibba*); Goudhurst KNT (*Gūtha*); Wadhurst SSX (*Wada*).

With tree-species: Ashurst KNT, SSX; Buckhurst ESX ('beech'); Ewhurst HMP, SSX, SUR ('yew'); Holdenhurst HMP ('holly'); Lindhurst NTT and Lyndhurst HMP ('lime'); Salehurst SSX ('sallow willow').

With reference to other vegetation: Fernhurst SSX.

With references to wild creatures: Brockhurst SHR (in Wem), SSX, WAR ('badger'); Crowhurst SSX, SUR; Deerhurst GLO; Hartest SFK; Hartshurst (in Wotton) SUR; Hawkhurst KNT and Haycrust SHR (both 'hawk'); Henhurst KNT ('birds').

THE LANDSCAPE OF PLACE-NAMES

With references to domestic creatures: Bolnhurst BDF ('bullocks'); Gathurst LNC, Gayhurst BUC, Goathurst SOM (all 'goat'); Lamberhurst KNT ('lambs'); Rossett YOW ('horse'); Ticehurst SSX and Tickenhurst KNT ('kids').

With descriptive terms: Brockenhurst HMP ('broken'); Chislehurst GTL ('gravel'); Longhirst NTB; Sandhurst BRK, GLO, KNT; Stonyhurst LNC.

With references to products: Bredhurst KNT ('boards'); Nuthurst SSX, WAR; Staplehurst KNT ('posts').

With references to industrial processes: Kilnhurst YOW; Tilehurst BRK.

With earlier place-names: Chithurst SSX (probably cēd); Limehurst LNC (Lyme Forest).

With a reference to topography: Madehurst SSX, Medhurst in Edenbridge KNT (**mǣd**).

With another woodland term: Gravenhurst BDF (**grāfa**).

With a reference to a meeting-place: Fingest BUC (*thing*).

With -inga- compounds: Bringhurst LEI ('Bryni's people'); Doddinghurst ESX ('Dudda's people'); Sissinghurst KNT ('Seaxa's people'); Warminghurst SSX ('*Wyrm's people').

Midhurst SSX is 'place in the middle of a *hyrst*' or 'place among *hyrsts*'.

With obscure or ambiguous qualifiers:

Bellhurst SSX. This name occurs three times in SSX, in Etchingham, Wartling and Beckley, the last example being a DB manor. The triple occurrence suggests a significant term, rather than a personal name, as qualifier. The early spellings for all three suggest *-l-* rather than *-ll-*. See the discussion of Baycliffe, under **clif**. Bellhurst could contain *bēl* 'fire' or Ekwall's **bel* 'glade'.

Penhurst SSX. Early forms have *Pene-*, as do those for Pembridge (listed under **brycg**). Derivation from the gen. pl. of *penn* 'enclosure' is rendered uncertain by the absence of *-nn-* spellings.

Penshurst KNT. There is a good deal of variety in the spellings, but *Pens-* and *Penes-* are the commonest. This looks like the gen.sg. of the unidentified word found in Penhurst.

Shadoxhurst KNT. The early spellings require an OE **sceattoc* which is unexplained and appears to be unique to this name.

Speldhurst KNT. The qualifier is *speld* 'splinter', but the precise meaning of the compound is uncertain.

Wheatenhurst GLO. The first element could be 'white' or *Hwīta*, personal name.

WOODS AND CLEARINGS

lēah 'forest, wood, glade, clearing', later 'pasture, meadow'. This word has a frequency among the topographical terms used in place-name formation which is comparable to that of *tūn* among habitative words. The order of frequency with which these words occur, which sets them apart from other place-name elements and renders comprehensive analysis impracticable, is, however, in marked contrast to the poor showing which both of them make in the corpus of earliest-recorded names. The analysis in Cox 1976 of the 224 names recorded between A.D. 672 and 731 yielded only six examples of *tūn* and seven of *lēah*. This indicates that both words came into common use in the mid-8th century. They may have been less frequently employed after about the mid-10th century, when both had developed meanings ('pasture, meadow' in the case of *lēah* and 'estate' in the case of *tūn*) which made them less useful in run-of-the-mill place-naming. Most names containing *tūn* and *lēah* were probably coined between c.750 and c.950.

In areas where *tūn* and *lēah* predominate over all other place-name elements it is instructive to map them together. They are mutually exclusive to a remarkable extent, and it can be demonstrated that *lēah* was the usual term for settlements in heavily wooded country and *tūn* for those in land from which most trees had long since been cleared. In Gelling 1974 this was shown to be the case in a large area with Birmingham as its central point. The map which accompanied this article, which achieved a modest fame among students of landscape history, is here reproduced as Fig. 44. A similar mapping exercise for DRH was subsequently carried out by Victor Watts (1976). Such maps repay close scrutiny: attention should be paid not only to the distribution of the symbols but also to the nature of the place-names in well-settled areas where neither *tūn* nor *lēah* names occur.

It may be regarded as established that *lēah* is an indicator of woodland which was in existence and regarded as ancient when English speakers arrived in any region. Rackham 1976 (p. 56) says that 'names, especially village names, ending in -ley or -leigh imply that some of the Wildwood remained at the time they were formed'. In Gelling 1974 it was suggested that isolated names containing *lēah* are likely to refer to woods in generally open country, where they would be jealously preserved. A clear example is Elmley Castle WOR, which appears on Fig. 44 as a single item interrupting the ring of *tūn* names which surrounds Bredon Hill. Where there are clusters of *lēah* names, such as that in north WAR, the meaning 'clearing' is the relevant one, and here the word has a quasi-habitative use denoting settlements in a wooded environment, the economy of which would, of course, be very different from that of settlements in south WAR. Many of the *lēah* names in the Arden region probably denote settlements which were established before the Anglo-Saxons arrived, but in other areas they may represent the breaking-in of new arable on the edges of ancient forest. The cluster of *lēah* names shown on Fig. 44

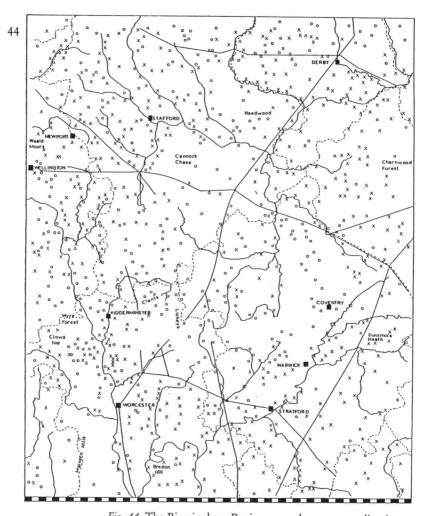

Fig. 44: The Birmingham Region; x = place-name ending in *-tūn*;
o = place-name ending in *-lēah*, or simplex name from *lēah*.

north of Wyre Forest in WOR, for instance, may indicate that the forest was much larger before an expansion of settlement in the Middle Saxon period.

Students of settlement history in areas where there are *lēah* names should consider the possible meanings in the light of all that is known about the early history of the region. Names which contain *lēah* in its late sense of 'pasture, meadow' can often be identified from the topography, or from the nature of the first element – for instance, a word such as **wisc**, the name of a domestic animal, or a reference to good pasture, as in Butterley. The sense-development was to some extent a natural one from 'clearing', but NED points out (under *lea*) that 'the sense has been influenced by confusion with LEAS sb.¹ (OE *lǣs*) which seems often to have been mistaken for a plural, and also with LEA sb.².' (NED's LEA sb.² is the elliptical use of OE *lǣge* adj. 'fallow, untilled').

238

The 'pasture' sense is a late development. The other main senses 'wood' and 'woodland clearing' probably overlapped. An Old English speaker would find no awkwardness in using the word to mean 'wood' in Elmley WOR and 'clearing' in Stoneleigh WAR, since the geographical contexts were so clearly distinct.

In Gelling 1984 there is a regional survey of the country showing in which counties *lēah* is more or less frequent. The results of this may be summarised here. The area in which *lēah* is a predominant place-name element extends in a broad band from NTB and DRH, in the north east, to DEV, in the south west. In the south midlands the band broadens so that it extends from the Welsh border to the ESX coast. The south-western sweep leaves YOW, LIN, NFK AND SFK outside the area where *lēah* is very frequent. In far north-western and south-eastern counties *lēah* is well-evidenced but does not predominate. In the north west it begins to be very common in LNC south of the river Ribble, and in the south east its frequency declines from SUR to KNT. In the counties where *lēah* is very common occurrences are not evenly distributed, but are clustered in a manner which defines the likely extent of woodland in the Middle Saxon period. It is interesting to note that some areas which had the legal status of 'forest' under Norman law are shown by place-name evidence not to have been wooded: Wirral in CHE is an example of this. The distribution of *lēah* names in CHE, STF, SHR, WAR and HRE is studied in detail in Gelling 1992.

Comprehensive analysis of compound *lēah* names is not practicable in the present study, but a large sample of compounds which make multiple appearances is presented in the reference section. Some of these names are outstandingly frequent, the compound with *hēah* 'high' (dat. *hēan*) being possibly the commonest of all if the tree-name formations are excluded. Some of the settlements with 'high wood' names are not themselves on high ground: Henley OXF and WAR and Hanley (Castle and Swan) WOR may be instanced. It is possible that such places were named from woods on higher ground overlooking the settlement, which would be a striking visual marker for travellers. Bradley and Langley are also very frequent.

The range of qualifiers is much what might be expected for a word predominantly associated with woodland, but allowance must be made for the late sense 'pasture, meadow'.

The comparative rarity of *-inga-* compounds is a noteworthy feature. Those noted in the reference section are probably not the whole corpus, but it seems clear that *-ingalēah* occurs very rarely in the context of the whole massive bulk of *lēah* names.

Reference section for **lēah**
This list contains a selection, based mainly on recurrent compounds in Ekwall 1960.

THE LANDSCAPE OF PLACE-NAMES

As simplex name: Lea, Lee, Leigh, Lye *passim*; Leece LNC, Leese CHE, Lees LNC (plural); Leam DRH, NTB (dat. plural).

As first element: there is no certain example.

As generic with personal names: this enormous category has not been quantified or analysed. Examples among places which have become towns include the following: Barnsley YOW (*Beorn*, several other Barnsleys contain different personal names); Bletchley BUC (*Blecca*); Dudley WOR (*Dudda*); Hinckley LEI (*Hȳnca*); Keighley YOW (**Cyhha*); Otley YOW (**Otta*).

As generic with tree-names:

'apple': Apley IOW, LIN, SHR, SOM, Apperley GLO, NTB, YOW, Appley LNC:

'ash': Ashley *passim*:

'aspen': Aspley BDF, STF, WAR, Espley NTB:

'birch': Bartley HMP, WOR, Berkeley GLO, Berkley SOM:

'box': Bexley GTL, Boxley KNT:

'elm': Almeley HRE, Elmley KNT, WOR(2):

'hawthorn': Hatherleigh DEV, Hatherley GLO, Thorley HRT, IOW, Thornley DRH, LNC, and indirect references in Hagley SHR, SOM, STF, WOR:

'hazel': Haseley IOW, OXF, WAR, and indirect references in Notley BUC, ESX:

'oak': Acklam (dat.pl.) YOE, YON, Akeley BUC, Oakle GLO, Oakleigh KNT, Oakley *passim*, Ocle HRE:

'willow': Sall NFK, Saul GLO, Sawley YOW(2) (*salh*); Weeley ESX, Willey CHE, HRE, SHR, WAR (*welig*); Widley HMP, Willey DEV, Withiel SOM (*wīthig*):

With references to other vegetation: Bentley *passim* ('bent grass'); Bramley and Bromley *passim* ('broom'); Fairlight SSX(2), Farleigh and Farley *passim*, Farnley YOW(3) ('fern'); Hadleigh ESX, SFK, Hadley HRT, SHR, Headlam DRH (dat.pl.), Headley HMP, SUR, WOR, YOW, Heatley CHE, Hedley DRH(2), NTB, Hoathly SSX(2) (*hæth, hāth*); Riseley BDF, BRK, Risley DRB, LNC ('brushwood'). Bitterley SHR, Butterleigh DEV and Butterley CHE, DRB, HRE are thought to refer obliquely to high-quality pasture.

With descriptive terms: Bradle DOR, Bradley *passim* ('broad'); Hardley HMP, IOW; Langley *passim* ('long'); Radley BRK, Raleigh DEV ('red'); Ripley DRB, HMP, SUR ('strip-shaped'); Rowley DEV, DRH, STF, YOE, YOW ('rough'); Softley DRH, NTB; Stanley *passim*, Stoneleigh WAR, Stonely HNT ('stone'); Whitleigh DEV, Whitley *passim* ('white'). Shirley DRB, HMP, SUR, WAR probably means 'bright clearing', but 'wood belonging to the shire' is a possible alternative.

With references to wild creatures: Areley WOR, Arley CHE, LNC(2), WAR, WOR, Earley BRK, Earnley SSX (*earn* 'eagle', in Gelling 1987 it is argued that these are fish eagles); Bagley BRK, SHR, SOM, YOW, Baguley CHE ('badgers'); Beeleigh ESX,

WOODS AND CLEARINGS

Beoley WOR ('bees'); Catley HRE, LIN; Crawley BUC, ESX, HMP, OXF, SSX ('crows'); Darley DRB(2), Durleigh SOM, Durley HMP ('deer'); Everley WLT, YON, Eversley HMP, Yearsley YON (*eofor* 'boar'); Finchley GTL, Finkley HMP ('finches'); Foxley NFK, NTP, WLT; Hartley BRK, DOR, HMP(3), KNT(2); Hiendley YOW, Hindley LNC; Martley SFK, WOR ('martens'); Midgley YOW(2), Migley DRH ('midges'); Moseley WOR, Mowsley LEI ('mice'); Purleigh ESX, Purley BRK (*pūr*, a bird-name); Wigley DRB, HMP ('beetles'); Woolley BRK(2), HNT, YOW ('wolves'); Wormley HRT, Wormsley HRE ('snakes').

With references to domestic creatures: Bulley GLO; Callaly NTB, Calveley CHE, Calverley YOW, Chawleigh DEV ('calves'); Cowley GLO; Gateley NFK ('goats'); Horsley DRB, GLO, NTB, STF, SUR, Hursley HMP; Lambley NTB, NTT, Lamesley DRH; Oxley STF; Shipley DRB, DRH, NTB, SHR, SSX, YOW ('sheep'); Stirchley SHR ('stirk'); Stoodleigh DEV, Studley OXF, WAR, WLT, YOW ('stud').

With references to topographical features: Bearley WAR, Birley HRE, Burghley NTP, Burley *passim* (*burh* 'fort', frequently referring to prehistoric fortifications); Dickley ESX, Ditchley OXF ('ditch'); Hurley BRK, WAR (*hyrne* 'corner'); Moreleigh DEV, Morley DRB, DRH, NFK, YOW (**mōr**); Selly WOR, Shelley ESX, SFK, YOW (**scelf**); Streatlam DRH (dat.pl.), Streatley BDF, BRK, Streetley ESX, Streetly CAM, WAR, Strelley NTT (**strǣt**).

With references to direction: Astley LNC, SHR(2), WAR, WOR ('east'); Norley CHE, Nordley SHR(2) ('north'); Soudley and Sudeley GLO ('south'); Westley CAM, SFK.

With other references to position: Handley CHE, DOR, DRB(2), NTP, Hanley STF, WOR(2), Healaugh YON, YOW, Healey LNC(2), NTB(3), YON, YOW, Heeley YOW, Henley OXF, SFK, SOM(2), SUR, WAR, Highleigh (in Sidlesham) SSX (all 'high'); Marley DEV, DRH, Mearley LNC ('boundary').

With references to classes of people: Broseley SHR and Burwardsley CHE (possibly 'fort-keeper'); Childerley CAM ('young noblemen'); Chorley CHE(2), LNC, STF ('peasant farmers'); Kingsley CHE, HMP, STF; Knightley STF ('retainers').

With references to woodland products: Bordesley WAR, WOR, Bordley YOW ('board'); Cowley DRB(2) ('charcoal'); Pilley HMP, YOW ('stakes'); Stapeley CHE, Stapely HMP ('posts'); Staveley DRB, LNC, WML, YOW ('staves'); Yardley ESX, NTP(2), WOR, Yarley SOM ('rods').

With references to crops: Barley LNC, YOW, Barlow DRB ('barley'); Flaxley GLO, YOW and Lindley LEI, YOW(2), Linley SHR(2) ('flax'); Wheatley ESX, LNC(2), NTT, OXF, YOW(2).

With -inga- compounds: Anningsley (in Chertsey) SUR ('Anna's people'); Billingley YOW ('*Billa's people'); Bingley YOW (probably 'Bynna's people'); Blechingley SUR ('Blæcca's people'); Cottingley YOW (*Cotta's people'); Fillongley WAR ('*Fygla's people'); Finningley NTT ('fen dwellers'); Gringley on the Hill NTT (first element uncertain); Headingley YOW ('Headda's people');

THE LANDSCAPE OF PLACE-NAMES

Hellingly SSX (possibly 'dwellers in a tongue of land'); Knottingley YOW (*Cnotta's people'); Madingley CAM (*Māda's people'); Oddingley WOR ('Odda's people'); Steppingley BDF ('Stæpa's people'); Washingley HNT (probably '*Wassa's people').

lundr ON 'wood'. Smith 1956 stresses the Scandinavian use for a sacred grove, but although a 12th-century English writer apparently shows awareness of this in rendering the second element of Plumbland CMB as *nemus paci donatum*, there is little reason to suspect anything more than a topographical meaning for the term in place-names in England.

The examples listed in the reference section include some well-recorded minor names. The element is best represented in Yorkshire, WML, NTT, NTP and DRB.

A few instances, such as Loatland Wood NTP and Lund Forest YON, are names of woods, not settlements, and two – Aveland LIN and Framland LEI – are wapentake meeting-places, but the corpus consists mainly of settlement-names. Sixteen of these settlements had some degree of administrative status, being DB manors and/or townships or parishes. This suggests that a wood called *lundr* was likely to be of economic value. A detailed study would probably show that many of the settlements are in areas where woodland was scarce at the time of the Viking wars. Timberland LIN, for instance, overlooks the fens by the river Witham in an area where there are few other woodland names, and the wood here was probably an important resource. Timberland was a considerable estate in 1086. Some of the parishes have shapes which suggest late formation: the parish of Lound NTT, e.g., has a boundary with Sutton which suggests the division of an open field area, and the parish of Lund YOE is a narrow strip. Some of the places with names in *lundr* probably represent infill settlement in the late 9th/early 10th centuries.

The high proportion of simplex names is consistent with the hypothesis that *lundr* refers to woods of some economic significance, as is the use of the word to qualify settlement-terms in Londonthorpe LIN and Lumby YOW.

The EPNS NTT survey states (p. 287) that *lund* was a common noun in that county until the late 12th century, so some of the names which are not recorded in DB may have been coined after the Norman Conquest, and by English speakers. In Birklands NTT, *Birkelund* (1250) was substituted for, or perhaps coexisted with, the synonymous English *Birchwude* (1188). There is, however, a preponderance of ON qualifiers in *lundr* compounds, so many of them probably arose in a Scandinavian-speech context. Eight of the 10 personal name compounds are wholly Scandinavian.

Except in simplex names, *lundr* is normally represented by -land in modern forms.

WOODS AND CLEARINGS

Reference section for ON **lundr**

As simplex name: Lound LIN, NTT(2), SFK, Lund LNC, YOE(2), YOW, Lunt LNC. Heath DRB was formerley *Lunt*. Craiselound and Eastlound LIN are *Lund et alter Lund* in DB.

As first element: Londonthorpe LIN (gen. *lundar*), Lumby YOW.

As generic with personal names: Autherlands in Ilton YOW (*Aldwulf*); Aveland LIN (ON *Afi*); Boyland NFK (*Boia*); Cowsland in South Leverton NTT (ON *Kollr*); Framland Hundred LEI (ON *Fræna*); Osland in Perlethorpe cum Budby NTT (ON *Authr*); Snelland LIN (ON *Sniallr*); Sutherland in Cropton YON (ON *Sútari*); Swanland YOE (ON *Sveinn*); Toseland HNT (ON *Tóli*).

With tree-names: Birkland Barrow in Bolton-le-Sands LNC and Birklands NTT (ON *birki* 'birch'); Eckland in Braybrooke NTP (ON *eik* 'oak'); Hasland DRB ('hazel'); Plumbland CMB ('plum'). Lund Wood in Wharram le Street YOE is *Thornlund* 1154 and later.

With references to wild creatures: Owlands in Marrick YON (ON *ulfr* 'wolf'); Rockland NFK(2) and Ruckland LIN (OE *hrōc* or ON *hrókr* 'rook').

With references to structures: Holland in Ripton HNT (OE *haga* or ON *hagi* 'enclosure'); Kirkland LNC (ON *kirkju* 'church').

With references to topography: Loatland Wood in Harrington NTP (ON *laut* 'hollow'); Morland WML (**mōr**).

With references to clearing: Stockland in Ebberston YON (ON *stokkr* or OE *stocc* 'tree stump'); Swithland LEI (ON *svitha* 'land cleared by burning').

With references to a product: Timberland LIN (OE *timber* or ODan *timbær*).

With an earlier place-name: Upsland YON (*Upsale* DB, *Upselund* 1280, ON *Upsalir* 'high dwelling').

With ambiguous qualifiers: Natland WML (various etymologies have been suggested: Fellows-Jensen 1985, pp. 149–50, favours ON **nata* 'nettle'); Rowland DRB (either ON *rá* 'roe-deer' or ON *rá* 'boundary'); Shirland DRB (*scīr* 'bright' or *scīr* 'shire').

rodu 'clearing'. This is the probable OE form of the word discussed in Smith 1956 under the head-forms **rod*, **rodu*. The asterisk is unnecessary as the word is well evidenced in charter boundaries.

rodu is predominantly a minor-name and field-name element. It is only common in LNC and YOW, but there are widespread occasional appearances elsewhere. In LNC and YOW it survived as dialect *royd*, and many of the occurrences in these counties will be post-Norman-Conquest. A study of Almondbury YOW (Redmonds 1983) says that there were approximately 40 *royd* names in the township, and that most of them probably date from 'between 1150 and 1350, a period when the population reached a high point and great stretches of the so-called waste were converted to arable and pasture'. There are, however, a significant number of settlement-names containing *rodu* in counties as far apart as SOM, HRE, NTP, CHE and NTB, so its pre-Conquest use was not limited to minor names or to the north of England.

As a simplex name for ancient settlements *rodu* has become Roade NTP, Rodd HRE, Rode, now North Rode and Odd Rode CHE and Road SOM. Roddam NTB is from the dative plural. The CHE, NTP and SOM places are in DB, the first two as small estates, but Road SOM with a tax assessment of 10 hides. Compound settlement-names in *-rodu* include Blackrod, Heyrod, Huntroyde and Ormerod LNC and Mytholmroyd YOW. In such names as these *rodu* presumably referred to a large clearing, suitable for a settlement-site. In charter boundaries, however, where there may be a later, developed, meaning, *rodu* is consistently used for linear clearings, and Kitson (forthcoming) considers that in this sense it is the etymon of modern *road*.

Analysis of compounds is impracticable on account of most of the material being minor names and field-names. In the few instances listed above the qualifiers are 'black', 'high', 'hunter', the ON personal name *Ormr*, and OE *(ge)mȳthum* 'river-junctions'.

*roth 'clearing'. This word has cognates in related Germanic languages but is not recorded in Old English. Roe Green and Wood in Sandon HRT appear as *Rothe* in a charter granting land to St Paul's minster, but as the charter is a forgery this does not constitute pre-Conquest documentation. The word is, however, well evidenced in place-names. It is the probable first element of Rothley LEI, NTB and Rothwell LIN, NTP, YOW. Rothend in Ashdon ESX was originally a simplex *roth name, like Roe Green HRT and some other minor names in that county.

There are several other 'clearing' words with initial *r-* which appear mainly in minor names and field-names, but are very occasionally used for settlement-sites. Here may be instanced:

*ryding, mistakenly indexed as *hryding* in Vols I–XVI of the EPNS series. *ryding is commoner in the north of England than in the south. In minor names and field-names it gives forms such as Riding, Ridding, Reading, Redding. It is very rarely used in ancient settlement-names, Armetridding ('hermit clearing') in Leyland LNC being possibly the only example.

*ryden and its Kentish form ***reden**. This is particularly common in minor names and field-names in ESX and SUR.

*rȳd, *rīed, *rēod, *ryde, *rede. These are the dialect forms given in Smith 1956 for an element which occurs, mostly in field-names, in south-eastern counties. It is probably the generic in Brandred and Coldred KNT ('burnt' and 'charcoal clearing').

Gillian Fellows-Jensen (1972, p. 102) suggests that Routh YOE derives from an Old Danish *rūth, or *rutha, 'clearing'.

sceaga 'small wood'. The word is related to ON **skógr**.

Literary references and occurrences in charter boundaries imply that a *sceaga* was a wood of limited extent, and the use of corresponding words in other Germanic languages suggests that it was likely to be projecting. In modern English, from the 16th to the 19th centuries, *shaw* is recorded in the sense 'strip of wood or underwood forming the border of a field'. The sense 'strip of wood' was probably established by the 9th century. It suits some charter-boundary instances, particularly one of 849 in the bounds of Alvechurch WOR – *in hæthleage sceagan thær he thynnest is* – and one of 1043 in the bounds of Sevington WLT – *betweox morsceagan*. The latter phrase, 'between the marsh-shaws', gives the earliest form for Moorshall WLT. Its occurrence in a number of ancient settlement-names suggests, however, that there was an earlier use of *sceaga* for more substantial woods.

Occasional names containing *sceaga* are sufficiently widespread to show that the term was generally available in the place-name-forming vocabulary; but the overall distribution is strikingly uneven. It is only common in a group of contiguous counties comprising CHE, DRB, LNC, WML and YOW, and detailed plotting in these counties would probably show that there are smaller areas in which the word is used freely in all categories of place-names. In DRB there is a marked concentration in High Peak Hundred, and the word is well represented in the adjacent part of YOW. A. H. Smith's study of the distribution in YOW (EPNS survey Part 7, p. 281) led to the conclusion that 'it seems very often to be peripheral to the main woodland areas'. The names included in the reference section are plotted on Fig. 45.

sceaga is much more common in minor than in major names, and a selection of minor names is included in the reference section in order to give a good impression of types of qualifiers. Among the descriptive terms, 'broad' is noteworthy, as suggesting that most shaws were narrow. Birds and small mammals figure prominently, and a number of shaws were characterised by a particular tree-species. Many of the names listed are likely to be of late OE or ME origin. The rare personal names are predominantly dithematic.

Shaw developed to *Shay* in some names in LNC and YOW, and there is an outlier in north SHR, where an area called The Shays lies between Loppington and Nonely. A paper by Dr Mary Higham in *Nomina* 12 (1989) reported an investigation of these names, and presented evidence that the areas so called were conspicuously unwooded, being in fact tracts of agriculturally low-grade land, some in upland areas, some on valley floors, which were often used for communal grazing. Dr Higham felt that the 'small wood' etymology was so inappropriate as to suggest that *shay* was a different word from *shaw*, but an appendix to the paper by M. Gelling showed that the evidence for interchange between the two forms points clearly to an identical origin. Discussion of the phonological development *shaw > shay* in this appendix was corrected, in some respects, by Victor Watts in *Nomina* 13, but that is less important for the

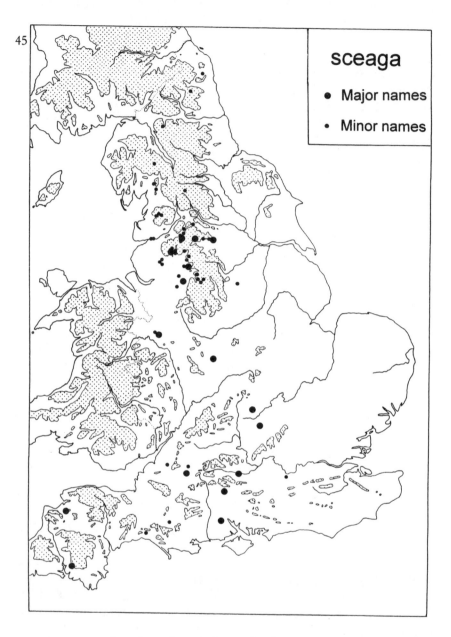

present purpose than the enigmatical use of a derivative of *sceaga* for areas of the type described by Dr Higham. M. Gelling's suggestion was that in areas where *shay* was a variant of *shaw*, *shay* was preferred for naming places where there had been no woodland for a long time, in order to differentiate these from places where *shaw* was still appropriate in its woodland sense. The occurrence of a term denoting woodland in an area long treeless may be compared with the use of **wald**.

WOODS AND CLEARINGS

Reference section for **sceaga**

As simplex name: Shaugh DEV, Shaw BRK, LNC, WLT, also several well recorded minor names in DRB and WLT. Shave Cross DOR is *La Schaghe* 1325.

As first element: Shawbury SHR (*burh* 'manor house'); Shawdon NTB (**denu**).

As generic with descriptive terms: Blackshaw YOW and Blashaw in Penwortham LNC ('black'); Bradshaw DRB, LNC, YOW ('broad', there are other instances in minor names); Fulshaw CHE, LNC ('foul'); Lanshaw in Shap WML ('long'); Lightshaw in Winwick LNC; Openshaw near Manchester LNC; Sansaw SHR ('sand'); Strumpshaw NFK ('stumps'); Wheatshaw in Bolton-le-Moors LNC ('wet').

With references to wild creatures: Bagshaw DRB ('badger'); Catshaw LNC; Cranshaw in Prescot and Cronkshaw in Whalley LNC ('crane'); Crawshaw LNC ('crow', also several minor names in YOW); Dunkenshaw near Lancaster and Dunnockshaw in Padiham LNC ('dunnock'); Earnshaw in Bradfield YOW ('eagle'); Evershaw in Biddlesden BUC ('boar'); Icornshaw in Lupton WML, Ickornshaw YOW (ON *ikorni* 'squirrel'); Marshaw near Lancaster LNC ('marten'); Ottershaw SUR.

With reference to a domestic creature: Hogshaw BUC.

With references to trees: Appleshaw HMP; Aspenshaw in New Mills DRB; Birkenshaw YOW, Birtenshaw in Bolton-le-Moors LNC and Brisco near Carlisle CMB ('birch'); Hathershaw in Oldham LNC ('hazel'); Oakenshaw YOW; Ollerenshaw in Chapel en le Frith DRB ('alder'); Wythenshawe CHE ('willow'); Nutshaw in Penwortham and Prickshaw in Whitworth LNC refer by implication to hazels and thorn trees.

With personal names: Audenshaw LNC (*Aldwine*); Barnshaw CHE (*Beornwulf*); Ellishaw NTB (**Illa*); Occleshaw in Wigan LNC (*Ācwulf*); Oughtershaw YOW (*Uhtrēd*); Renishaw in Eckington DRB (ME *Reynold*).

With references to topography: Moorshall in Kington St Michael WLT (**mōr**); Walshaw in Whalley LNC (**wælle**); Wishaw WAR (*wiht* 'bend', here probably 'curved hollow').

With river-names: Brunshaw in Whalley LNC (river Brun); Frenchay (earlier *Fromeshawe*) GLO (river Frome).

With references to products: Bickershaw LNC ('bee-keeper'); Huntshaw DEV (*Hunishauwe* 1238, 'honey').

With ambiguous qualifiers: Bramshaw HMP (early spellings vary, some suggesting 'bramble', others suggesting 'broom'); Buckshaw in Holwell DOR (*Buggechage* 1194, the first element could be the fem. personal name *Bucge* or ME *bugge* 'hobgoblin').

THE LANDSCAPE OF PLACE-NAMES

skógr ON 'wood', cognate with **sceaga**. This element is best represented in minor names in CMB, LNC, WML and YOW. Outside these counties it is rare in all categories of names.

Very few places with names in -*skógr* have a special administrative status. Aiskew YON and Thurnscoe YOW are DB manors and have respectively township and parish status, DB mentions a number of holdings in Haddiscoe NFK, Middlesceugh CMB is a parish with Braithwaite; the majority, however, are minor names, though frequently well recorded in ME sources. *skógr* was adopted into ME, and although NED, *s.v. scogh*, describes it as rare in literary sources, place-name evidence suggests that the ME word was widely familiar in northern counties. Interchange with OE **sceaga** is evidenced in the spellings for Barnskew WML and Roscoe LNC. It is probable that some of the names in the reference section are of post-Conquest origin, though those with unequivocal ON qualifiers will have been coined by Norse speakers.

Analysis of qualifiers suggests that *skógr* had a fairly general meaning. Tree species are well evidenced but do not predominate as they do with **holt**.

Reference section for ON **skógr**

As simplex name: minor names Sceugh, Scow, Skeaf, Skewfe, Skews, Skuff in CMB, WML, YOW are more likely to be from ME *scogh*, a reflex of *skógr*, than to be of true ON origin.

As first element: Skewkirk YOW, Skewsby YON. Scorborough YOE is *Scogerbud* DB, which suggests a compound of *skógr* and *búth* 'booth': the replacement of the generic by OE *burh* 'fort' has not been satisfactorily explained.

As generic with tree-names: Aiskew YON (ON *eik* 'oak'); Aldersceugh near Blennerhasset CMB; Brisco(2) and Busco YON (ON *birki* 'birch'); Ellershaw in Boltons CMB (ON *elri* 'alder'); Hessleskew YOE (ON *hesli* 'hazel'); Thurnscoe YOW (ON *thyrnir* 'thorn').

With references to structures: Burscough LNC (OE *burh*); Gascow in Ulverston LNC (ON *garthr* 'enclosure'); Scalescough near Carlisle CMB (ON *skáli* 'shieling-hut'). Loscoe DRB(2), YOW(2) is a recurring compound of ON *lopt* 'upper chamber' and *skógr*. Another instance occurs as a field-name in YOW, *Loftschoghe* 13th century. Loskay in Kirkby Moorside YON has a spelling, *Loftischo* 1282, which indicates derivation from a phrase *lopt i skógr* 'loft-house in a wood'. These names may, as suggested for Woodhouse and Woodhall, refer to buildings with some special function concerned with woodland management.

With references to position: Middlesceugh CMB; Northsceugh in Cumwhitton CMB and Noska in Skipton YOW; Susscarrs in Wheldrake YOE (*Southscogh* 1235); Wescoehill in Harewood YOW.

With personal names: Barnskew in Crosby Ravensworth WML (ON *Bjarni*); Haddiscoe NFK (ON *Haddr*); Huddlesceugh in Kirkoswald CMB (ME *Hudde*);

WOODS AND CLEARINGS

Sarscow in Eccleston LNC (ON *Sæfari*); Tarlscough in Ormskirk LNC (ON *Tharaldr*).

With descriptive terms: Flaska in Matterdale and Fluskew in Dacre CMB (also a lost *Flaskew* in Dufton WML, probably ON *flatr* 'level'); Greenscoe in Dalton LNC (ON *grǽnn* or OE *grēne*); Myerscough in Lancaster (**mýrr**).

With references to wild creatures: Haresceugh in Kirkoswald CMB ('hares', ODan *haræ* or OE *hara*); Roscoe in Standish LNC ('roe deer', ON *rá* or OE *rā*).

With reference to domestic creatures: Lambsceugh in Hesket in the Forest CMB (OE or ON *lamb*); Swinscoe STF (OE *swīn* or ON *svín*).

With references to owners: Briscoe in St John Beckermet CMB (*Breta* 'of the Britons'); Cunscough in Halsall LNC ('king', ON *konungr*).

With a reference to vegetation: Humblescough in Garstang LNC (probably ON *humli* 'hop plant').

With ambiguous qualifiers:

Blainscough in Standish LNC. Gillian Fellows-Jensen (1985, p. 105) says that ON *Blǽingr* 'the dark one' could be a personal name or a stream name.

Scalderskew in St Bridget Beckermet CMB and Skelderskew YON. Perhaps the ON personal name *Skjǫldr*, but the double occurrence supports Gillian Fellows-Jensen's suggestion (1985, pp. 156-7) that the qualifier is *skjaldari* 'shield-maker', referring to a specialised use of timber.

thveit ON 'clearing, meadow, paddock'. This element belongs to a group of words which share the general sense of cutting or being cut. Smith 1956 takes the view that the root meaning of *thveit* was 'something cut down' and that there was a sense-development from 'a felled tree' to 'a clearing'. In English names the translation 'clearing' is usually acceptable, but allowance must be made for the later senses of 'enclosed pasture' and 'meadow'. The last sense is clearly represented in the only two instances of *thveit* in CHE, Morphany (earlier *Morthwait*), south of Daresbury, and a cluster of field-names in Wirral, near Bidston (Gelling 1995, pp. 192-3). This later development resembles that of OE **lēah**, but for *thveit* there is no reason to suspect an earlier meaning 'wood'.

There are occasional occurrences of *thveit* in most areas where there was substantial Norse settlement, but it is most frequent in the north west, particularly so in CMB. It is also common in the part of south west Scotland adjacent to CMB, and examples there are mapped in Nicolaisen 1976 (p. 106). Since the word is, to quote Professor Nicolaisen, 'associated with the secondary development of less promising ground, usually on a higher level', allowance should be made for its being sometimes a borrowed appellative used

THE LANDSCAPE OF PLACE-NAMES

by ME speakers. Compounds with ON qualifiers are, however, likely to be names coined by ON speakers for areas brought into cultivation in the early stages of Norse settlement, and a high proportion of the names included in the reference section are of this nature.

About 60 *thveit* names are included in Ekwall 1960, and since this gives a large enough corpus for analysis no attempt has been made at a thorough study of the numerous minor names and field-names containing *thveit* in CMB, LNC, WML and Yorkshire. Only a few names have been included from sources other than Ekwall 1960. Some of the remaining names are discussed in Fellows-Jensen 1985, which includes examples from DMF.

The qualifiers used in the *thveit* compounds which are analysed in the reference section support the concept of new settlements in marginal land. Some *thveit*s were 'rough', 'sour' or 'stony', the dominant vegetation in some was bracken, broom, reeds or sedge, and one made a suitable retreat for a hermit. Buildings in some of the clearings were booths, shielings or storehouses. There are two references to wolves. A comprehensive study which included all field and minor names would probably reinforce this message. The list of *thveit* names in the EPNS CMB survey (p. 494), for instance, includes compounds with qualifiers meaning 'blaeberry', 'burnt', 'rough grass', 'marsh', 'moss', and 'thistle'. Three Thackthwaites in CMB provided thatching material and Burthwaite Forest is named from a 'briar *thveit*'.

The five compounds with 'cross', presumably the Christian symbol, present a problem of interpretation. Perhaps settlements of the sort to be found in late clearings were remote from early churches and this created a need for preaching crosses.

Reference section for ON **thveit**

As simplex name: Thwaite NKF(2), SFK, also minor names in CMB, WML, YOE, YON, YOW.

As generic with descriptive terms: Braithwaite CMB(2), YON (also Braffords in Swanland YOW, 'broad', ON *breithr*); Graythwaite LNC ('grey', OE *græg*); Langthwaite LNC, YON, YOW, Longthwaite CMB ('long', OE or ON *lang*); Micklethwaite YOW ('big', ON *mikill*, also several minor names and field-names in CMB and YOW); Ruthwaite CMB ('rough', OE *rūh*); Southwaite in Mallerstang WML (*Sourthwatt* 1324, ON *saurr* 'sour ground'); Stainfield near Haconby LIN ('stone', ON *steinn*); Swinithwaite YON ('place cleared by burning', ON *svithningr*).

With personal names: Allithwaite LNC (ON *Eilifr*); Austhwaite CMB (ON *Afast*); Bassenthwaite CMB (French *Baston*); Finsthwaite LNC (ON *Finnr*); Gunthwaite YOW (ON feminine *Gunhild*); Hampsthwaite YOW (ON *Hamall*); Yockenthwaite YOW (*Yoghan*, ON adaptation of Irish *Eoghan*).

WOODS AND CLEARINGS

With references to buildings: Bouthwaite YOW ('storehouse' ON *búr*); Castlethwaite WML (French *castel*); Curthwaite CMB ('church', ON *kirkja*); Husthwaite YON ('house', ON *hús*); Satterthwaite LNC ('shieling', ON *sætr*); Waberthwaite CMB (possibly ON **veithi-búth* 'hunting or fishing booth').

With ON kros *or late OE* cros 'the Cross': Crosthwaite CMB, WML, YON, Crostwick and Crostwight NFK.

With references to crops: Haverthwaite LNC ('oats', ON *hafri*); Heathwaite LNC, YON ('hay', OE *hēg* or ON *hey*, also Haithwaite in Nicholforest and a lost *Haythwayt* in Castle Sowerby CMB); Linethwaite CMB and Linthwaite YOW ('flax', OE *līn* or ON *lín*).

With earlier place-names: Guestwick (earlier *Geistweit*) NFK (Guist); Radmanthwaite (earlier *Rodmertheit*) NTT (probably a place-name from OE **hrēod-mere* 'reed pool'); Subberthwaite LNC (lost place-name *Sulby*); Tilberthwaite LNC (lost place-name *Tillesburc*).

With references to trees: Applethwaite CMB; Thornthwaite CMB(2), WML, YOW (also Thorniethwaite DMF, OE or ON *thorn*).

With references to other vegetation: Brackenthwaite CMB(3) (probably ON **brækni*); Branthwaite CMB(2) (probably OE *brōm* 'broom'); Reathwaite in Westward CMB ('reed', OE *hrēod*); Seathwaite CMB ('sedge', ON *sef*).

With reference to a domestic animal: Calthwaite CMB ('calf', OE *calf* or ON *kalfr*).

With reference to a wild creature: Outhwaite in Roeburndale LNC and Ullthwaite in Hugill WML ('wolf', ON *ulfr*).

With a reference to topography: Huthwaite NTT, YON (hōh); Seathwaite LNC (sǣ).

With a reference to an inhabitant: Armathwaite CMB(2) ('hermit', French *ermite*).

With a direction: Easthwaite in Irton with Santon CMB (ON *austr* replaced by English *east*) and Eastwood (earlier *Estthweyt*) NTT.

With obscure or ambiguous qualifiers:

Hunderthwaite YON. The first element could be an ON personal name *Húnrøthr*, but Gillian Fellows-Jensen (1972, pp. 97–8) argues for the OE administrative term *hundred*.

Lowthwaite in Uldale CMB (*Loftthwayt* 1399). ON *lopt* 'upper chamber' could refer to a two-storey building, or it could be used in a transferred sense 'hill'. This last use is evidenced in Norway but is not certain to occur in England. This compound may be analogous to that in Loscoe, Loskay, with **skógr**, this being another reference to a special building in a woodland setting.

THE LANDSCAPE OF PLACE-NAMES

Nibthwaite LNC (*Thornubytueith* 1208, *Neubethayt* 1236, *Neburthwait* 1336). Gillian Fellows-Jensen (1985, p. 150) considers this to be either 'new booth clearing' or 'new storehouse clearing' with ON *forn* 'old, former' prefixed in the earliest references. The spellings are not sufficiently consistent for certainty.

Roundthwaite WML (*Rounerthwayt* 1294). EPNS survey (2, p. 51) and Ekwall 1960 ascribe the first element to an ON **raun*, gen. **raunar*, 'mountain ash'. Gillian Fellows-Jensen (1985, pp. 154–5) doubts the existence of this word, and suggests ON *raun* 'test, trial', comparing the appellative *raunarvöllr* 'place where a trial takes place'. This is probably to be preferred, but neither the recorded nor the unrecorded word *raun* has been certainly noted in other place-names in England. Cf., however, Dodgson in EPNS CHE Survey 4, p. 256, on a field-name Ransel.

Slaithwaite YOW. Gillian Fellows-Jensen (1985, pp. 161–2) equates this with Slethat in Ruthwell DMF and endorses the suggestion in EPNS survey 2, pp. 307–8, that the qualifier is ON *slag* 'blow' in a sense such as 'cutting down'. As in the previous name, this would be the only instance noted of such a term in a settlement name.

vithr ON 'wood'. This word, which is cognate with **wudu**, is a rare element in England, and it does not occur in all areas where there was Scandinavian settlement. It has not been noted in DRB, NTP, NTT, for all of which there are detailed surveys, and there are no certain occurrences in CHE. The 21 names analysed in the reference section include some minor names from YOE, YON, YOW and WML, but a number of items listed in the elements sections of the EPNS survey for WML and YOW have been rejected because the documentation is inconclusive or very late. Three names, Witherslack WML, Widdale (in Aysgarth) and Wydale (in Brompton) YON, have been omitted as they could equally well contain ON *vith* 'willow'. It seems clear that *vithr* should not lightly be adduced to account for -with in minor names or field-names, as it was, for instance, in EPNS CHE survey 4, p. 179, in a comment on a Tithe Award name, *Nugwith*.

In spite of its rarity in place-names, *vithr* seems to have been understood by English speakers and recognised as equivalent to **wudu**. Several compounds have OE qualifiers, and in Beckwith YOW *vithr* can be shown to have replaced **wudu**, since the name is recorded *c.*972 as *Bec Wudu*.

Reference section for ON **vithr**

As generic with descriptive terms: Blawith LNC ('dark', ON *blár*); Mewith Forest in Betham YOW ('narrow', ON *mjór*); Smawith in Rufforth YOW ('narrow', OE *smæl*); Waithwith (in Catterick) YON ('wet', ON *vátr*); Yanwath WML ('level', ON *jafn* replacing OE *efn*).

With tree-names: Askwith YOW ('ash', ON *askr*); Beckwith in Pannal YOW ('beech', OE *bēce*); Birkwith in Horton with Ribblesdale YOW ('birch', ON *birki*); Eppleworth (earlier *Eppelwyth*) in Cottingham YOE ('apple', ON *epli*). Nutwith in Mashamshire YON, in which 'nut' may be OE or ON, implies hazels. Bramwith YOW contains OE *brōm* 'broom'.

With personal names: Atterwith in Long Marston YOW (ON *Hattr*); Bubwith YOE (OE *Bubba*); Tockwith YOW (ON *Toki*).

With references to wild creatures: Hartwith YOW ('hart', OE *heorot*); Rookwith YON ('rook', either OE *hrōc* or ON *hrókr*).

With references to ownership or use: Menwith YOW ('common', OE *gemǣne*); Skirwith CMB ('shire', OE *scīr*).

With other qualifiers: Lockwood (*Locwyt* 1273) in Skelton YON (OE *loca* 'enclosure'); Strutforth in Soulby WML ('plantation', ON *storth*); Westworth (earlier *Westwith*) in Guisborough YON. Fulwith in Pannal YOW could contain OE *fūl* 'foul' or OE *fugol* 'bird'.

wald, Kentish and WSaxon **weald**, 'forest', later 'open high ground'.

The sense 'forest' is predominant in English literary sources until *c.*1200, but this is not the only sense of cognates in other Germanic languages, where 'wilderness' seems to be equally well represented. The ON cognate *vǫllr* means 'untilled land, plain'. It is not clear to what extent 'high ground' is part of the ancient Germanic sense, but it seems likely that in England most forests called *wald* had high situations, and it is the 'high ground' sense which persisted in English when the association with woodland had been lost. Modern English *wold* does not mean 'forest'. The complex problem of the use of the variant forms *wald* and *weald* will be considered later. For the immediate purpose of establishing the meaning in place-names the two forms will be considered synonymous.

Place-name specialists have traditionally explained the change of meaning from 'forest' to 'open high ground' by saying that the term was generally applied to forests in high situations, and that when the trees had been cleared the survival in district-names such as Cotswolds caused the term to be newly interpreted. This appears to be broadly correct, but recent studies by historical geographers have emphasised that all areas called *wald* or *weald* in ancient place-names were at one time likely to have been heavily wooded. The later meaning probably does not appear in district-names or ancient settlement names, even if 'forest' is wholly inappropriate to the modern landscape. In the great Wealden Forest which occupied much of Kent and Sussex and parts of Hampshire and Surrey the transformation of the landscape from forest to open land never occurred, and this may be the only area in which the association of forest with *wald* or *weald* names is clearly apparent in the modern landscape. Earlier names for The Weald are discussed under **wudu**.

wald was accorded a lengthy discussion in Gelling 1984. This included a summary of the debate among landscape historians which took place in the late 1970s concerning the significance of *wald* in place-names. Since 1984 there have been two important discussions, one by Professor Alan Everitt in his book on the evolution of landscape and settlement in Kent (Everitt 1986), the other by Dr Harold Fox in a paper about some wold areas in the East Midlands (Fox 1989). As regards place-name usage the main point to emerge from these studies is that *wald* was a term used for districts, rather than for single woods or for the immediate environment of a single settlement, a conclusion which was tentatively advanced in Gelling 1984. Everitt and Fox both consider the occurrence of *wald* in place-names to be vital evidence for the precise definition of the areas of wold districts.

Some wolds are clearly defined by the continuing use of the district-name: here belong the Lincolnshire and Yorkshire Wolds and the Cotswolds. Others, however, as Dr Fox observes (p. 79) 'have never been described as such except in the faint language of early place-names'. The great forest of Bromswold, which lay in NTP, HNT and BDF, is accorded a literary reference of the early 12th century as the refuge of Hereward the Wake: its limits, however, are established by the use of Bromswold as an affix to the settlement names Newton and Leighton, the prefixing of *Wold* to the name which became Old Weston, and the occurrence of *wald* in two BDF names, Wold, in the parish of Odell, and Harrold. Two other wolds in LEI and NTP are defined by Dr Fox entirely from place-name evidence. In CAM there was a part of the Isle of Ely which was *The Wold*, and another area called *Weald* or *Wild* on the border with HNT: both are defined by place-name evidence which is set out in EPNS CAM survey pp. 246 and 54. In KNT, Professor Everitt has defined a belt of territory characterised by *wald* names (including Walderchain, Waldershare, Womenswold, Sibertswold, Ringwould, Studdal and Hammill) which he calles 'the wold region or Downland' (1977, p. 1). In his 1986 book he discussed the characteristics of this region. It is a part of the county which was deforested early, contrasting with The Weald, where permanent settlement probably only became common in the 10th century.

The portrait of wold country which emerges from the studies of Everitt and Fox is of regions where early Anglo-Saxon wood clearance often left stands of trees which served as woodland pasture for largely pastoral communities. Most of these regions were elevated, though, as pointed out in Fox 1989, p. 82, this is not a universal characteristic. Fox instances Southwold SFK and Bede's *Deirewald*, in which lay Beverley YOW, as 'low-lying in the extreme'. Wolds do not provide the best quality land, and in the east midlands they are characterised by deserted medieval villages and by Scandinavian place-names, the latter indicating relatively late settlement by agrarian communities.

The treatment of *wald* names in the reference section is not altogether satisfactory because the term probably never refers to a particular wood which was the defining factor in the environment and economy of a particular settlement, and this makes the classification by qualifiers less meaningful than it is for other woodland terms. The suggestion made in Gelling 1984 (p. 227) that compounds of *wald* with personal names should be interpreted as 'Swīthbeorht's/Stígr's/ Wigmund's part of the district called Wold' rather than 'Swīthbeorht's wood' etc. is supported by subsequent studies. Even apparently isolated *wald* names, like Southwold SFK, are probably surviving relics of vanished wold areas. Weald by Bampton OXF may have been thought of as the south-eastern edge of the vanished Cotswold forest.

The geographical distribution of the variant forms *wald* and *weald* is puzzling, and has not been satisfactorily explained. In terms of OE dialect, *wald* is the Anglian form, *weald* the Kentish and West Saxon, but in practice both forms seem to have been in use in all areas, and they frequently interchange in the spellings for individual names. Gospatric's Writ, a Cumberland document of the mid-11th century, uses the form *weald*. Place-names well outside the sphere of West Saxon influence, such as Wild in Syresham NTP, derive their modern form from *weald*, with -*wald* and -*weld* interchanging in the spellings. One of the CAM wold areas, that in the Isle of Ely, has predominantly -*wald*, -*wold* spellings but with occasional -*weld*; the other, Croydon Wilds, derives its modern form from *weald*, while having -*wald*, -*wold*, -*weald* and -*weld* in ME spellings. A major inconsistency is found in KNT, where the names in Everitt's 'wold region or Downland' derive from 'Anglian' *wald*. In Gelling 1984 the suggestion was hazarded that the Kentish names crystallised earlier than the West Saxon phenomenon of breaking which caused *wald* to become *weald*. At all events the widespread interchange between these two forms can hardly be accounted for in terms of dialect, and it requires study by a better qualified philologist than the present author.

wald and *weald* are most frequently evidenced in simplex names. The commonest use as first element is in the compound with *hām*. This gives the recurring name Waltham, which has been of special interest since 1975, when Rhona Huggins published a paper suggesting that it was a term for royal administrative centres in forest areas. The use of *wald* for group names in Waldingfield, Woldingham, Walderchain and Waldershare emphasises the 'district' meaning. There are a few unequivocal personal-name compounds, and a few with descriptive terms. The absence of any certain references to wild creatures is surprising. The high proportion of compounds with qualifiers which defy explanation is consistent with this being an early OE place-name term.

wald is not evidenced in all counties for which there are detailed surveys. It has not been noted in DEV, IOW, LNC, RUT, WLT, and it is absent from BRK except in the compound Waltham. CHE has only two field-names, both uncertain, and WML one field-name and an uncertain minor name.

Reference section for **wald, weald**

As simplex name: Southam Holt (*Le Waude* 1262) WAR, Old NTP, Weald BUC, HNT, OXF, The Weald SSX/SUR, North and South Weald ESX, Harrow Weald GTL, Wield HMP, Wild in Aldenham HRT, Wild in Syresham NTP, Westbury Wild BUC, Croydon Wilds CAM, Wold in Odell BDF, The Wolds DRB, LIN, NOTT, YOE, Burton Wold NTP, Upton Wold WOR, Wolvey Wolds WAR.

As first element: Walderslade in Chatham KNT (*Waldeslade* 1190, **slæd**); Waldron SSX (*ærn* 'building'); Waltham BRK, ESX(2), HMP(2), KNT, LEI, LIN, SSX(2) (*hām*). Waldingfield SFK and Woldingham SUR contain **feld** and *hām* with a group-name *Waldingas*, 'wold-people'. Walderchain in Barham and Waldershare KNT are 'narrow valley' and 'land-division' of the *Waldwaru*, 'wold-dwellers'.

As generic with personal names: Sibertswold KNT (*Swīthbeorht*); Stixwould LIN (ON *Stígr*); Wiggold GLO (*Wicga*); Wymeswold LEI (*Wīgmund*).

With descriptive terms: Fairwood in Well YON (ON *fagr*/ME *fair*); Hammill KNT (*hamel* 'misshapen'); Harrold BDF ('hoar'); Longhold in Clipston NTP (cf. *Longewold* 13th century in the Isle of Ely, part of the Wold described in EPNS CAM survey, p. 246).

With -inga- *compounds*: Easingwold YON ('Ēsa's people'); Horninghold LEI ('people of a horn-shaped piece of land'); Womenswold KNT (*Wimlincga wald* 824, *Wimlingas* is a folk-name of obscure etymology).

With references to position: Methwold NFK (OE *middel* replaced by ON *methel*); Northwold NFK; Southwold SFK.

With a reference to ownership: Prestwold LEI ('priests').

With a reference to vegetation: Hockwold NFK ('mallow').

With a reference to domestic creatures: Studdall KNT ('horse-stud').

With obscure, uncertain or ambiguous qualifiers:
Bromswold NTP/HNT/BDF. Early spellings for this extensive forest, e.g. *Brouneswald* 1286, are appropriate to derivation from the OE personal name *Brūn*. There are, however, two other occurrrences of the place-name, and this casts doubt on a personal-name etymology. There is a lost *Brun(n)eswold* in the Isle of Ely (EPNS CAM survey 246), and *Bromeswold* 1317 is cited without location in EPNS DRB survey 754. Even without this recurrence it is doubtful whether a personal name is acceptable as the qualifier for the name of such a great forest, a consideration which applies also to Cotswold. *Brūn* 'brown

one' might be an OE forest-name, but for this and for Cotswold a pre-English district-name would seem more appropriate.

Cotswold GLO/OXF/WOR. The earliest spelling is *Codesuualt* 12th century. An OE personal name *Cōd has been inferred from this and Cutsdean, at the southern end of the massif. An earlier district-name would seem more appropriate, but no suggestions are available for the origins of such a hypothetical toponym.

Coxwold YON, Cuxwold LIN. Professor Richard Coates (1995b) argues cogently that these two names are identical and that a third instance of the compound is Cucket Nook in Lythe YON, which was *Cukewald* 1279. Cuxwold is *Cucuwald c.*1115. Professor Coates interprets this name as 'cuckoo forest', arguing that *cuckoo* may well have had an OE antecedent rather than being a borrowing from French. Apart from this, however, no compounds have been noted of *wald* with words for wild creatures. There appear to be no 'crow' or 'rook' names in the corpus, nor (remarkably) any references to wolves or deer. This may be another instance, like Bromswold, of a recurrent *wald* name with a mysterious qualifier.

Grimswolds Fm in Snitterfield WAR. The spellings, e.g. *Grumeswold* 1227, *Groswold* 1280, *Grusselwold* 1306, are too varied for safe etymology.

Hawold YOE. The early spellings *Houwald* could indicate **haugr** or **hōh**.

Ringwould KNT. The OE spelling *Roedligwealda*, from a charter of 861, makes it doubtful whether the etymology proposed in Ekwall 1960 from *Hrēthlingas* 'Hrethel's people' can be accepted, though a 13th-century form, *Rudelingewealde*, is consistent with it.

Womenswold KNT. This has been classified above as an *-inga-* name, but base of the group-name *Wimlingas* is obscure.

wudu, earlier **widu**, 'wood'. Apart from **lēah**, this is the commonest of the 'wood' terms. It was used for large stretches of woodland: forests with *wudu* names are Bernwood BUC, Charnwood LEI, Inglewood CMB, Selwood SOM, Sherwood NTT, Whittlewood NTP and Wychwood OXF. The Wealden forest is spoken of in the Anglo-Saxon Chronicle under the year 893 as the *micla wudu* called *Andred* and also as the *weald*. In the annal for 477 it is the *wudu* called *Andredesleage*, and in a document of 1018 it is *Andredesweald*. (*Andred* is the OE form of *Anderita*, the Romano-British name of the fort at Pevensey). These references suggest that *wudu*, **lēah** and **wald** could all mean 'forest', but *wudu* probably kept that meaning after it had become obsolescent for the other terms.

The use of *wudu* in place-name formation differs notably from that of other woodland terms in its extreme rarity as a simplex name and its very high frequency as a qualifier. As a first element it is extremely common with habitative terms. The 'grove' words are also frequent as qualifiers for settlement terms, but the Grafhams and Graftons are not such a large proportion of the whole corpus as are the Woottons, Woodhams and Woodcotes.

It is here assumed that in these habitative names *wudu* is a topographical term, not a reference to building material, since, at the time when most of these names arose, wood must have been almost universal as a building material in areas where supplies of timber were available. Woodbridge SFK, Woodchurch CHE, KNT and Woodkirk YOW have, however, been omitted from the reference section, as the 'building material' option seems more appropriate to these. The recurrent names Woodhall and Woodhouse may refer to buildings which housed people who had functions connected with the management of woodland. These names are likely to be mostly post-Conquest; OE *hall* and *hūs* are rare in pre-Conquest names, though ON *hús* was frequently employed. The two YOW names, Woodsome and Wothersome, which derive from the dative plural of *wuduhūs*, are in an area where ON influence was strong.

Since *tūn* was very seldom applied to settlements within wooded areas, it is likely that Wootton is a 'functional' name, used for settlements close to woodland which played a special role in the handling of timber.

The predominance of compounds in which *wudu* is the first element leaves a comparatively small corpus in which the word has a defining qualifier. The position of the wood is frequently stated, and there is fairly frequent use of earlier place-names, several of which are pre-English. Compounds with descriptive terms are fewer than might have been expected, in view of the colourless nature of the word. Tree-names are rare, perhaps indicating that these woods contained a variety of species. References to other sorts of vegetation are even rarer, and there is a surprising absence of animals. To judge by the scarcity of personal names it was unusual for a *wudu* to belong to an individual.

Reference section for **wudu**

As simplex name: Woodham DRH (dative plural).

As first element with habitative terms:

tūn: Witton DRH, NFK(2), NTB(2), YON(2); Woodton NFK; Wootton BDF, BRK, DOR(3), HMP(2), IOW, KNT, LIN, NFK, NTP, OXF, SHR(4), SOM(2), STF(2), WAR(3), WLT(2); Wotton BUC, GLO(2), SUR.

cot: Woodcote HMP, OXF, SHR(2), SSX, SUR(2), WAR; Woodcott CHE, HMP.

hall: Ekwall 1960 includes Woodhall HRT, LIN, YOE, YOW with a statement that there are 'several' other examples. In fact this and Woodhouse are very frequent as minor names, and no attempt has been made to assemble a complete list.

WOODS AND CLEARINGS

hūs: Ekwall 1960 includes Woodhouse LEI, YOW(3), acknowledging that there are many others. Eight SHR examples are listed in Part 1 of EPNS survey of SHR, p. 322. Woodsome and Wothersome YOW are from the dative plural, *wuduhūsum*.

Woodstock OXF contains *stoc* 'secondary settlement' and Woodthorpe DRB, LIN contains ON *thorp* with the same meaning.

As first element with other generics: Woodbarrow SOM and Woodborough WLT (**beorg**); Woodborough NTT and Woodbury DEV (*burh* 'ancient fort'); Woodburn NTB (**burna**); Woodchester GLO ('Roman site'); Woodcroft NTP; Woodford CHE(2), CNW, ESX, NTP(2); Woodgarston HMP (*gærstūn* 'paddock'); Woodhay BRK, HMP ('enclosure'); Woodhorn NTB ('horn' referring to a coastal promontory); Woodland DEV, Woodlands DOR, KNT, SOM (**land**); Woodleigh DEV (**lēah**); Woodrow WLT, WOR; Woodsetts YOW ((*ge*)*set* 'fold').

As generic with references to position: Astwood BUC, WOR and Eastwood ESX; Heywood LNC ('high'); Marwood DEV ('boundary'); Melwood LIN and Middlewood HRE; Northwood GTL, IOW, Norwood GTL(2), SUR; Ringwood HMP (*Rimucwuda* 955, probably 'boundary'); Southwood NFK; Upwood HNT; Westwood WAR, WLT, WOR.

With earlier place-names or PrWelsh toponyms: Brewood STF (**breʒ**); Charnwood LEI (Welsh *carn* 'rock'); Chetwode BUC (*cēd*); Crewood in Crowton CHE (probably Welsh *cryw* 'fishing-weir'); Halewood LNC (Hale); Hockerwood NTT (an OE hill-name, also found in Hockerill HRT, discussed under **hyll**); Kentwood BRK (river Kennet); Stockwood DOR (probably an earlier *Stoke*, the settlement is *Stokes Sancti Edwoldi* in 1238); Wicklewood NFK (probably an earlier **Wiclēah* 'wych-elm wood or clearing').

With descriptive terms: Blackwood YOE, YOW; Brandwood and Brantwood LNC, Brentwood ESX, Burntwood STF (all 'burnt'); Broadwood DEV(2), SOM; Evenwood DRH ('level'); Fulwood LNC, NTT ('foul'); Horwood BUC, DEV ('dirt'); Marshwood DOR; Stobswood NTB ('characterised by stubs'); Stowood OXF ('stone'); Verwood DOR ('fair'); Weetwood NTB and Wetwood STF ('wet'); Whitwood YOW ('white'). Calverleigh DEV, *Calewudelege* 1194, has a name meaning 'bare wood' as the first part of the compound.

With topographical terms: Brookwood SUR (**brōc**); Downwood HRE and Dunwood HMP, STF (**dūn**); Ewood LNC(2) (**ēa**); Holmwood SUR (**hamm**); Hopwood LNC, WOR (**hop**); Lingwood NFK (**hlinc**); Lythwood SHR (**hlith**).

With references to ownership or occupancy: Bushwood WAR ('bishop'); Charlwood SUR ('farmers'); Henwood WAR ('religious community'); Inglewood Forest CMB ('Angles'); Kingswood GLO(2), SUR, WAR; Portswood HMP ('town'); Ratchwood NTB (*wrecca* 'outcast'); Sherwood NTT ('shire');

THE LANDSCAPE OF PLACE-NAMES

Wychwood OXF (Hwicce). Threapwood CHE was a 'disputed' wood, partly in CHE and partly in FLI. Callingwood STF contains French *calenge* 'challenge'.

With personal names: Botwood SHR (*Botta*); Goodwood SSX (fem. *Gōdgiefu*); Intwood NFK (*Inta*); Simonswood LNC (*Sigemund*); Whittlewood NTP (*Witela*); Woolstanwood CHE (*Wulfstān*).

With references to trees: Ashwood STF; Iwode HMP ('yew'); Linwood HMP, LIN(2) ('lime'); Selwood SOM ('willow').

With references to other vegetation: Bestwood NTT ('bent grass'); Marchwood HMP (*merece* 'smallage').

With reference to birds: Arnewood HMP and Earnwood SHR ('eagles'); Cawood LNC, YOW ('jackdaw'); Cornwood DEV ('cranes').

With terms for enclosures: Hackwood HMP (Coates 1989 suggests that this is a 13th-century name containing ME *hake* 'door-fastening'); Haywood HRE, STF and Heywood SHR(2), WLT ((*ge*)*hæg*); Lockwood YOW (*loca*); Ningwood IOW (**niming*).

With terms indicating a function: Borthwood IOW ('board'); Needwood STF ('need', perhaps referring to supplies of timber, or to provision of refuge); Saltwood KNT (perhaps providing timber for salt production); Walkwood WOR (*weorc* 'construction').

In phrases: Bowood DOR, WLT (earlier *Bovewode*, 'above the wood'); Westwood KNT (*Bewestanwudan* 805, 'west of the wood'). Underwood DRB, NTT probably means 'place at the edge of the wood'.

Names with ambiguous, obscure or uncertain qualifiers:
Barnwood GLO. On the evidence of the spellings, which are post-Conquest, this could be 'Beorna's wood', 'wood of the warriors' or 'wood with a barn'.

Bernwood BUC. This survives in Bernwood Fm in East Claydon, but it was earlier the name of a royal forest which lay to the southwest, taking in part of the Chilterns, encompassing Brill, Oakley and Boarstall. The Anglo-Saxon Chronicle under the year 917 refers to *Byrnewuda*, probably meaning a district rather than a settlement. Etymology must depend on this spelling. EPNS BUC survey suggested that *Byrne-* was a metathesised development of a name from PrWelsh **brinn**, and this is reasonable. It does not suit the area round Bernwood Fm, but it could refer to Brill or Muswell Hill, or it could have been a reference to the whole Chiltern range. A hybrid of this sort would be analogous to Brewood and Charnwood. Ekwall 1960 gives OE *byrgen* 'burial mound' as the qualifier, but this seems less appropriate for the defining feature of a forest. The post-Conquest forms for Bernwood closely resemble those for Barnwood GLO, but it is probable that this is a different name.

Bowood DEV. This compound of *boga* 'bow' with **wudu** occurs 13 times in DEV, with variant modern forms Boode, Bowda, Buda, Bude. *boga* is extremely common as a first element in DEV, occurring 17 times with dūn, twice with **clif** and occasionally with other topographical generics, in addition to this group of names with *wudu*. The reference must for the most part be to shape, but in the *wudu* compound the alternative of materials for making bows cannot be excluded.

Garswood LNC. Only two forms are available, *Grateswode* 1367 and *Gartiswode* 1479, and these do not point to an obvious etymology.

Gaywood NFK (see discussion of Gaydon WAR, under dūn).

Hanwood SHR. The first element could be *hana* 'cock', *Hana* personal name or *hān* 'rock'.

Harewood HMP, HRE, YOW, Harwood LNC(2), NTB(2), YON. These names are not necessarily identical. Possible qualifiers are *hār* 'hoar', *hara* 'hare' and **hær* 'rock', but *hara* is unlikely as animals are not usually referred to in *wudu* compounds. A recent suggestion (Coates 1997b) is that all these names mean 'grey wood', and that OE *se hara wudu* was a translation of the recurrent British name recorded as *Letocetum*, which is discussed under cęd. Professor Coates further suggests that both the British and the OE names were semi-technical terms for surviving patches of wildwood. Harraton DEV is *Harewodeton* 1244.

Lipwood in Haydon Bridge NTB. This has been derived from *hlīep* 'leap', used of a steep slope. Another possibility is *lippa* 'lip', which might have been used of the edge of a slope.

Marwood DRH. This is recorded as *Marawuda* c.1050, and derivation from OE *māra* 'greater' has been suggested. There does not seem to be another instance of this word in place-names, however.

Packwood WAR. The same first element is found in Pakefield and Pakenham SFK, and in three instances of Packington, in LEI, STF and WAR. There is also a minor name Packmores in Warwick. A personal name **Pac(c)a* inferred from these place-names seems a likely solution, but there is no evidence for it outside place-names.

Scrainwood NTB. The first element is the gen.pl. of *scrēawa* 'shrew mouse', but ME *schrewe* also means 'villain', and the Old High German equivalent could mean 'dwarf goblin'. Such meanings may have been current in OE.

CHAPTER 7

Ploughland, Meadow and Pasture

This chapter contains discussions of some of the Old English terms which were used in place-names for the three categories of land – arable, meadow and pasture – required for subsistence farming.

A major theme of the history of land-use from the Neolithic period onwards is that of increasing demand for arable, and if the significance suggested here for **æcer, feld** and **land** be accepted place-name study can make a useful contribution to our understanding of the progressive encroachment of plough on pasture, and the processes of reclamation of marsh and heathland. These three words are the 'land-use' terms which make a substantial appearance in settlement-names, and it is hoped that the discussions of these, together with the distribution maps in Figs 46, 48 and 49, will be accepted as evidence for settlement expansion beginning in the immediate post-Roman period and continuing into the decades which followed the Viking wars.

Other aspects of land-use are less well represented in settlement-names. Meadow is not often referred to: its presence could be taken for granted in ancient settlements, and since meadowland was not subject to conversion for arable **mæd** was not a likely generic to be used in the naming of new farming units.

Pastureland could be of various types, including marshland in summer, and all kinds of woodland. Many of the terms discussed in Chapters 2 and 6 would be applied to land the main use of which was for pasturing animals. This chapter is relatively short because there were not many terms for which a specialised 'pasture' sense was the main one. Woodland pasture had some specialised terms, however, and two of these, **bær** and **denn**, are placed in this chapter because they refer directly to the feeding of stock. In the Wealden areas of Kent and Sussex many pastures called **denn** evolved into farming settlements, but these were more of the 'minor' than the 'major' kind. When outlying pastures evolved into parishes they were more likely to have names in **lēah**, the most important of the 'woodland' terms.

Rough grassland, of the type misleadingly called 'waste' in medieval records, was of vital importance for grazing, and this function is probably reflected in the use of **hæth** as a qualifier for habitative generics.

OE **æcer**, ON **akr** 'plot of cultivated land, measure of land which a yoke of oxen could plough in one day' (Fig. 46). It is the first sense, 'cultivated land', which is relevant to the study of settlement-names. Modern *acre*, as found in field-names, is a development from the second sense.

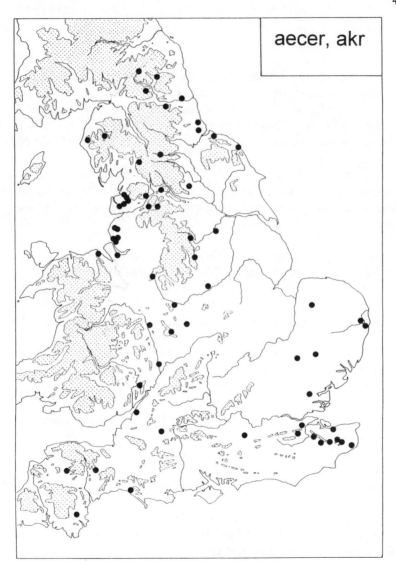

46

aecer, akr

It is obvious that a word meaning 'cultivated land' must have had some additional connotation when used as a generic in a settlement-name. Study of relevant names in their geographical setting combined with an analysis of qualifying elements used with *æcer* and *akr* leads to the conclusion that the additional connotation is 'newly broken-in'. This was a term which, like **land**, was used for farms on the edge of cultivation, probably for the most part new establishments of the Anglo-Saxon or Viking periods. When used in settlement-names both words can be translated 'piece of marginally cultivable land of limited extent'. Many places with names in *æcer* or *akr* can be seen, even in the modern landscape, to have a close relationship to heath, high moorland or marsh.

Acre NFK is the name of three settlements, distinguished as Castle, South and West, which lie beside the river Nar with wide expanses of heath to north and south. Alsager CHE is almost surrounded by heath. Bicknacre ESX is adjacent to a large area of rough pasture. Gatacre SHR is in heathland on the SHR/STF border, and Ridgeacre WOR, now in Birmingham, is a notoriously bleak high area. Hasilacre Hill in Pirbright SUR is in heathland, and Nepicar KNT is on the edge of Wrotham Heath.

High moorland names include Acreland in Sawley and Crumacre in Gargrave YOW, the two Dillicars in WML and YOW, and Innsacre in Shipton Gorge DOR. Onesacre in Bradfield YOW is on the lower slopes of Onesmoor. Muker YON is by the Pennine Way and Cliviger LNC is in high moorland southeast of Burnley.

Names with a clear relationship to marsh include Bessacar YOW, in the marshes south-east of Doncaster, Renacres and Shurlacres LNC, both on the edge of Halsall Moss, Rivacre CHE, by the Mersey estuary, and Handsacre STF, by the marshes of the Trent Valley.

Analysis of the qualifying elements used with *æcer* and *akr* shows that the newly-broken-in marginal land was most frequently used for growing flax (OE *līn*) or beans. The plot might be 'crooked', 'narrow' or 'rough', and the soil sandy or stony. Vegetation mentioned includes bent grass, fern and weeds. The only domestic creature associated with these plots is the goat. There are seven compounds with personal names, four of which are Norse.

A number of the names included in the reference section belong to single farms. Compounds with numerals, such as Ramacre Wood in Castle Hedingham ESX ('three') and Tinacre in Clawton DEV ('ten') have been omitted, as in these the meaning 'measure of land' is clearly indicated. Halnaker SSX has been left aside, as the forms do not point conclusively to OE **healfan æcer*, and the meaning 'half a strip of ploughed land' or even 'half an area of newly-broken-in land' seems inappropriate to this Domesday manor and important feudal centre. The area is, admittedly, heathland, as Hilaire Belloc's poem *Hannaker Mill* suggests.

Reference section for **æcer**

As simplex name: Acre NFK.

As first element: Acreland (in Sawley) YOW (**land**).

PLOUGHLAND, MEADOW AND PASTURE

As generic with references to crops: Beanacre WLT, Benacre KNT(2), SFK ('beans'); Dillicar WML, YOW ('vetches'); Lenacre KNT(2), Linacre DRB, LNC, Linacres NTB, WOR, The Linegar GLO, Llinegar FLI ('flax'), there is a lost *Linacre* in Horseheath CAM; Nepicar (in Wrotham) KNT ('turnip'); Overacres (in Elsdon) NTB (late-recorded, probably 'oats'); Renacres LNC ('rye'); Wheatacre NFK, Whiteacre KNT ('wheat').

With personal names: Alsager CHE (*Ælle*); Edder Acres (in Shotton) DRH (*Æthelrēd*); Innsacre (in Shipton Gorge) DOR (*Ine*); Melliker (in Meopham) KNT (fem. *Milde*); Onesacre (in Bradfield) YOW (ON *Ánn*); Stainsacre (in Whitby) YON (ON *Stein*); Stirzacre (in Garstang) LNC (ON *Styrr*); Susacres (lost in Brearton) YOW (ON *Sótr*).

With descriptive terms: Crumacre (in Gargrave) YOW, Crumbacre (in Kirkleatham) YON ('crooked'); Muker YON (ON *miór* 'narrow'); Oldacres (in Sedgefield) DRH; Rivacre (in Hooton) CHE ('rough'); Whitacre WAR, Whitaker (in Padiham) LNC 'white').

With references to vegetation: Bessacar YOW ('bent grass'); Farnacres (in Whickham) NTB ('fern'); Hasilacre (in Pirbright) SUR ('hazel'); Weddicar CMB and Woodacre LNC ('weed').

With topographical terms: Cliviger LNC (**clif**); Fenacre (in Burlescombe) DEV (**fenn**); Marshacre (in Elberton) GLO; Ridgeacre WOR; Roseacre LNC (ON *hreysi* 'cairn'); Shurlacres LNC ('clear spring'). Minsteracres DRH ('mill-stones') may be broadly classified here, as alluding to the sort of rock from which mill-stones could be made.

With reference to a domestic animal: Gatacre SHR, Goatacre WLT, Goodacre (in Bridestowe) DEV (all 'goat').

With reference to a wild creature: Tarnacre LNC (ON *trani* 'crane').

With references to soil: Chadacre SFK (*cert* 'rough ground'); Sandiacre DRB ('sandy'); Stoneacre (in Otham) KNT.

With reference to position: Fazakerley LNC ('border acre' with **lēah**).

With reference to a building: Uzzicar CMB ('house').

With uncertain or ambiguous qualifiers:

Bicknacre ESX. As with other Bick- names the choice lies between **bica* 'point', which would place Bicknacre in the descriptive terms category, and the personal name *Bica*.

Fishacre (in Broadhempston) DEV. The first element is 'fish', perhaps with reference to the use of fish-waste as manure.

Greenacre (in Elham) KNT. This is *Grimeshaker* 1275. OE *Grīm* is considered to have been an early pseudonym for the god Woden, later a vaguer term for a supernatural being. *Grīm* names are discussed in Gelling 1988, pp. 148–50.

The superstitious sense is appropriate when a name refers, as most *Grimes-* names do, to prehistoric earthworks, but it is not obviously suitable in this compound with *æcer*. It is possible that this is a late OE name containing an English borrowing of the ON personal name *Grímr*.

Handsacre STF. Ekwall 1960 conjectures a personal name *Hand on the analogy of *Fōt*, which occurs as a nickname in DB.

Waddicar (in Melling) LNC. The first element could be either 'woad' or 'weed'.

bǣr 'pasture', especially 'woodland pasture for swine'. In Anglo-Saxon charters which grant land in southern counties *bǣr* is a term for an area of pasture, usually outlying, which belongs to an estate. Compounds such as *den-bǣr*, *weald-bǣr* and *wudu-bǣr* are sometimes used, and one Kentish charter refers to *pascua porcorum que nostra linguam Saxhonica denbera nominamus*. In ME spellings *bǣr* would be indistinguishable from **bearu**, but it seems certain that *bǣr* is much the less frequent of the two.

bǣr is probably the source of the name Bere Forest, applied to two woodland areas in HMP. It may be found in Bere Regis DOR, where there was a royal forest, but apart from this there is no likely instance in settlement-names. Dr Paul Cullen informs me that Bere near Canterbury is more likely to derive from *bȳre* 'byre, cowshed'.

brēc Anglian, Kentish, **brǣc** WSaxon 'breach'. This term for newly-broken-in ploughland is manifested mainly in field-names, most of which probably derive from ME *breche*, rather than being of OE origin. The OE word is, however, found as the first element of a few settlement-names. Ekwall 1960 gives this derivation as certain for Bircham and Breckles NFK and for Bratton DEV(2), SOM, WLT, as probable for Bretton DRB, YOW(2), and as possible for Braxted ESX. The Bretton names are interpreted in EPNS surveys as 'settlement of Britons', however, and Braxted is derived in the ESX survey from *bracu* 'fern', so these names are best left aside. The others seem convincing examples of OE *brēc*, *brǣc*.

At Bircham NFK, northeast of Kings Lynn, the name belonged in the first instance to Great Bircham. The nearby villages of Bircham Newton and Bircham Tofts were respectively called *Niwetuna* in DB and *Toftes* in 1205, and both names suggest that expansion of settlement was still taking place at the end of the Anglo-Saxon period. Bircham itself, however, since the generic is *hām* 'village', is an early name, and the settlement may be a new, relatively early, Anglo-Saxon foundation in heathy country. Breckles NFK, west of Attleborough, lies at the edge of a large area of heath. The generic in this name is *lǣs*.

The two DEV Brattons, Bratton Clovelly southwest of Okehampton and Bratton Fleming northeast of Barnstaple, and Bratton west of Minehead SOM, do not have situations markedly different from those of their neighbours, though in all three areas it is easy to believe that there was ground to be broken in from moorland. Bratton WLT, northeast of Westbury, is easily understood as a settlement formed by encroachment on downland pasture.

denn 'woodland pasture, especially for swine'. This is the usual source of -den in place-names in SSX and KNT, confusion with **denu** being to some extent obviated by the frequent development of **denu** to -dean in those counties. There are a few minor names containing *denn* in SUR, but, as explained in Gelling 1984, it is very unlikely that the element occurs in ESX. It need not be considered as a possible element in settlement-names outside the Weald of KNT and SSX and areas bordering that forest. Within that region it is mainly found in minor names.

There are two important discussions of *denn*: Brandon 1978 (*passim*) and Everitt 1979 (pp. 103–6). The use of *denn* in settlement-names implies that swine pastures belonging to estates outside the forest sometimes evolved into agricultural establishments and acquired independent status. Everitt 1979 says 'They might be situated at almost any distance from the parental settlement, and in a few cases were as much as 45 miles away, as at Tenterden, the pasture of the men of Thanet, though usually between about eight miles and 20 miles'.

Only a small number of names containing *denn* belong to settlements of sufficiently high status to appear in Ekwall 1960. These include, for SSX, Danehill, which was earlier a simplex *Denne*, Iden ('yew'), Playden ('sport', either animal or human), and for KNT Benenden (pers.n. *Bionna+ing*), Chillenden (pers.n. *Ciolla*), Cowden, Pinden (probably 'enclosure'), Smarden ('butter') and Tenterden (*Tenetwaru* 'people of Thanet').

Analysis of the many minor names in KNT and SSX which have *denn* as generic is outside the scope of the present study. Cursory examination of the SSX corpus suggests that the most frequent qualifiers are personal names, sometimes with connective *-ing-*, and trees (e.g. Birchden in Withyham, Haselden in Dallington, Maplesden in Ticehurst, Thornden in Ashburnham). There are occasional descriptive terms, e.g. 'round' in Rounden in Brightling and *rūm* 'spacious' in Rumsden in Rotherfield. There is another Cowden in Wartling, and a rare reference to industry in Hammerden in Ticehurst.

ersc 'ploughed land', used for arable land in areas which are not entirely suitable for crop growing. This place-name element is the subject of a recent investigation (Cole forthcoming) in which all known instances are studied except for those in field-names. The suggestion put forward in Gelling 1984, that *ersc* might have referred to the stubble left by a single season's cropping, must be discarded. The word survived in modern dialect as *arrish* and *earsh* with the meaning 'stubble field', but this cannot account for the use of the term in enduring place-names. *ersc* is best seen as a term, like **æcer**, **feld** and **land**, which records the expansion of arable during the Anglo-Saxon period.

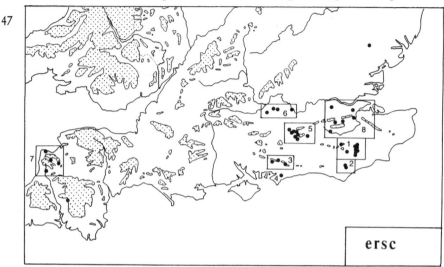

47

As regards distribution, *ersc* is limited to southern England, with Pebmarsh ESX and *Oakhurst* in Shenley HRT as outliers north of the Thames. *Oakhurst*, last heard of in 1828, was *Acersce* c.945. Clustering is a very marked characteristic, as can be seen from Fig. 47. The groups numbered 1–8 on that map are each discussed in Cole forthcoming in relation to the geology and soil conditions of the areas concerned. She concludes that in all these regions there would be serious problems for the would-be cultivator, and that 'the distribution, difficult soil conditions of places named *ersc* along with evidence from surrounding place-names are consistent with it meaning occasional ploughland amid woodland, pasture, marsh or moorland and representing early attempts to cultivate later-settled areas of England'.

Some 40 names are discussed in Cole forthcoming. Six of these – Burwash SSX, Pebmarsh ESX, Ryarsh KNT, Sundridge KNT, Winnersh BRK and Wonersh SUR – have the 'major name' status loosely used as a criterion for study in this book. All six are parish-names and Pebmarsh, Ryarsh and Sundridge are also Domesday manors; and *Acersce*, later *Oakhurst*, in Shenley HRT, was an esate in the 10th century. This is a far from negligible proportion of the corpus, showing that it was possible for a settlement to thrive in the sort of situation to which *ersc* refers.

As regards the nature of the compounds in these major names, Burwash SSX is named from proximity to an undiscovered fortification (*burh*). Pebmarsh ESX is 'Pybba's ploughed land', an association with a particular person which is echoed in some of the minor names, such as Battenhurst in Ticehurst SSX (*Bēta*), Billeshurst in Lingfield SUR (*Bill*), Locksash in Marden SSX

PLOUGHLAND, MEADOW AND PASTURE

(Locc), Orznash in Rotherfield SSX *(Ōsa)*, Turzes in Burwash SSX *(*Tirheard)* and Willinghurst in Wonersh SUR *(Willa)*. Ryarsh KNT is one of several *ersc* names with 'rye' as qualifier, others being Ryehurst in Binfield and Ryeish in Shinfield BRK, and Ryersh in Capel SUR. Rye is a suitable crop for an *ersc*, as it is autumn sown and more tolerant of poor conditions than other grains. Barley is mentioned in the heavily disguised name Bush, in Cuxton KNT, which is *Bererse* in 12th-century references, and EPNS GLO survey 4, p. 123, notes a field-name Oatersh. Flax was grown at Linersh in Bramley SUR. Sundridge KNT was *sundor*, 'in separate ownership', and this compound occurs again in KNT in Sundridge Park, Bromley. Winnersh BRK contains the rare place-name element **winn* 'pasture, meadow'. Wonersh SUR was 'crooked' *(wōh,* dative *wōgan)*, and descriptive terms are also found with *ersc* in minor names, e.g. Lagness in Pagham SSX ('long'), Mellersh in Compton SUR ('spotted'); Radge in Tavistock DEV ('red'), Socknersh in Brightling SSX ('saturated'). The compound with a tree-name found in Oakhurst HRT is paralleled by Hazelhurst in Ticehurst SSX.

It will be seen from the examples cited that *ersc* is particularly liable to confusion with commoner elements, perhaps most frequently with **hyrst**.

The place-name element *edisc*, which is not treated in this book, may be etymologically related to *ersc*. The ultimate source of both words has not yet been ascertained. There is no detailed study of names containing *edisc*, but it is clear that the distribution is distinct from that of *ersc*. Most examples are in ESX and East Anglia (Bendysh, Brockdish, Brundish, Cavendish, Edgefield, Thornage) or in CHE and LNC (Edgeley, Standish, Wornish). There are outliers (including another Bendish in HRT), but there appear to be no examples south of the Thames. *edisc* differs from *ersc* also in being used as a first element in Edgefield and Edgeley and as a simplex in Escombe DRH, which is the dative plural *ediscum*.

Literary occurrences of *edisc*, particularly its use in glosses for Latin words, are suggestive of man-made enclosures, so *edisc* has been omitted from this study as being 'habitative' rather than 'topographical'. Other 'enclosure' words, such as *haga, (ge)hæg, pearroc*, are also omitted.

feld 'open country'.
Very little of the text of the present work has been taken directly from the 1984 book *Place-Names in the Landscape*, which was its forerunner. An exception is made here, however, in respect of the first five paragraphs of the 1984 article on *feld*, because these make a suggestion about the sense-development of the word – from 'open land' to 'communally-cultivated arable' – which seems worth repeating without summary or paraphrase. The 1984 text is:

> The word is used in literary texts to describe unencumbered ground, which might be land without trees as opposed to forest, level ground

as opposed to hills, or land without buildings. In many references there is a contrast between *feld* and areas which are difficult of access or passage. The contrast most often recorded is that with woodland, but contrast with hills is also well evidenced, and one text, an agreement dated 852 about land at Sempringham and Sleaford LIN, says that the estate is given 'mid felda and mid wudu and mid fenne', which suggests that *feld* might be contrasted with marsh as well as with woodland and hills. In charter boundaries the phrase 'ut on thone feld' is quite frequent on the edge of woodland. The general tenor of the passages in which *feld* occurs in OE texts clearly indicates that for most of the pre-Conquest period it was used indifferently of land which might or might not be under the plough. It certainly had no special connotation of arable.

It seems likely that *feld* came to mean arable land, particularly the communally-cultivated arable which characterised the open-field farming system, in the second half of the 10th century (which is the date at which the open-field system is now thought likely to have arisen). There is evidence for this use of *feld* in WOR charters dated 966 and 974 (Hooke 1981, p. 175; Fox 1981, p. 84), and the new sense appears in literature at the end of the century. Ælfric's colloquies, written about then, include a statement by a ploughman that he drives his oxen 'to felda' at dawn, and yokes them to the plough. The natural interpretation is that the *feld* is the area being ploughed, the implement having been left there at the close of the previous day's work. But at an earlier date, and at least until the middle decades of the 10th century, Anglo-Saxon charters use a term variously spelt *irthland*, *yrthland*, *ierthland*, *earthland*, *eorthland* in this sense, whereas *feld* is used of common pasture. The two words occur in a BRK charter dated 961 which has instead of boundary clause the statement: 'Thas nigon hida licggead on gemang othran gedal lande feldlæs gemane and mæda gemane and yrthland gemæne' (These nine hides lie intermingled with other lands held in shares: open pasture common, and meadow common, and plough-land common). Here the arable land of the community is *yrthland*, 'ploughland', while *feldlæs*, 'field leasow', is generally taken to refer to land used for pasturing stock. The use of *yrthland* is well illustrated by some bounds of Sotwell BRK, which are attached to a charter of 957. These run 'swa be Maccaniges wirthland swa swa oxa went' (so by the ploughland of Mackney just as the ox turns).

There is one text in which *yrthland* glosses Latin *agros*, and it might have been expected that these English and Latin terms, *irthland* and *ager*, would be used in post-Conquest surveys for the open fields.

PLOUGHLAND, MEADOW AND PASTURE

But in fact the terms which emerge in regular use after the Conquest are *feld* and *campus*.

The semantic change in *feld* is dramatic, and one would expect it to be due to a sudden change in the pattern of farming. It is argued below that there must have been an early phase of Anglo-Saxon encroachment on pasture land which gave rise to numerous ancient settlement-names in *-feld*. This did not cause the word to develop a new meaning, however, and the next likely period for such a development is at the end of the ninth and the beginning of the tenth centuries, when the influx of Scandinavian farmers must have led to a great increase of arable land in the east and north of England. As the rough pasture was converted to arable, the grazing available on the old arable – on fallow fields and balks – would become even more important, so there could have been a period when the plough was commonly to be seen on the *feld*, and the cattle were grazing on parts of the *irthland*. By the middle of the tenth century there were some areas, notably north-west BRK, where a number of estates had very little or no rough grazing or woodland inside their borders, so the only types of open land to be seen must have been meadow and ploughland, and *feld* 'open land' may have come to seem the appropriate term for the larger of these two divisions.

It is clear at any rate that the sense 'arable land' need not be reckoned with in ancient settlement-names. As a term employed in the naming of villages *feld* probably means 'open land previously used for pasture', and it may be an indicator of areas which were converted from rough pasture to arable in the Anglo-Saxon period. In some instances the contrast implied may be partly between pasture and arable. Some isolated names, such as Clanfield OXF and Watchfield BRK, are in areas where woodland is likely to have disappeared at an early date in the Anglo-Saxon period, and it might be suggested as a hypothesis that the land covered by these parishes was at a very early period of English name-giving reserved for communal pasture by the people of the surrounding villages. A farmer who studied the layout of the open fields at Clanfield OXF (Pocock 1968) thought that he could detect the breaking-in of the furlongs by five groups of people working from different directions. It is possible that what he observed was the taking in of former common pasture by the communities of the surrounding villages.

One passage in that discussion seems by hindsight unsatisfactory: this is the second half of the fourth paragraph, where a connection seems to be implied between Scandinavian settlement in eastern and northern areas and a shortage of rough pasture in Berkshire. These are, of course, two separate matters. Dense Scandinavian settlement in some areas is likely to have led to expansion of ploughland at the expense of pasture, and, at about the same time or a little later, developments in land use in some areas outside the range of Scandinavian settlement may have had a similar effect. More documentation survives for the non-Scandinavianised areas, so it is there, particularly in charters relating to estates in BRK and WOR, that we have explicit evidence for the change in English and Latin terminology for arable land. This change may plausibly be considered to have been more widespread than the recorded evidence for it.

The region-by-region survey of the frequency of *feld* which was included in Gelling 1984 need not be repeated here. Attention was drawn to the occurrence of bands of parishes with *feld* names. A notable example is found in BRK, on the western edge of Windsor Forest. Here there are eight contiguous parishes – Bradfield, Englefield, Burghfield, Shinfield, Wokefield, Stratfield, Swallowfield and Arborfield; Heckfield, another Stratfield, and Sherfield adjoin these on the HMP side of the county boundary. In SFK a wide belt of *feld* names runs from north to south through the county, from Homersfield and Ringsfield on the NFK border to Westerfield near Ipswich. It was also noted that in the northern half of England many places with *feld* names lie close to the 500ft contour, where *feld* meets upland: this can be observed on Fig. 48. A contrast between *feld* and marsh is evidenced in YOE, where Cavil, Kelfield and Duffield lie on the north bank of the Humber.

It is here suggested that the encroachment on arable commemorated by the use of *feld* in settlement-names took place mostly in the 6th and 7th centuries. Driffield YOE is known from archaeological evidence to have been the focal point of one of the major 6th-century settlement areas on the southern side of the Yorkshire Wolds.

feld was certainly in use as a place-name-forming generic in the early Anglo-Saxon period: there are 10 instances among names recorded by A.D. 730 (Cox 1976). Not all of these refer to settlements. They include sites of battles and synods, but at least four of them refer to places where estates with settlements had been established by the 7th century. The occasional use of *feld* into the 10th century is suggested by Scrafield LIN, a compound with an ON qualifier, but it is clear that **land** was the preferred term for new arable in the period of the Viking settlements.

About 260 names are analysed in the reference section. No attempt has been made to collect all early-recorded minor names, and the recurrence of Heathfield, Merryfield and Whitfield in minor names in the southwest has been ignored for counting purposes. Descriptive terms make up the largest category of qualifiers, some of these – 'broad', 'clean', 'heath' and 'white' – being repeated in various counties. Aesthetic qualities, rarely mentioned in place-names in general, are well evidenced with *feld*. Personal names are the next largest category, with a fairly high proportion of dithematic items, one

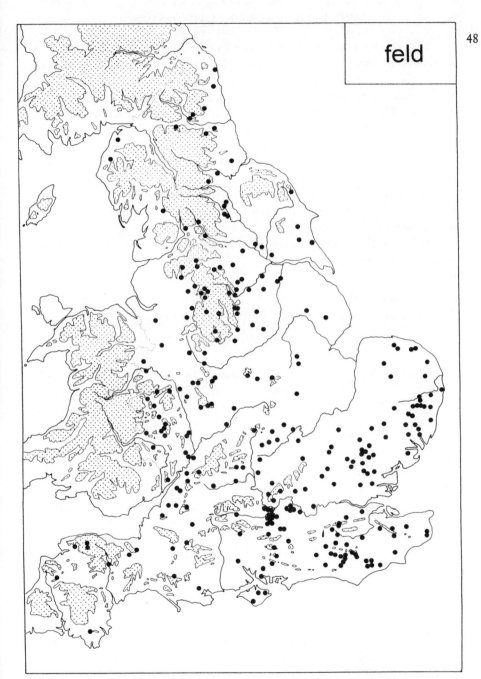

of which is feminine. There are also nine compounds of *feld* with a personal name + *-inga-*. These factors are compatible with the suggested 6th- and 7th-century date as the main period for the use of *feld* in settlement-names.

THE LANDSCAPE OF PLACE-NAMES

Wild creatures are mentioned more frequently than domestic ones, and no references have been noted to crops. Bent grass and broom are the most frequently mentioned vegetation. Some areas of *feld* were characterised either by conspicuous individual trees or by stands of a particular tree-species.

A notable group of *feld* settlements were named from adjacent rivers, and in these, as in the group named from other pre-existing place-names, there is a high proportion of pre-English items. In general the qualifying elements used with *feld* support the hypothesis that areas of grazing land were the original referents.

The number of *feld* compounds with etymologies so obscure or ambiguous as to prevent classification is comparatively low in relation to the size of the corpus.

This is a place-name element of oustanding interest to the student of settlement history.

Reference section for feld

As simplex name: Feldom YON, Feldon WAR (dat.pl.); Field STF; Leafield OXF (French definite article prefixed). Byfield NTP is 'place beside the *feld*'.

As first element: Felbridge SUR; Felsted ESX; Feltham MDX; Felton HRE, NTB SHR(2), SOM.

As generic with personal names: Alstonfield STF (*Ælfstān*); Arborfield BRK (probably fem. *Hereburh*); Bayfield NFK (*Bǣga*); Beauxfield KNT (*Bēaw*); Bedfield SFK (*Bēda*); Bitchfield LIN (*Bill*); Brimpsfield GLO (*Brēme*); Brushfield DRB (*Beorhtrīc*); Canfield ESX (*Cana*); Catfield NFK (ON *Kati*); Caversfield OXF (*Cāfhere*); Chelsfield KNT (*Cēol*); Cockfield SFK (*Cohha*); Cratfield SFK (**Crǣta*); Ellisfield HMP (**Ielfsa*); Elsfield OXF (*Elesa*); Englefield SUR (*Inga*); Haresfield GLO (**Hersa*); Hellifield YOW (probably ON *Helgi*); Hemingfield YOW (**Hymel*, probably with connective *-ing-*); Hewelsfield GLO (*Hygewald*); Homersfield SFK (*Hūnbeorht*); Isfield SSX (**Isa*); Laxfield SFK (*Leaxa*); Luffield BUC (*Lufa*); Macclesfield CHE (**Maccel*); Mangotsfield GLO (*Mangod*); Manfield YON (*Mana*); Matfield KNT (**Matta*); Mountfield SSX (**Munda*); Sedgefield DRH (*Cedd*); Tatsfield SUR (*Tātel*); Uckfield SSX (*Ucca*); Watchfield BRK (**Wæcel*); Wattisfield SFK (**Wacol*); Wethersfield ESX (*Wihthere*); Winkfield BRK (*Wineca*); Wingfield WLT (*Wina*); Wokefield BRK (**Weohha*).

With descriptive terms: Aldfield YOW ('old'); Bradfield BRK, ESX, NFK, SFK, YOW and Broadfield HRE, HRT ('broad'); Bredfield SFK (?*brǣdu* 'breadth'); Bromfield CMB ('brown'); Charfield GLO (probably 'rough surfaced'); Clanfield HMP, OXF, Clanville HMP, SOM and Glenfield LEI ('clean'); Driffield GLO, YOE ('dirt'); Enville STF ('level'); Fairfield DRB, KNT; Falfield GLO and Fallowfield LNC, NTB ('fallow coloured'); Hadfield DRB, Hatfield ESX(2), HRE, HRT, NTT, WOR, YOE, YOW, Heathfield DEV, SOM, SSX, Hothfield KNT and (with *tūn*) Hethfelton DOR ('heath', Heathfield is a common minor name in DEV); Heckfield HMP

274

PLOUGHLAND, MEADOW AND PASTURE

(probably 'high'); Horfield GLO ('dirt'); Kelfield LIN, YOE ('chalk'); Longville SHR(2) (referring to the same stretch of land); Marvel IOW, Merrivale DEV, Mirfield YOW (*myrig* 'pleasant', Merrifield and Merryfield are common minor names in DEV); Metfield SFK (**mǣd**); Micklefield YOW ('large'); Ninfield SSX ('newly broken in'); Nuffield OXF (earlier *Togfeld, Toufeld*, 'tough', cf. Tovil *infra*); Shenfield ESX ('beautiful'); Sherfield HMP(2) ('bright'); Stanfield NFK, Stainfield LIN ('stone'); Therfield HRT and Turville BUC (*thyrre* 'dry'); Tovil KNT ('tough'); Wheatfield OXF, Whitefield DEV, IOW, LNC, Whitfield DRB, KNT, NTB, NTP, Widefield DEV ('white', there are also several minor names in DEV); Wingfield DRB (*winn* 'pasture').

With references to topographical features: Brafield NTP and Brayfield BUC (*brægen* 'brain', perhaps used for a raised area); Broxfield NTB (**brōc**); Burghfield BRK (**beorg**); Chesterfield DRB, STF ('Roman ruin'); Crowfield SFK (*crōh* 'nook'); Cruchfield BRK (**crūc**); Cryfield WAR (**creowel* 'fork', i.e. 'river-junction'); Hilfield DOR (**hyll**); Holmesfield DRB (**holmr**); Scrafield LIN (ON *skreith* 'landslip'); Sarnesfield HRE (Welsh *sarn* 'road'); Stratfield BRK, HMP (**strǣt**); Toppesfield ESX ('top'); Walfield CHE ('spring'); Washfield DEV ('rushing water'); Wightfield GLO (*wiht* 'bend', but the referent has not been determined); Winchfield HMP (*wincel* 'nook', here a small recess in a contour).

With references to vegetation: Benfieldside DRH, Bentfield ESX, Binfield BRK ('bent grass'); Bitchfield NTB ('beech trees'); Bramfield SFK, Brimfield HRE, Bromfield SHR, Broomfield ESX, KNT, SOM ('broom'); Dockenfield HMP ('docks'); Farnsfield NTT ('fern'); Fersfield NFK ('furze'); Hasfield GLO ('hazel trees'); Ifield KNT, SSX ('yew trees'); Lindfield SSX ('lime trees'); Mayfield STF ('madder'); Nutfield SUR.

With river-names: Blithfield STF (Blythe); Cantsfield LNC (Cant); Charsfield SFK (?*Char*); Ightfield SHR (*Giht*); Limpsfield SUR (*Limen*, a derivative of the British word for 'elm', perhaps the old name of river Eden); Mansfield NTT (Maun); Panfield ESX (Pant); Sheffield YOW (Sheaf); Swallowfield BRK (Swale); Tanfield DRH (Team); Worfield SHR (Worfe); Yarnfield WLT (?*Gerne*).

With earlier place-names other than river-names: Dogmersfield HMP (**Doccan-mere* 'dock pond'); Edenfield LNC (**Ēg-tūn* 'island settlement'); Hundersfield LNC (**Hūnan worth* 'Hūna's enclosure'); Lichfield STF (Romano-British *Letocetum* 'grey wood'); Mawfield HRE (Welsh *(Mais) Mail Lochou*); Nympsfield GLO (a British name meaning 'holy place').

With references to wild creatures: Catsfield SSX; Cavil YOE ('jackdaw'); Cranfield BDF ('crane'); Cuckfield SSX ('cuckoo'); Darfield YOW ('deer'); Dronfield DRB ('drones'); Duffield DRB, YOE ('doves'); Dukinfield CHE ('ducks'); Finchfield STF; Froxfield HMP, WLT ('frogs'); Marefield LEI ('martens'); Netherfield SSX ('adders'); Padfield DRB ('toads'); Ravenfield YOW ('raven'); Snitterfield WAR ('snipe'); Yarnfield STF ('eagles').

With references to domestic creatures: Enfield MDX (**ēan* 'lambs'); Fairfield WOR ('hogs'); Gosfield ESX ('geese'); Houndsfield WOR; Rotherfield HMP, OXF, SSX ('cattle'); Sheffield SSX ('sheep'); Titchfield HMP ('kids'); Withersfield SFK ('wether').

With -inga- compounds: Atherfield IOW (Æthelhere's people'); Bedingfield SFK ('Bēda's people'); Benefield NTP (*Beringafeld c.*970, 'Bera's people'); Bingfield NTB ('Bynna's people'); Finchingfield ESX ('Finc's people'); Haslingfield CAM (probably **Hæsela*'s people'); Huntingfield SFK ('Hunta's people'); Itchingfield SSX ('Ecci's people'); Shinfield BRK (**Scīena*'s people'); Waldingfield SFK ('wold dwellers'); Waldringfield SFK ('Waldhere's people'); Wingfield SFK (probably 'Wiga's people').

With ethnic names: Englefield BRK ('Angles'); Markfield LEI and Markingfield YOW ('Mercians').

With references to man-made structures: Austerfield YOW ('sheep fold'); Beaconsfield BUC ('beacon'); Edgefield NFK (*edisc* 'enclosure'); Hoofield CHE ('shed'); Hurdsfield CHE ('hurdle'); Sheffield BRK (possibly **scēo* 'shelter'); Stalisfield KNT (*steall*, here perhaps 'stable'); Warfield BRK ('weir').

With references to a product or activity: Coldfield WAR ('charcoal'); Madresfield WOR ('mower'); Threshfield YOW ('threshing'); Wakefield NTP, YON ('festivities'); Whatfield SFK ('wheat').

With references to position: Marshfield GLO ('boundary'); Northfield WOR; Suffield NFK, YON ('south'); Westerfield SFK ('more western'); Westfield NFK, SSX.

With references to religious practices: Eaglesfield CMB and Ecclesfield YOW (**eclēs* 'Christian community'); Wednesfield STF (Woden).

Names with obscure or ambiguous qualifiers:

Bardfield ESX. The spellings suggest an OE **byrde*, which Ekwall 1960 takes to be a derivative of *bord* 'border'. In Stibbard NFK an identical word is used as generic. It is not advisable to look for a precise meaning without more evidence than is provided by these two names.

Blofield NFK. The first element could be **blāw* 'blue' (though the evidence for this word is not conclusive) or a derivative of *blāwan* 'to blow' denoting an exposed place.

Bramfield HRT. This has been explained as 'steep' from *brant*, which suits the spellings, but Watts (forthcoming) points out that the land has only a gentle slope. He suggests *brænde* 'burnt'.

Eldersfield WOR. There is a 972 spelling *Yldresfeld*. The qualifier may be **hyldre*, ME *hilder*, 'elder tree', but in other names ascribed to that word *H-* is preserved.

Framfield SSX. A good case can be made for the existence of a personal name **Fremma*. Coates 1996b suggests, however, OE *fremede* 'foreign', which might refer to an administrative status at odds with geographical position. There is a spelling *Fremedfeld* 1265 which supports this.

Harefield MDX. The first element is *here* 'army' but the significance is unclear. OE sources use the word *here* for Viking armies, but it is not likely that temporary use by such a body would produce a permanent place-name.

Hayfield DRB. The forms are inconsistent: some indicate *hǣthfeld*, others suggest 'hay'.

Henfield SSX. The earliest spelling, *Hanefeld* 770, strongly suggests 'rocks'; it would be difficult to confirm this by fieldwork in this heavily developed area.

Howfield KNT. Perhaps a personal name **Huhha*, as suggested for Hughenden (under **denu**).

Huddersfield YOW. Possibly the personal name *Hudrǣd*, but an OE **hūder* 'shelter' has been suggested for this and some other names including Hothersall (discussed under **halh**).

Leconfield YOE. An obscure element **lecing*.

Lingfield SUR. In the will of Ælfred Dux, 871–9, *Leangafeld* occurs three times, *Læncanfelda* once. The two spellings occur in successive sentences. This inconsistency, which is highly unusual, renders any etymology doubtful.

Morville SHR. The spellings require a first element *mamer*, which is obscure.

Nesfield YOW. Possibly *nēates-* 'of cattle', with, as Fellows-Jensen 1972, 160–1, points out, some Scandinavian influence apparent in ME spellings. The alternative suggestion, ON *nezti* 'lowest', is also possible, however. The place is on the 400ft contour, at the edge of high moorland.

Nosterfield CAM, YON. Probably identical names. Suggestions for Noster- have assumed that N- is a transference from *atten* 'at the' or *in*, but it is at least as likely to be original. This would rule out *ēowestre* 'sheepfold' (which in any case has given markedly different spellings in Austerfield YOW) and also a hypothetical **ōster* 'hillock'. Nosterfield is unexplained.

Pakefield SFK. This has the same first element as Packwood, discussed under **wudu**.

Petersfield HMP. The church is dedicated to St Peter, and this has been accepted as the reason for the name. Coates 1989 points out, however, that there are well known instances of church dedications being suggested by place-names, and that *Peteresfield* 1181 may represent an OE name 'Peohthere's open land'.

Ringsfield SFK. This could be 'open land of the circle', but it is not apparent what *hring* would refer to. Alternatively **Hring* could be a personal name: *Hring-* is well evidenced in compound personal names.

Shedfield HMP. The first element is the genitive plural of *scīd* 'piece of wood sliced thin'. The significance is not clear.

Springfield ESX. The spellings suggest an *-inga-* compound based on *spring*, which is a rare but well evidenced place-name element. The OE word meant 'place where water issues from the ground'. The sense 'coppice' appears in ME, and 'plantation of young trees' is evidenced in early modern sources. The 'woodland' senses may have been current in OE, but neither these nor the 'water' sense make a satisfactory base for a group name **Springingas* combined with *feld*. The matter is best left open.

Sternfield SFK. The first element may be the mysterious word **sterne*, found as a generic in Sewstern LEI and possibly in Tansterne YOE.

Stansfield SFK, YOW. The first element could be 'of the stone' or a personal name *Stān*.

Stonesfield OXF, earlier *Stuntesfeld*. *Stuntes-* is the first element of minor names in the same parish. These have been explained as containing a nickname *Stunt* from OE *stunt* 'foolish', but a pre-existing place-name is possible. The original meaning of *stunt* may have been 'short, truncated', and this could have been a hill-name.

Swafield NFK, Swayfield LIN. The first element appears to be *swæth* which in OE meant 'mark left by a moving body'. It is doubtful whether this can be legitimately taken to mean 'track' in the sense 'path'. The meanings 'space covered by the swing of a scythe' and 'measure of grassland' appear in ME. It is not clear what the term means in its rare occurrences in ancient settlement-names.

Tanfield YON. The first element may be OE *tān* 'twig, sprout, shoot', discussed under Tansor (**ofer**). If so, it is here in the genitive plural. The sense of the compound is obscure.

Tiffield NTP. OE *tig*, recorded in a few compounds such as *fore-tig*, 'market-place', may have had as one of its meanings 'meeting-place', and this would be a possible solution for Tiffield.

Warmfield YOW. The first element could be 'stallions' or 'wrens': OE **wrǣna* and *wrænna* would be indistinguishable in ME spellings.

Wingfield BDF. The spellings are too varied for safe interpretation. They include *Winfield* c.1200, *Wintfeld* 13th century, *Winefeld* 1249. One form, *Wynchefeld* 1276, has led to the suggestion that the first element is 'winch' (as in Winchendon), but this is from an Assize Roll, not the most trustworthy source of place-name spellings. The matter is best left open.

Woodsfield WOR, *Wordesfeld* 1275. A genitive of *worth* 'enclosure' is perhaps a possibility, but this common generic is rarely used as a qualifier, and is more likely to take a feminine than a masculine inflection.

hæth, KNT and SSX hāth, 'heather, tract of uncultivated land'. There is little to add to the discussion of this word in Gelling 1984, but it might, perhaps, have been asserted more strongly there that the reference in place-names will generally be primarily to the quality of the ground, rather than to the presence of the plant. The use of the term *hǣthfeld* in OE charter boundaries suggests that this was a compound appellative for the sort of land which later became a common.

There are some simplex settlement-names – Heath BDF, DRB, HRE, SHR, YOW, Hethe OXF, Hoath KNT – but much the most frequent use of *hǣth* in ancient names is as a first element. Among habitative names the usual compound is with *tūn*: Hatton occurs in CHE(2), DRB, GTL, LIN, SHR(4), STF and WAR. The other habitative compound noted is with *cot*: this gives Heathcote DRB and WAR, and there is another DRB example, Hearthcote Fm in Church Gresley. In topographical settlement-names *hǣth* is frequently compounded with **dūn**, **feld** and **lēah**, and examples are listed under those words. A 'one-off' compound, Heyshott SSX, has as generic a word meaning 'corner, angle, projecting piece of land'.

There are very few ancient settlement-names with *hǣth* as generic in a compound. Blackheath GTL (which recurs as a hundred-name in SUR and in some widespread late-recorded minor names) may be the only specimen. Horseheath CAM may be another, but the spellings are inconsistent: they include *-hede*, *-ete* and *-eye*, as well as those with *-hethe*, so 'horse heath' is not a certain etymology.

land, OE and ON, 'land, estate', also 'new arable area' (Fig. 49).
When used in ancient settlement-names this word must have had a more specific meaning than the modern descendant, and in Gelling 1984 it was suggested that the specific meaning was 'ground newly broken-in for arable farming'. There is a good deal of evidence for the use of *land* as one of the terms employed to denote new settlements of the 7th–12th centuries, established in areas colonised or reclaimed in response to an increasing need for arable. Most of this evidence is circumstantial, as is usually the case with deductions made from place-names, but the meaning is explicit in the recurrent name Newland(s). The other, less frequent, term with this specific meaning is OE **æcer**, ON **akr**. New settlements of the Anglo-Saxon period could also be given **feld** names, but this is a reference to the previous state, not to the new function.

The assembled corpus of 74 ancient settlement-names in *-land* includes five East Anglian *-inga-* compounds, which belong at the earlier end of the time scale, as well as a number of northern names with ON qualifiers, and some post-Conquest formations. This suggests that *land* in the sense 'new arable area' was used in place-name formation for a long period, and it is reasonable to regard the ON word as subsumed in the use of the OE term.

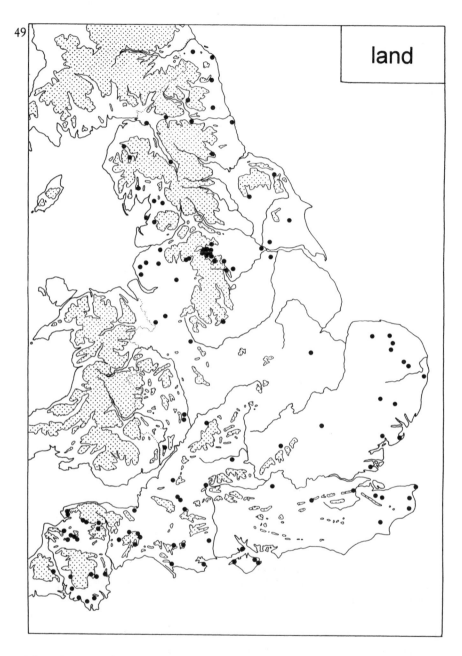

There is an overlap between *land* and **feld**, but **feld** appears to have been more commonly used in the earlier part of this long period of arable expansion, while *land* was the usual term in later centuries, certainly predominating during the decades of Norse settlement. A meaning 'estate' also has to be reckoned with in the later part of the period, and in this respect *land* overlaps with *tūn*. In two south-western names, Hartland DEV and Helstone near Camelford CNW, *-land* and *-ton* interchange in ME spellings. There are two parishes called Kirkland in CMB, and here 'estate belonging to a church' is the obvious etymology.

The use of *land* in minor names and field-names is different from that in settlement-names. In those categories it sometimes has the specialised sense 'strip in a field-system' (as, e.g. in the recurrent name Longlands), but usually it means the same as the modern word, 'ground, part of the earth's surface'. Another modern sense is found in district-names, which include those of some counties and, of course, England, Ireland, Scotland.

In the analysis offered in the reference section three uses of *land* are distinguished: that in ancient settlement-names, that in district-names, and the occurrence of *-land* as part of a compound appellative. It is the first category which is of special interest to students of settlement history.

In Gelling 1984 examples were cited of settlements in which *-land* names obviously relate to reclamation of heath, high moor and marsh. Relationship to wood is less common, but the recurrent name Woodland(s), when it belongs to an ancient settlement, probably means 'land cleared for cultivation on the edge of wooded country'. Attention was also drawn in 1984 to a remarkable concentration of *-land* names in YOW. Thirteen such names are shown on sheet 102 of the O.S. 1in. map, six of them bunched in the area between Halifax and Huddersfield. Several of these YOW names have ON qualifiers, and it seems reasonable to postulate late expansion of arable exploitation in this rugged terrain.

land is barely evidenced as the first element in a compound name. Lambrook SOM is *Landbroc* in 1065, and may mean 'boundary brook'. There are a few names from *landgemǣre* 'boundary': these include Landermere Hall in Thorpe-le-Soken ESX. Landmoth YON has been variously interpreted: it could be ON *landamót* 'meeting of the lands', i.e. 'boundary', or it could be OE **landagemōt* 'district meeting-place'.

Compounds with *land* as generic show an uneven distribution into categories of qualifier. The largest groups consist of those with personal names and those with references to a topographical feature. Descriptive terms are well represented, and include a number ('cleared', 'swamp', 'heavy', 'stone', 'wild') which accord well with the interpretation of *land* advocated here. No references have been noted to wild fauna or flora, so presumably neither of these seemed important in the circumstances in which these names were coined. There are a few references to domestic animals, but none to crops. The *-inga-* compounds are all in East Anglia, where other evidence for groups of people known as 'followers of x' or 'dwellers at x' is abundant. These *-ingaland* names may date from a relatively early period when such groups were the basis of social organisation. In the seven instances of Newland(s) the reference to newly-broken-in arable is explicit.

THE LANDSCAPE OF PLACE-NAMES

Reference section for **land**

As generic with personal names: Allisland DEV (*Ælli*); Barkisland YOW (ON *Bǫrkr*); Buteland NTB (*Bōta*); Byland YON (*Bēaga*); Cadlands HMP (*Cada*); Cooksland STF (**Cuc*); Donyland ESX (*Dunna* + *-ing-*); Dotland NTB (ON *Dot*); Gilsland CMB (ME *Gille*); Goathland YON (*Gōda*); Kidland DEV (*Cydda*); Leaveland KNT (*Lēofa*); Rusland LNC (ON *Hrōlf*); Sisland NFK (*Sige*); Snodland KNT (**Snodd* + *-ing-*); Thurgoland YOW (ON *Thorgeirr*); Thurland LNC (ON *Thorolf*); Thurstonland YOW (ON *Thorsteinn*).

With descriptive terms: Blackland WLT; Greetland YOW (either OE *grēot* 'gravel' or ON *grjót* 'rock'); Leyland LNC ('uncultivated'); Redland GLO ('cleared'); Soyland YOW ('swamp'); Spotland LNC ('small piece'); Stainland YOW ('stone'); Swarland NTB ('heavy'); Willand DEV ('wild'); Woolland DOR (**winn* 'meadow, pasture'); Yealand LNC ('high').

With reference to a topographical feature: Brookland KNT; Ealand LIN, Elland YOW (**ēa**: Ponteland NTB is *ēa-land* with the river-name prefixed); Holland ESX, LNC, Hoyland YOW(3), Hulland DRB (**hōh**); Litherland LNC(2) ('of the concave hillside', gen. of ON *hlíth*, discussed under **hlith**); Marland DEV, LNC (**mere**); Pilland DEV (**pyll**); Shelland SFK (**scelf**); Totland IOW ('look out place'); Woodland DEV, Woodlands DOR, KNT, SOM.

With -inga- compounds: Haveringland NFK (**Hæfer*'s people'); Irmingland NFK ('Eorma's people'); Kessingland SFK ('Cyssi's people'); Poringland NFK (first element obscure); Ringland NFK (first element obscure).

With references to structures: Burland YOE ('byre'); Crosland YOW (ON *cros* 'cross'); Shopland ESX ('shed'); Stockland DEV, SOM (*stoc*, 'outlying farmstead'); Thorpland NFK (ON *thorp* 'hamlet').

With references to domestic animals: Strickland WML(2) ('stirks'); Studland DOR ('horses'); Swilland SFK ('swine'); Yaverland IOW ('boars').

With earlier place-names: Curland SOM (Curry); Membland DEV (probably a British stream-name *Mimid*); Welland WOR (possibly a British stream-name).

With reference to ownership: Burland CHE ('peasant').

With references to direction: Norland YOW ('north'); North Sunderland NTB (early spellings indicate that this is 'southern land').

With 'new': Newland BRK, GLO, LNC, WOR, YOW. Newlands NTB and as a minor name several times in CMB probably refers to post-Conquest settlements.

Names with obscure or ambiguous qualifiers:
Climsland CNW. Climson, a nearby settlement, is *Clymestun* c.970, *Clim-* has not been noted elsewhere and it is unexplained.

PLOUGHLAND, MEADOW AND PASTURE

Crowland LIN. Ekwall 1960 proposes an OE *crūw 'bend', which would be related to *crēowel, assumed to occur in Cryfield. This would be appropriate if the sharp angle of the county boundary here represents an old river-course.

Dowland DEV. Adjacent Dolton has the same first element. The spellings suggest *duhel, which is unexplained.

Molland DEV. Molton, 4 miles west, has the same first element. Possibly a late adaptation of a Welsh hill-name, a form of the word found in Malvern.

Oldland GLO. 'Old' is generally ambiguous in place-names: it could mean 'disused' or 'in use for a long time'.

Tolland SOM. The spelling *táá land* in a document of the late 11th century contradicts Ekwall's derivation from the river-name Tone. OE *tā* 'toe' is possible, referring to the narrow tongue of land between streams. This word may also be found in Dumbleton GLO (*Dumeltan* in some OE spellings), where it would refer to the toe-like ridges on one side of Dumbleton Hill.

Wackland IOW. OE *wacu* could mean 'look out' or 'festivities'. It has been assumed to have the latter sense in Wakefield.

Wigland CHE. Either a reference to beetles, or *Wicga* used as a personal name.

District-names in -land:

Bowland Forest LNC/YOW. First element 'bows', probably referring to curving valleys.

Cleveland YON. First element genitive plural of **clif**.

Cumberland, 'land of the Welsh'.

Hartland DEV. The ME spellings indicate *heorot-īeg 'hart island' as first element. The history of the name is complicated, but it is probable that it was a district-name for the country inland from Hartland Point, rather than originally a settlement-name.

Holland LIN. Early spellings are *Hoiland*, as are some of those for Holland ESX, but the derivation from hōh, which is acceptable for the ESX name, seems impossible for this fen district.

Kidland Forest NTB. Apparently identical with the settlement-name Kidland DEV, from the personal name *Cydda*.

Northumberland, 'land of the Northumbrians'.

Portland, Isle of, DOR, 'land by the harbour'.

Rutland, '*Rōta*'s land'.

Threapland CMB, 'disputed land'. Threapland is a settlement, but the name may originally have been that of an area on the west boundary of Bothel and Threapland parish.

Westmorland, 'land of the people west of the moors'.

Compound appellatives with -land:

Buckland. This recurrent name derives from OE *bōcland* 'estate granted by royal charter'. The possession of a royal charter conferred important privileges on the holders of estates. There are 29 Bucklands whose existence was recorded in the Domesday Survey or earlier: these are in the counties of BRK, BUC, DEV(13), DOR(2), GLO, HMP, HRT, KNT(4), MDX, SOM(3) and SUR. A study of these Bucklands by Dr A. R. Rumble (Rumble 1987) suggests that the more precise sense of the place-name, as opposed to the appellative, was 'estate created by an Anglo-Saxon royal diploma'. There were, of course, a great many estates with *bōcland* tenure, but, since the possession of a royal charter only occasionally gave rise to a new settlement-name, Dr Rumble suggests that the appellative was only used as a place-name when the estate in question was a new formation, created either by division of an earlier administrative unit or by the amalgamation of two earlier ones.

Copeland CMB, DRH, Coupland NTB, WML, from ON *kaupaland* 'purchased estate'.

Faulkland SOM, from OE *folcland*. This appellative is the contrasting term to *bōcland*, denoting an estate which does not enjoy the freedom from legal restrictions and tax dues conferred by a royal charter. Faulkland is near Buckland Dinham, and the contrast in tenure presumably gave rise to the name.

Foreland KNT, first recorded 1326. The ME and modern term *foreland*, 'cape, headland', may have had an OE antecedent.

Sinderland CHE, Sunderland CMB, DRH, LNC, YOW, Sunderlandwick YOE: also in minor names and field-names. One of the OE instances of *sundorland* occurs in the OE translation of Bede's *Ecclesiastical History*. Here, when dealing with Bede's statement that he was born 'in territorio eiusdem monasterii', the translator renders *in territorio* as *on sundorlonde*. This has led to claims that Sunderland DRH was Bede's birthplace, but the statement is likely to be a reference to the tenurial status of one of the monastery's estates, not necessarily to one which bore that name. The meaning is 'separate land' and various interpretations can be put on this. It is perhaps more likely to refer to tenurial status than to geographical position.

mǣd 'meadow'. The declension of OE nouns to which *mǣd* belongs takes *-w-* in oblique cases, and Modern English *meadow* is from the dative *mǣdwe*. The word is related to *mow*, and it denotes land from which hay crops could be obtained. This was the most valuable category of land to early farmers, and the rarity of *mǣd* in ancient settlement-names is probably due to the presence of meadow being taken for granted as a condition of serttlement. *mǣd* is, of course, common in minor names and field-names.

About half of the small corpus of settlement-names in which *mǣd* occurs have it as first element. Here belong Madehurst SSX and Medhurst KNT (**hyrst**), Meaburn WML and Medbourne LEI, WLT (**burna**), Meddon DEV (**dūn**), Metcombe DEV (**cumb**), Metfield SSX (**feld**). The only *mǣd*- compound not to have a topographical generic is Maydencroft in Ippollitts HRT (*Medcroft* 1269).

There are a few compound settlement-names with *mǣd* as generic. Breightmet LNC and Hormead HRT have descriptive qualifiers, 'bright' and 'dirty' respectively. Denmead HMP is in a valley (**denu**). Chilmead in Nutfield SUR furnishes a clear instance of the rare term **cille*, identified by Dr Gillis Kristensson (1973) as the first element of Childrey BRK; it means 'spring', perhaps specifically 'spring in a gulley'. A stream rises by Chilmead Fm. Shipmeadow SFK refers, rather surprisingly, to sheep. Whiteoxmead SOM is *Witochesmede* in Domesday Book, so 'Hwittuc's meadow'. Bushmead in Eaton Socon BDF, well recorded from 1227, is 'bishop's meadow', but as no episcopal connection is known *Biscop* is here perhaps a by-name. Hardmead BUC has a confusing variety of early spellings, and the etymology is best left open.

There are two neighbouring *mǣd* names on the Welsh border which deserve careful discussion. Presteigne RAD and Kinsham HRE are roughly three miles apart, on either side of the boundary between England and Wales. Between them lies a great triangle of meadow land through which a large tributary stream flows to its confluence with the river Lugg. The boundary makes a detour round Presteigne, dividing this meadow between the two settlements.

Presteigne is *Presthemed* 1137–9, Kinsham is *Kingeshemede* 1216. It is clear that this area of meadow land, which is surrounded by hills and is a very striking landscape feature, was called *Hemm-mǣd*, 'border meadow', and that the settlements at its east and west ends were named respectively from priestly and royal ownership of this valuable resource. Presteigne is probably the estate called *Humet* which appears in the Domesday Survey among the estates of the lord of Richards Castle. *Humet* is a garbled version of the name *Hemm-mǣd* without the prefix.

The etymology of Presteigne was correctly (though tentatively) put forward by B. G. Charles in his 1938 book *Non-Celtic Place-Names in Wales*. The spellings set out there illustrate the irregular development. Mis-transcription of -*m*- as -*in*- gave such forms as *Prestheinede* in the 13th and 14th centuries, and the misunderstanding was probably perpetuated by the obscurity of the name to Welsh speakers. B. G. Charles was not aware of the connection between Presteigne and Kinsham, and he makes no reference to the landscape feature, but the credit for the 'border meadow' etymology must go to him.

The HRE name Kinsham is, curiously, omitted from Ekwall 1960. It is possible that he failed to recognise ME spellings such as *Kingeshemede* as belonging to this place. He does, however, include Presteigne, on account of a mistaken ascription to SHR, and his fanciful etymology for Presteigne has misled later scholars and is proving difficult to dislodge from modern consciousness. Ekwall's conjecture was that *-hemed* in *Presthemed* was OE *hǣmed* 'marriage, sexual intercourse' used in an unrecorded sense 'household'. This unlikely suggestion led to the inclusion of *hǣmed* in Smith 1956, and it was then taken as an established place-name element by J. McN. Dodgson, who cited it in EPNS CHE survey (Part 1, p. 78) as the first element of Henbury, with the observation that in Presteigne and Henbury 'there may be indelicate connotations'. This ghost element should be eliminated once and for all from place-name studies.

A correct account of the 'border meadow' at Presteigne was given in Gelling 1984 (p. 250), but Kinsham was not included in that discussion. In Coplestone Crow 1989 (pp. 117-18) Presteigne and Kinsham are discussed together, and a full range of spellings is supplied for Kinsham, showing that it is certainly 'the king's portion of the border meadow'. There are traces of the mistaken *-in-* for *-m-*, giving some spellings like *Kyngesheinde* 1292, but the ultimate development is due to the dropping of the final syllable in a three-syllable name, a phenomenon characteristic of the Welsh borderlands. This has played a part in the development of *Presthemed* to Presteigne, but in that name the *-in-* corruption has also persisted.

wisc, *wisse 'marshy meadow'. The first form is recorded in a charter of A.D. 898, which grants land at Farleigh KNT, specifying that six acres of meadow on the *miclan wisce* belong to the estate. This, together with the survival of the word as *wish* in SSX dialect, establishes the meaning. Place-name evidence suggests the existence of an OE **wisse*, either a variant form or a related word, with the same meaning. Names with spellings from *wisc* and **wisse* are treated together here.

wisc is primarily a southern place-name term, but it certainly occurs in Cranwich NFK and there are two instances in CHE. **wisse* is probably found in Hautbois NFK and in Barwise WML, so the terms were widely available, though sparingly used. Sledwich near Barnard's Castle DRH is another possible north-country instance.

wisc is best evidenced in SSX, where EPNS survey (p. 562) says it is 'very common indeed' in field-names. There are several simplex minor names in SSX, such as The Wish in East Dean and Wish Tower in Eastbourne. **wisse* is probably the first element of the major name Wisborough (**beorg**). As generic it occurs in the minor names Glydwish in Burwash, with *glede* 'kite', and Buckwish in Henfield, with *bōc* 'beech tree'. Buckwish Fm is situated at the edge of a large marshy area which has Rye Fm ('at the island') at the centre, and this conjunction offers good examples of the meanings of both *wisc* and īeg.

There is a recurrent compound of *wisc* and lēah which must be considered an appellative, using lēah in its late OE sense and meaning something like 'marshy meadow pasture'. This is the source of Whistley BRK, Westley Fm in Mamhead DEV, Whistley Fm in Syresham NTP, Wisley SUR, Wistlers Wood in Woldingham SUR and Whistley Fm in Potterne WLT. Wistley Hill in Coberley GLO is *Uuisleag* in an original charter of 759, so may contain **wisse*. There is a charter boundary mark *wiscleageat* in HMP, and field-names from *wisclēah* have been noted in CHE and WLT.

The major names Cranwich NFK, Dulwich GTL and Hautbois (earlier *Hobwiss*) NFK have as qualifiers 'crane', 'dill' and **hobb* 'tussock of grass'. Hautbois is probably from **wisse*, and this word is the likeliest candidate for first element of Wisbech CAM (bæc). There are a few minor names (in addition to those already mentioned) which have *wisc* or **wisse* as first element: these include Wiscombe Park in Southleigh DEV and Wishanger in Miserden GLO.

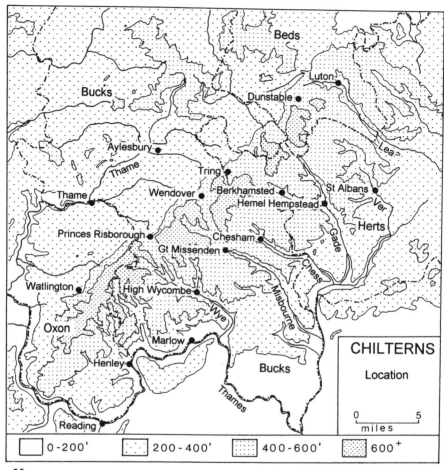

50

Appendix: the Chilterns, a Case Study

Ann Cole

Previous chapters have shown that the Anglo-Saxons had a wide range of landscape terms in their place-name-forming vocabulary, each with a very precise meaning. The way in which these terms were used in practice can be illustrated by looking at the toponymy of the Chilterns southwest of the Lea Valley and their immediate surroundings. To understand the choice of terms, the landscape must be visualised as the Anglo-Saxons originally saw it; first the physical landscape and then such human imprints as the Anglo-Saxons' predecessors may have left.

The massif called the Chilterns is a chalk escarpment running northeast–southwest. The scarp slope on the northwest side rises steeply for some 400ft to the scarp crest, the highest points of which are over 800 feet OD. Thence the land dips gently southeast towards the Lower Thames and London Basin. Cut into the scarp slope is a series of short, rounded valleys (coombes), while the dip slope is dissected by a series of long dry valleys, each with a gentle gradient, some with tributary valleys, and a few which cut right through the escarpment as wind gaps. The interfluves between these valleys vary in extent: in the southern part, for instance, they are wide plateau-like features, whereas northwest of Chesham they have been reduced to a series of narrow ridges.

Some account of the geology is necessary for understanding the conditions faced by the early settlers of the region, particularly with regard to water supply and soils. The Lower Chalk forms a gently rising belt of country along the foot of the escarpment in the Vale of Aylesbury. The Middle and Upper Chalk form outcrops on the scarp slope and along the sides of the dip-slope valleys. Elsewhere on the hills there are extensive areas of superficial deposits. On the higher parts of the scarp crest and interfluves is the Clay with Flints producing a deeper, if stonier, soil than is found on the Chalk. It is less liable to dry out and supports a different plant community. Lower down the dip slope but still on the interfluves are deposits of Older River Gravels (formerly called Plateau Gravel), and soils here are very stony and free draining. In addition there are small patches of Reading Beds (sands and clays) on high areas, and two very small outliers of London Clay at Nettlebed and Lane End. On the valley floors a narrow strip of Coombe Rock occurs. This is material which slumped off the valley sides and is mainly chalk debris. Beneath the Chalk and outcropping to the northwest in the Vale of

APPENDIX: THE CHILTERNS

Aylesbury is a belt of Upper Greensand, and beneath that the Gault Clay, and here the countryside takes on a very different aspect. To the south and east along the Thames Valley are deposits of river gravels and alluvium making up terraces and a flood plain, a strong contrast to the Chiltern chalklands.

It is important to note that although the Chilterns are a range of chalk hills the surface is far from uniform. Higher, flatter parts are covered by Clay with Flints and the Chalk is most evident on the slopes. This produces subtle changes in the soils and natural vegetation and has a profound effect on the hydrology and hence the availability of water for man and beast.

Since the Chalk is permeable rainfall infiltrates very readily and moves down to the water table. This rises steadily in the winter months, reaching a peak in the spring, and falls to its lowest level in the autumn. Springs occur where the water table intersects the land surface. There is a marked spring line along the scarp foot at the junction of the Chalk with the Upper Greensand and Gault, giving rise to streams which flow towards the Thame. These are reliable springs with a good flow of water throughout the year. Other springs occur partway down the dip-slope valleys. The lowest of these are perennial and feed permanently flowing streams. Others break out higher up their valleys in springtime only and nourish a temporary stream or winterbourne. Thus only the lower reaches of the valleys have a perennial water-supply, such as the rivers Wye, Misbourne, Chess, Gade, Ver and Lea. In recent decades even these have had their flow severely reduced because so much ground water has been extracted for London's use. Some of the springs no longer flow. Elsewhere in the Chilterns water is often scarce. On the Chalk there are no ditches round the fields and no network of tributaries to feed the rivers, but on the Clay with Flints and the London Clay damp hollows occur and puddles form readily along trackways, encouraging man to construct ponds in many parts of the hills. Water-supply in small quantities is not as great a problem in the Chilterns as in some other chalk downlands.

The human impact on the Chilterns in pre-Saxon times was limited (Fig. 51). Most of the barrows occur in eye-catching positions on the scarp crest, as at Ivinghoe Beacon, and Iron Age hill-forts are usually on prominent points, as at Maiden's Bower, Ivinghoe Beacon, Boddington and Pulpit Hills. Three more overlook the Thames, and two the Wye. Cholesbury and Wyfold are the only substantial hill forts on the plateau. There are a number of bank and ditch features: a series of short stretches runs from the Icknield Way to the scarp crest in the southern Chilterns, and there are three long stretches, one running straight from Mongewell to Nuffield, and two, each with distinct changes of alignment, in the plateau of the central Chilterns. These last are probably of Iron Age date, and they may have separated cattle from arable country: each is called Grim's Ditch. The Icknield Way, an important trade route, ran along the foot of the escarpment, linking Norfolk to the heart of southern England, and many of the scarp crest features are visible from it.

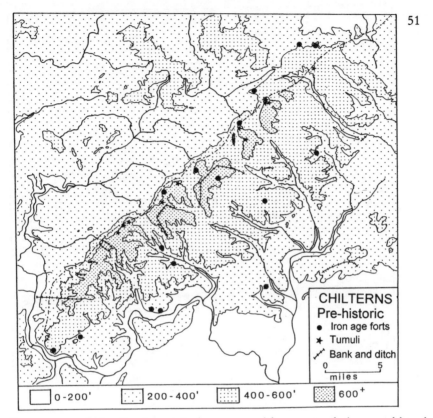

Large areas of the dip slope are, however, without any obvious prehistoric remains.

Fig. 52 shows that several Roman villa sites occur along the foot of the escarpment. Others occur in the dip-slope valleys, mainly towards the lower end where there were perennial streams, and particularly in the Chess and Gade valleys and near Verulamium, the Chilterns' first town. There would have been open farmland around these settlements. Two main roads, Akeman Street and Watling Street, link Verulamium to the English Midlands, crossing the hills. A small town, *Durocobrivis* (Dunstable), grew up at the junction of Watling Street and the Icknield Way. Numerous other minor roads developed near Verulamium and on the margins of the area, including one from Dorchester across the hills to a Roman bridge at Henley. At some early stage the track developed which is known in part as Knightsbridge Lane, running from Henley through Assendon to Pyrton and eventually to Oxford. Even by the end of the Roman period, however, man had left little imprint in terms of lasting features such as roads and substantial buildings in the heart of the Chilterns; he was merely nibbling at the edges. But there must surely have been Romano-British peasants living in the hills, even if their traces are

APPENDIX: THE CHILTERNS

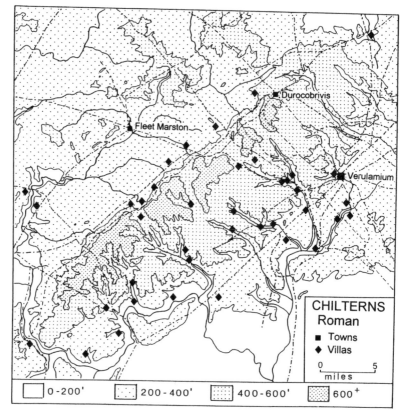

52

invisible to us, before the arrival of the Saxons with their different language and new place-name-forming vocabulary. Chiltern place-names contain very few references to man-made features of pre-Saxon origin, and there are not many Old English habitative terms. There is correspondingly a great wealth of topographical terms.

In the Chilterns it is the valleys which offer the most desirable settlement-sites. The contrast between the long dip-slope valleys and the short scarp-slope coombes is reflected in the use of **denu** and **cumb** as shown on Fig. 21. Almost all the **denu** names are in the dip-slope valleys, and the majority are in the upper stretches which lack flowing water. Great and Little Gaddesden are on either side of the source of the Gade near Milebarn. Great Missenden is close to the source of the Misbourne, and Little Missenden is a short distance downstream. The valleys at Great and Little Hampden, Dunsden, Nettleden and Frithsden are dry. Harpsden, although near the mouth of its long valley, is where springs emerge and flow sufficiently to maintain a short winterbourne in about one year in four at the present day. The long Assendon valley has three settlements of that name, Lower, Middle and Upper, the last a farm in Stonor village. In exceptionally wet years a bourne rises between Middle and Upper Assendon. It used to flow along the Fairmile to reach the Thames at Henley, but is now hidden from view. Hambleden and Yewden are in a valley where the bourne flows more often, but not now every year. The qualifier of Hambleden, **hamol*, means 'maimed' or 'misshapen' referring to the sharp bend near Fingest (Fig. 21) and/or the irregular valley sides: it contrasts with other *denu*s, which are straight or gently curving. Where the scarp slope is longer and less steep valleys, sufficiently long to be called **denu**, can develop, examples being Gangsdown, Dean and Ipsden. The last, 'upland valley', can be seen leading into the heart of the hills from the Icknield Way which crosses the parish near the church.

There are two places where the landform and the name appear to be mismatched – an unusual phenomenon anywhere. One is Checkendon, which lies in a slight depression in rolling upland, about one mile southeast of a magnificent **denu**-shaped valley which is nameless but contains Ouseley Barn and Bottom Farm. The second is Elvendon Priory, a modern building on the site of an old farm in an excellent **denu**-type valley, yet with a name whose early spellings suggest dūn, not **denu**. In the great majority of cases **denu** in the Chilterns is used of the long stretches of valley usually on the dip slope, often near or above the highest point at which springs rise and winterbournes flow, and which are large enough for several settlements of the same name. For these settlements to have flourished the problem of water supply must have been solved. Possibly with a higher water table a shallow well was sufficient to tap supplies.

The other valley term widely used in the Chilterns is **cumb** (Fig. 21). The scarp slope provided four steep-sided bowl-shaped valleys which the Anglo-Saxons named Coombe (near Ellesborough), Huntercombe, Swyncombe and Watcombe. Present day Watcombe Manor is in Watlington, a mile from the scarp face, but its forerunner was in the present day hamlet of Howe at the mouth of a fine curving **cumb** which had Watcombe Wood, now Howe Wood, on its steep slopes. Swyncombe's little Saxon church and Swyncombe House nestle in a hollow among the beechwoods: this is a much shrunken settlement illustrating the cut-off, disadvantageous situations of settlements in **cumb** sites. Huntercombe is at the head of Gangsdown – a **denu** name, despite the modern form. A long **denu**-type valley may well end in a **cumb**-like feature, although this is the only one in the Chilterns to have a **cumb** settlement at its head. Occasionally **cumb** is used to describe bowl-shaped embayments in the middle of dip slope **denu**s: Warmscombe Firs and Lane represent one such in the Assendon valley, and Coombe's Farm is another in the Hughenden valley. Other *combe* names appear on maps of the Chilterns in **cumb**-type valleys, but without early spellings it is not certain that these are old. It may be that **cumb** is under-represented on Fig. 21.

APPENDIX: THE CHILTERNS

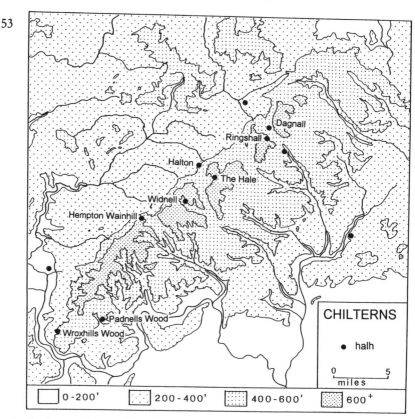

53

halh is another valley term occurring in the Chilterns where the topography is appropriate. Referring to a wider, shallower, less distinct valley shape than **cumb**, it is used where wind-gaps pierce the escarpment producing funnel-shaped embayments along the line of the scarp foot (Fig. 53). Dagnall lies at the centre of the funnel which continues on the dip slopes as the Gade Valley. Nearby Ringshall Farm overlooks a hollow, not bowl-shaped enough to be a **cumb**, in the side of the main **halh**. The pattern is repeated at Halton which lies to one side of the funnel-shaped gap leading through to the Missenden valley, where the prime site, Wendover, has a Celtic name. The Hale is in a substantial, irregularly shaped side valley. The next wind-gap, at Princes Risborough, has two *dūn*-named settlements in it, the hills being more definite landmarks by which to identify the settlements than a **halh**; but a small side valley of irregular shape at Hempton Wainhill has a **halh** name. On the dip slope **halh** is sometimes used for the head of a **denu** if it is sufficiently open and gently sloping. Widnell in Ellesborough occurs at the head of the Hampden **denu**, which is wider and less steep than those of most such valleys, so that **halh** was felt to be more appropriate than **cumb**. Padnells Wood in Rotherfield is similarly placed at the head of a dip-slope valley. Wroxhills

Wood is on the slope around a small funnel-shaped hollow opening onto the Thames near Goring. Gales channelled up the valley have brought down a tangle of beech trees at its head, but in more clement conditions wind may provide a good updraught for the birds of prey which gave Wroxhills its name.

There is a single example of **slæd**, in Exlade Street. This is quite a deep, steep-sided valley, although its shape is now masked by the embankment of the A4074 which cuts across it. The hamlet is on chalk with a veneer of Boulder Clay, but there is no stream, the valley floor is not particularly moist, and no gullies lead into it, so it is not clear why **slæd** was chosen in preference to **denu** here.

A very common modern valley term in the Chilterns is *bottom*. Stretches of the floors of many valleys are called -bottom, and there are examples of Bottom Farm and Bottom House Farm in some of them. Others are named from the village in whose territory they lie, as in Pednor and Flaunden Bottom, others have descriptive qualifiers, as in Stony Bottom or, more fancifully, Pigtrough Bottom and Drunkards Bottom. These features do not have the specialised shape of the early-recorded bottoms discussed in Chapter 4. The age of the names is unknown, but it seems likely that they are comparatively modern. Lock's Bottom and Bottom Waltons in Burnham are shown on Fig. 17 together with **botm** names likely to be of pre-Conquest origin, but it is possible that they should be included in this grouping in spite of the Bottom element being recorded in the 13th century.

The word **dell** is used in the Chilterns to describe pits or quarries, and it usually occurs in the names of woods and farms which appear in the records in the late 13th and the 14th centuries. Although many were undoubtedly used merely to obtain chalk for liming the fields, a number bear such a close relationship to Roman roads that it is probable that they provided material for road construction in the first place and for road maintenance as well as for liming the fields later. The Viatores have drawn attention to a series of 17 pits beside Margary 21a (St Albans - Welwyn - Braughing) which they suggest were used in the construction of the road. This road, at one point, curves round the rim of a large pit opposite Chalkdell Farm in Sandridge. Two of the other four Chalkdells, Chalkdell Wood in St Peters and Chalkdell Wood in Knebworth, are adjacent to Roman roads, as are Waterdale Farm, Bride Hall (earlier *Bridelle*) and The Dell in Wheathampstead. This concurs well with Kitson's findings, mentioned in Chapter 4.

Other valley terms are very rare or absent in the Chilterns. With no ravines or gorges terms such as **corf**, **clof** and **clōh** were not needed, and **hop** was not used as Chiltern valleys do not have the remoteness which that term implies.

Hill terms are used just as selectively. Fig. 32 shows the large numbers of **dūns** in the lowlands north and west of the escarpment, but only six within the hills. The three at the mouth of the Princes Risborough gap, Horsenden, Saunderton and Calverton, are best considered as outliers of **dūn** country. Only one of the six, Averingdown, a lost village represented by a church on a bump at the end of a long isolated ridge near High Wycombe, could be considered to have a typical **dūn** site. The others, Hayden and Bromsden Farms and Caddington, high up on the Chiltern plateau where the ground slopes gently away on three sides, and Bovingdon high up yet in a slight hollow, illustrate the use of **dūn** in its 'upland expanse' sense. Elvendon remains a puzzle.

Desborough in High Wycombe is the only **beorg** within the Chilterns: it refers to a hill crowned by an Iron Age fort. Ellesborough and Edlesborough refer to small rounded hills at the foot of the escarpment, both crowned by churches: the one at Edlesborough with its scarped churchyard is shown on Fig. 28a. Risborough (Monks and Princes) refers to Pulpit and Whiteleaf Hills, a rounded pair clearly visible from the Vale.

The specialised term for a burial mound, **hlāw**, occurs in two major names: Taplow with its great tumulus in the old churchyard commanding a fine view over the Thames, and Bledlow at the scarp foot, overlooked by the tumuli on Bledlow Cop and Wain Hill, also fine viewpoints. Whittles Farm in Mapledurham, overlooking the Goring gap, is the only other instance of **hlāw** in the hills. No early spellings are available for Bledlow Cop, but Turville Hill was formerly known as Copson Hill, and this very narrow, steep-sided hill is a possible example of a **copp**.

hlith occurs in The Lyde, Bledlow, a deep hollow with a powerful spring, and at West Leith, Tring, where the farm is tucked into a little valley at the scarp foot.

hyll is widely used but in minor names of only local significance. Some of the Hill and Hillend Farms first evidenced in the late 13th century appear on the 1:25,000 maps, but many **hyll** names are too minor even for this distinction. The exceptions are Coleshill, on a hilltop, and Pishill, whose tiny church looks down on the cluster of houses around the road along the valley floor.

A more important hill-term in the Chilterns is **hrycg**, which is shown on Fig. 54. Hawridge, Asheridge and Chartridge lie strung out along the narrow steep-sided ridges which converge on Chesham, and a second Ashridge, near Gaddesden, is similarly placed, as are Bledlow Ridge, Andridge, Towerage and another Ashridge on ridges which converge on High Wycombe. The hamlets of Badgemore (earlier *Baggeridge*) and Witheridge, and the woods called Shambridge, Groveridge and Kildridge in the southern Chilterns are on smaller, less steep-sided spurs; this countryside is less deeply dissected, but the landforms are similar. The **hrycg** names are all on the dip slope. The corresponding scarp face landforms are described by the terms **hōh** and **ōra**. Some slopes are steep enough for the term **clif**, but these are often part of

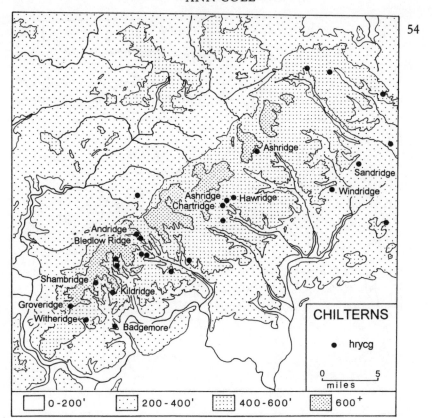

features called **hōh** and **ōra**, so **clif** is not used here. The scarp face is rarely as long and straight as those near Clevancy or Swallowcliffe in WLT, where the steepness of the slope is more prominent than the spurs or coombes along it. There is, however, an example of a river **clif** where the Thames has cut up hard against a chalk bank at Cliveden, from whose gardens there is a dramatic drop to the river.

Sharpenhoe, Houghton Regis, Totternhoe and Ivinghoe are four classic **hōh**s on the northern half of the scarp slope (Fig. 55). At each there is a dramatic, eye-catching spur projecting into the plain. Fig. 36 shows Ivinghoe Beacon from the north-north-east. Viewed from other directions it exhibits a more irregular profile, like Totternhoe and Sharpenhoe, but still with the characteristic gentle slope up to the summit and a steeper, concave slope down to the Vale of Aylesbury. On the dip slope in the northern Chilterns **hōh** is used in its less exact sense to denote blunt spurs along the sides of the valleys, as at Luton Hoo and The Hoo in Great Gaddesden.

The alternative hill-spur term, **ōra**, is more characteristic of the southern half of the scarp slope, where level-topped headlands with the typically rounded shoulders jut out into the Vale. The Chinnor **ōra** is shown on Fig.

APPENDIX: THE CHILTERNS

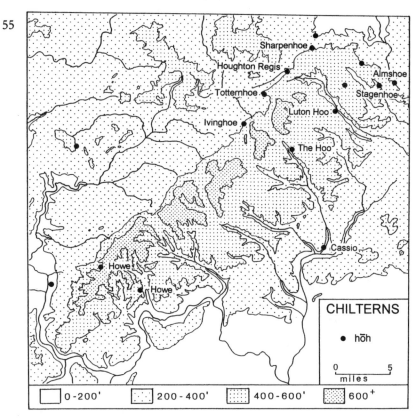

40, and all the Chiltern ōras are marked on Fig. 56. Aldbury Nowers and the ridge which was called *Buckleshore* near Whipsnade have a similar shape. The characteristic ōra shape of Bixmoor can be seen well from the north and west, and that of Chalk Wood (earlier *Chalkore*) from the south. Stonor, which has now supplanted Upper Assendon as a village name, refers to the hill north of Stonor House which is shown on Fig. 40. Launder's Farm can be thought of as the other end of the Stonor feature; it does not have a separate ōra. There are five dip-slope ōras in the heart of the Chilterns: Courns Wood, Denner Hill, Honor End Farm, Ballinger and Pednor, all in wooded countryside, deeply dissected into hills and valleys, which means that their outlines are difficult to see and that there are few distant views of them. This suggests that the countryside was more open at the time of name-giving. Their function as 'signposts' is discussed *infra*.

The hills and valleys are the most obvious features of the Chilterns to anyone passing through, but for earlier inhabitants water was of crucial importance, particularly in an area of permeable rocks where surface water is rarely found. Many Chiltern place-names comment on water sources.

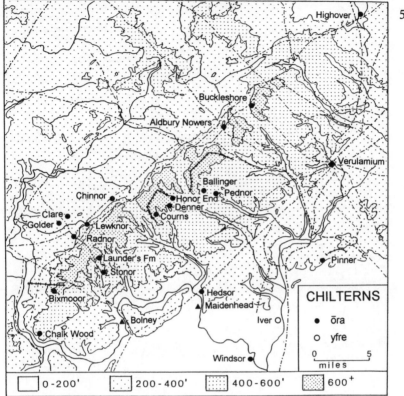

Three different terms for springs occur (Fig. 57), the commonest being **well**. The scarp foot has a series of good, reliable springs feeding the streams tributary to the Thame and Thames. Some are thrown out by the Chalk Marl and Totternhoe Stone, and a few lie further into the Vale of Aylesbury, thrown out by the Upper Greensand and Gault. Many of the spring-line settlements have **well** names: Mongewell, Britwell Salome, Brightwell Baldwin, Crowell, Adwell, Miswell, Sewell and Bidwell. Two exceptionally prolific springs merit a special term, at Ewelme, an æwylm which is the source of a river which rises in a pool below the church, flows through an erstwhile trout farm, and then into old cress beds; and, further north, at Oughton (earlier *Alton*) Head, where the spring is a source of the river Hiz. This is an example of æwyll contracted to Al- as in Alton DOR, HMP, WLT. Both Ewelme and Oughton Head are close to ancient routes and were no doubt a welcome source of clear, cold water to hot, thirsty travellers. Some spring-line settlements, such as Chinnor, Lewknor, Shirburn and Wendover, do not have **well** names.

Dip-slope springs are less frequently noted in settlement-names because they are less regular in flow. Extraction of ground water to supply the needs

APPENDIX: THE CHILTERNS

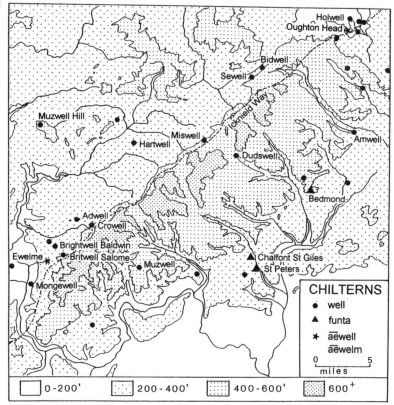

57

of London has resulted in a falling water table for over a hundred years, so that formerly reliable springs and the upper reaches of the streams are now dry. The Bulbourne normally rose at Dudswell, and three-quarters of a mile above that in very wet years. The Misbourne's highest perennial spring was at Great Missenden Abbey, and traces of a watercourse extend to Mobwell, for which no early spellings are available. Amwell in the Lea valley used to have good springs which have now run dry.

The springs at Muzwell Farm, Lane End, are not connected with the chalk aquifer. Here a miniature rift valley has down-faulted the Reading Beds and London Clay into a hollow measuring some half a mile each way at Moorend Common, trapping rainfall on an impermeable layer. Water oozes out of the ground in hollows well grown with juncus and willow scrub, squelchy underfoot. It is drained by a small deep-cut stream flowing into a swallet. Muzwell Farm is built on a steep rise overlooking seeps supplying a second stream. Its qualifier, **mēos**, is used in the sense 'boggy ground' here. At Miswell Farm near Tring, another *mēos-well*, a large pond occupies the site of the springs and one cannot tell whether this was once a boggy area too. The compound occurs again in Muswell Hill near Brill, which is awash with

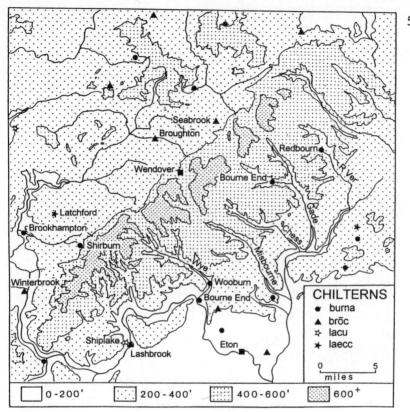

springs and seeps. The first element of Misbourne is probably a derivative of **mēos**. Between Great and Little Missenden the stream is bordered by wet ground growing with glyceria, where it is too squelchy to reach the bourne dry-shod.

The third spring term used in the Chilterns is **funta**. At Bedmond the water source was probably the spring near the school, whose water may have been made available in a trough, the **byden** of the name. Chalfont, 'calves' spring', is in the Missenden valley. Both of these places are on a Roman road leading to Verulamium and probably served to refresh travellers. Fig. 6 shows that other examples of **funta** occur close to Roman roads. The use of this special Latin-derived term at Bedmond and Chalfont may imply that the Anglo-Saxons knew the area before some Roman structures round the springs had fallen into disrepair.

Terms for streams are little used because of the lack of running water (Fig. 58). Only **burna** is used within the Chilterns since this area is within the zone where a distinction was made on geological grounds between **burna** and **brōc**. The clear, spring-fed, often stony-floored streams with their seasonal regimes are classic **burna**s. They are found in the lower stretches of the

APPENDIX: THE CHILTERNS

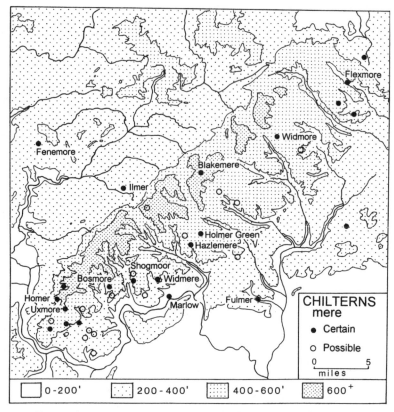

59

dip-slope valleys where streams are perennial, down-valley from the **denu** names. Redbourn is on the river Ver near its source. Bourne End is on the Bulbourn near its confluence with river Gade. Wooburn (from *wōh* 'crooked') lies on the great curve of the river Wye before its confluence with the Thames at a second Bourne End. Not only was the Ver once called the Redbourne, but the Chess was earlier called *Pittlesburne* and the Misbourne retains an old **burna** name, so it is clear that these chalk streams were regarded as **burna**s. Shirburn, 'bright stream', is a scarp foot spring-line settlement. Wendover is beside a clear stream flowing over white pebbles which arises from a scarp foot spring. This is a British name meaning 'white river', the Celtic equivalent of Shirburn.

Streams occasionally flow in the Hambleden, Assendon and Harpsden valleys, but they are seasonal and in some years do not rise at all. If they did not rise every year in Saxon times they would fail to qualify as winterbournes. The chalk aquifer of the Chilterns is too limited for winterbournes to develop, which differentiates this from the areas of Dorset and Salisbury Plain, where most of the Winterbourne place-names occur. The absence of regular perennial streams is the reason why villages in the Chiltern valleys have **denu** rather than **burna** names.

Chiltern place-names do not contain **brōc**, the term for a muddy, rain-fed stream, but once the Chalk is left behind, and streams flow over the Gault Clay or alluvium, **brōc** names are met with. Brookhampton, Broughton and Seabrook are on clay, and Winterbrook is on alluvium. At Lashbrook, a small backwater of the Thames, **læcc** 'bog' is combined with **brōc**, reinforcing the reference to the swampy, overgrown nature of the watercourse. Nearby the term **lacu** occurs in Shiplake, the sheep-washing stream, referring to one of the side channels on the Thames floodplain. Eton, 'settlement on a river', lies beyond the hills on the river Thames.

The pond term, **mere**, is very frequent in the Chilterns, but because only very small settlements, usually farms, are associated with these ponds the names are not all listed in the EPNS survey of OXF, BUC and HRT. Additionally it is not always easy to determine whether they are derived from **mere** 'pond', **(ge)mǣre** 'boundary' or **mōr** 'marsh'. A number of farms with fine old ponds and **mere**-like names have been included here on the assumption that they are most likely to contain **mere**. Fig. 59 shows that **mere** is uncommon in the lowlands beyond the Chilterns, where only Marlow, Fulmer, Ilmer and Fenemore's Farm are noted. Within the hills **mere** is very frequent on the interfluves, particularly in the southern Chilterns where these are wide, and where perennial streams are absent. Most **mere**s lie below the 600 foot contour. Their sites have been carefully chosen, first for the impermeability of the soil. One confirmed and two possible examples are on the Older River Gravels (Plateau Gravel), although this is so free-draining that it is usually avoided as a pond site. Only two are on the Reading Beds, sands and clays, with better water-holding properties, because the outcrops are so limited. The great majority, certainly 11 and up to 17, are wholly on Clay with Flints, which can be made relatively watertight. A further four, perhaps six, lie on the boundary of Clay with Flints and a more permeable rock. Where possible the pond is sited in a natural hollow, often enlarged by digging it out and using the spoil to build up a bank on the downhill side. To maintain the pond the overland flow after heavy rain is supplemented by channelling into it the run-off from the compacted soil of adjacent tracks, water-supply and access being thus neatly combined. The presence of such a pond allowed small permanent settlements to grow up, like Homer, Uxmore, Shogmoor and Bosmore Farms. There are at least seven examples of Widmer, -more, -mere, which could mean either 'wide' or 'willow pond'; the latter is more likely as willow (as opposed to sallow) trees are not common on chalk hills and could therefore be a useful identifying characteristic. The difficulty of securing a reliable water-supply on the Chiltern plateau is illustrated by the need to ration drinking water and postpone washdays indefinitely during dry

APPENDIX: THE CHILTERNS

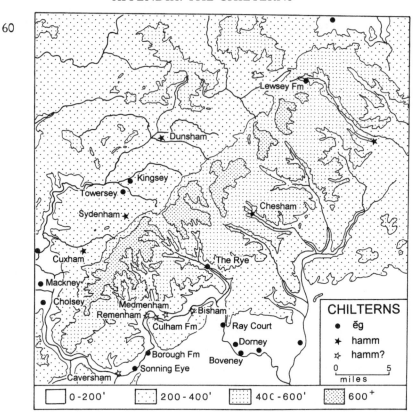

60

summers in Stoke Row, half a mile from Uxmore, until 1864 when a well 368 feet deep was sunk as a gift of the Maharajah of Benares.

Other water terms discussed in Chapter 1 are absent from the Chilterns. It is clear that those which do occur were chosen carefully and applied consistently. The unusual frequency of **mere** testifies to the vital importance of water supply in an area so lacking in surface water. There is a corresponding rarity of terms for wet ground discussed in Chapter 2.

The term **hamm** occurs only once in the hills, at Chesham, at the meeting point of several valleys where the river Chess rises. The level valley floor just below the town is well watered and lusher than the surrounding hills. This is an example of **hamm** meaning 'land hemmed in higher ground', but also with the connotations of good pasture found in the 'river meadow' sense. Beside the Thames, each in a river bend enclosing flood-plain meadows, lie several fine **hamm** candidates. Medmenham is thought to mean 'middle hamm', although the spellings do not point conclusively to **hamm**. The name implies that there are other **hamms** on either side. To the east this would be Bisham, to the west it is likely to be Remenham or perhaps Culham Farm.

304

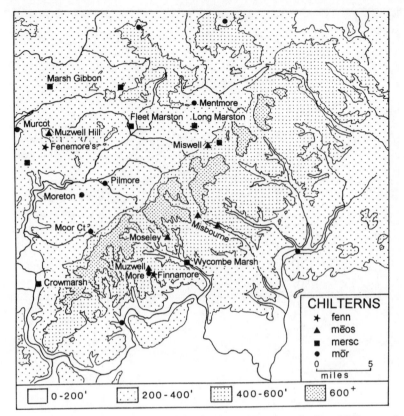

Caversham is another possible **hamm**. North of the Chilterns **hamm** reappears in the Middle Thames and Thame valleys.

The distribution of ēg is very similar (Fig. 60). The only two examples in the hills are The Rye in High Wycombe, a patch of undeveloped land between two channels of the river Wye in flat, flood-prone valley land, and Lewsey Farm near Luton, near the headwaters of the river Lea in a rather marshy area. Like **hamm**, ēg occurs beside the Thames, at Boveney, Dorney, Ray Court in Maidenhead (still plagued by flooding), Borough Farm (from *burh* and ēg) and Sonning Eye, where flood waters can lap up the riverside gardens almost to the cottage doors. Cholsey, Mackney, Towersey and Kingsey continue the series north of the Goring Gap.

Wycombe Marsh, just downstream from The Rye, is the only occurrence of **mersc** in the hills; otherwise **mersc** is used of sites in the well-watered lowlands to the north west. It is much more difficult to ascertain the distribution of **mōr** because of the difficulty of distinguishing it from **mere**. There are no major names, and such minor names as may contain **mōr** are mostly first evidenced after 1500. However, references to the chapel of St Mary le More near Moor Farm half a mile south of Lane End go back to the

APPENDIX: THE CHILTERNS

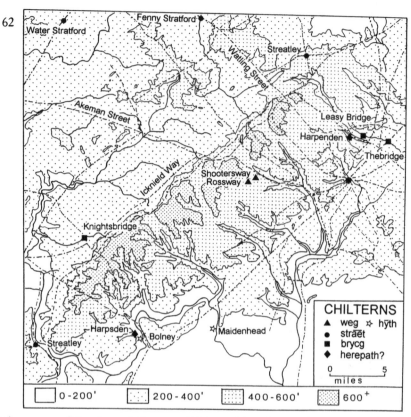

13th century (Hepple and Doggett 1992, p. 78) and must be connected with the cluster of Moor names in the wet-floored little rift valley described in connection with Muzwell *supra*. Finnamore Farm (now Bluey's Farm), the only name containing *fennig*, must also be related to this wet patch (Fig. 61).

The roads and tracks of the Chilterns are shown on Fig. 62. They include secondary and tertiary Roman roads around Verulamium identified by the Viatores, some of which are likely to be Romanised ridgeways, and the course of some may be open to debate.

References to roads are infrequent in Chiltern place-names (Fig. 62). The Roman-road term **strǣt** occurs in Streatley BRK, on the Dorchester to Silchester road as it passes through the Goring Gap, and in Streatley BDF, on Margary 170b (Limbury – Marston Moretaine), high on the Chiltern scarp, just before the steep descent to Sharpenhoe. It is also evidenced in the names of the two major Roman roads, Akeman Street and Watling Street.

The general term **weg** occurs only in two minor settlement-names. Rossway (from **roth** 'clearing') is probably the minor road from Chesham to near Wigginton which is followed by a parish boundary for most of its length. It is joined by Shootersway (from *scēacere* 'robber') before descending the

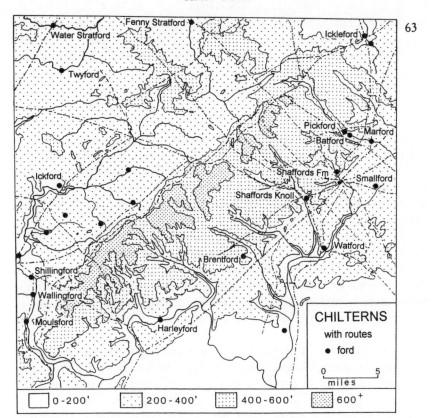

Fig. 63

scarp to join Akeman Street and then the Icknield Way. The Icknield Way, a long-distance prehistoric track linking Dorset to the north coast of Norfolk, has been referred to as a **weg** for over a thousand years. Kitson (forthcoming) suggests that it means 'way leading to the Iceni' (the Iron Age tribe inhabiting north-west Norfolk).

Harpenden and Harpsden (discussed in Chapter 4) may incorporate the term *herepæth* (discussed under **pæth**). There is a minor Roman road along the side of the Harpenden valley. The Harpsden valley is the natural route from Rotherfield ('cattle pasture') to Bolney ('bullocks' hythe'), and this might have been made use of by an early road. Other terms for tracks are absent from the Chilterns, and terms for river crossings are equally rare because with so few perennial streams the need for them did not arise.

Fig. 63 shows that **ford**, the main 'river-crossing' term, occurs fairly frequently along the Thames and its tributaries and occasionally on the perennial stretches of the Chiltern streams. Pickford and Batford are crossings of the Lea. Nearby Marford is where Margary 221 crosses the same river heading for Ayot St Lawrence; this is the only certain reference to a Roman ford. Shafford Farm refers to a crossing of the river Ver just north of

APPENDIX: THE CHILTERNS

Verulamium, but as Watling Street does not cross the Ver at this point the ford can only have been of local importance. Shafford's Knoll (a different name) in Abbots Langley is close to the point where Margary 163a crosses the Gade, and this may refer to the Roman ford. The original site of Brentford in Coleshill is now lost. As the Chiltern rivers are comparatively easy to cross there would have been little need to concentrate traffic on the fords used in Roman times or to maintain them. Equally there was no pressing need to build bridges, so there are few bridge names, just a Bridge Farm in St Stephens and two by the river Lea, namely Leasy Bridge near Wheathampstead and *Thebridge* ('thief bridge'), now Waterend Farm. There is now no bridge at Waterend Farm, but there is a ford where the Roman road crosses the river. Knightsbridge Lane takes its name from a bridge over a small stream in Pyrton.

The Chiltern rivers are too small for there to be any examples of **hȳth** or **stæth**, but hȳth occurs on the Thames at Maidenhead close to where the Verulamium – Silchester road crosses the river. This could act as a trans-shipment point or a base for a ferry. Bolney near Henley may have served the same function, because Knightsbridge Lane coming from Hythe Bridge in Oxford to Henley was an alternative overland route to the roundabout course down the Thames.

In the discussion of **ofer** in Chapter 4 it was noted that the landform in question is frequently found along routeways, the Droitwich saltways being instanced; and in the discussion of ōra it was suggested that sailors used this landform to help determine their landfall in many places round the coasts of Kent, Sussex and Hampshire. The Chiltern ōras also make good landmarks, and their position relative to routeways is shown on Fig. 56. On the Dorchester to Silchester road, views of Bixmoor from the north or Chalk Wood from the south indicate the approach to the Goring Gap and the important crossing with the Icknield Way. They also indicate the Way's important crossing over the Thames at Goring and Streatley. Knightsbridge Lane (Oxford–Henley) is well supplied with ōra/ofer landmarks, starting with Shotover near Oxford, from which can be seen the long ridge of the Chilterns with Chinnor and Lewknor at its foot, and the Greensand ridge on whose slopes stand Golder and Clare: all these names have ōra for generic. The feature at Clare is a modest ōra, but it comes into its own when it hides the Chilterns from view as the traveller approaches from the west. *Radnor*, now Pyrton, lies between the two ridges, and the original name referred to a small hill which marks the crossing of Knightsbridge Lane with a Roman road to Dorchester. This small and not-very-red-soiled hill is a poor example of an ōra, and this may account in part for the change of the settlement-name to Pyrton. Knightsbridge Lane climbs the slope to Christmas Common from which all these landmarks can be seen in the reverse direction. It then follows a ridge down into the Assendon valley, passing the Stonor ōra and Launder's

Farm as it does so. The Assendon valley, perhaps the **denu** of the pack asses, leads automatically to the Roman bridge in Henley and the hȳth at Bolney.

The traveller on the Icknield Way going northeast from Goring can chart his progress with a series of features called ōra and hōh. The Howe refers to Watlington Hill, Lewknor to Beacon Hill and Chinnor to The Cop; Aldbury Nowers survives as the name of the woodland at SP 9513, Ivinghoe's hōh is Ivinghoe Beacon, there is a lost *Buckleshore* in Studham which is probably the hill with the Whipsnade lion cut into its turf, and north of that there is Totternhoe. Travellers bound northeast or southwest would learn which landmark heralded the approach to Akeman Street, Watling Street or one of the minor Roman roads linking the Chilterns with the Vale of Aylesbury and crossing the Icknield Way. Features called ōra are quite often associated with lesser Roman roads or non-Romanised trackways rather than major Roman roads on which, since they were direct, well-paved routes, the traveller would be less in need of guiding landmarks. Aldbury Nowers and *Buckleshore* both have minor Roman roads crossing them. It is, however, difficult to see how the five dip-slope ōras could be used by travellers, because views are so limited and they are not near known routes. The only surviving pre-Saxon features of note nearby are the two Grim's Ditches. Honor End Farm is at one end of one ditch and Courns Wood is at the other, whilst Denner Hill is between the two. Pednor and Ballinger are in the same block of upland as the Grim's Ditch near Cholesbury. Pinner and Nower Hill MDX are by yet another Grim's Ditch. Hedsor, also on the dip slope, overlooks the point where the Verulamium – Silchester road crosses the Thames, and would be useful to the north-east-bound traveller, or perhaps to a boatman using the hȳth at Maidenhead.

The use of terms for woodland, ploughland and pasture can be better understood if the history of the Chiltern woodlands is summarised first.

In post-glacial times the Chilterns became thickly wooded with a variety of deciduous trees, including much more oak but less beech than today. Some clearance must have occurred in pre-Roman times, for instance on the scarp crest where barrows and hill forts are found, and on the dip slope in the vicinity of the Grim's Ditches and the Cholesbury and Wyfold hill forts. By the end of the Roman period the lower, well-watered ends of the valleys contained farmland, as evidenced by the presence of Roman villas, and Roman roads passed through the woodland. It is probable that small numbers of Romano-British people were living in clearings in the hills. At the coming of the Anglo-Saxons the Chilterns were still well wooded, but the woodland cover has waxed and waned through the centuries. Some woods have cores which probably go back to wildwood, as suggested by the survival of some rare plants. Individual woods may have increased or decreased in area from time to time. Some woods were destroyed and others created: those through which the Grim's Ditches run are not likely to be descended from wildwood.

APPENDIX: THE CHILTERNS

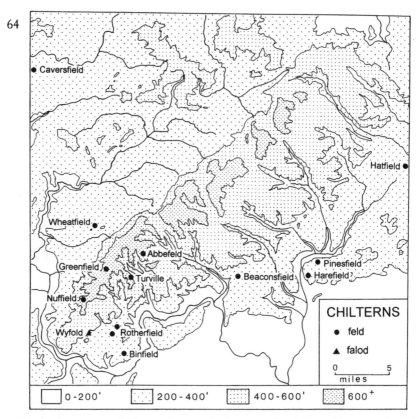

64

Both the area and the make-up of the woodland have fluctuated with man's needs. In Anglo-Saxon times the Chilterns were not, like the Weald, an area used wholly for woodland pasture: they also supplied timber and fuel to scarp-foot and valley-floor settlements. A number of religious houses were established in the Chilterns in the 12th and 13th centuries, and this fact, together with a general population increase resulted in assarting, an activity recorded in both documents and place-names. In medieval times, and increasingly in the 16th, 17th and 18th centuries, coppice wood was being supplied to London along the Thames for fuel. This was a valuable source of income which would have inhibited further clearance, especially in the south Chilterns, since coppice could produce more profit than crops; it also gave greater emphasis to careful woodland management. By the 19th century coal had supplanted wood as fuel, so, as the furniture industry developed, notably in High Wycombe, coppiced wood gave way to pure stands of beech which persist today (Hepple and Doggett 1992, Chs 5, 7, 10).

The earliest clearings are represented by the terms **feld** and **lēah**. Names containing **feld** in its 'rough, open, upland pasture' sense are shown on Fig. 64. The distribution is very uneven. Rotherfield, Binfield and Nuffield are on

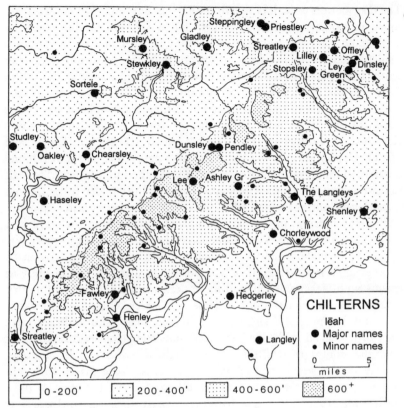

the southern part of the Chiltern plateau where it is at its flattest and least dissected. The name Turville, now that of a village in the Hambledon valley, probably referred originally to the upland now called Turville Heath. Greenfield was nearby on the scarp crest, and the lost *Abbefeld* was on the plateau roughly between Stokenchurch and Lane End (see Baines 1981 for a full discussion). The qualifiers of Nuffield, Turville and Binfield ('tough', 'dry' and 'bent grass') aptly describe the nature of this rough, upland pasture, and in the name Rotherfield there is a suggestion of the use to which it was put – cattle-rearing. The link between Rotherfield by way of the Harpsden valley to Bolney, the bulls' landing-place, has already been mentioned. Wyfold Grange is a further hint of cattle-rearing, as it probably refers to a cattle fold. It is noteworthy that place-name references to crops and domestic animals occur in the same general area as **feld** in the southern Chilterns, as at Swyncombe (swine), Studdridge (stud), Pishill (peas): they are absent from the central Chilterns but reappear north of the river Gade nearer the major **lēah** names, with Barley End, Studham, Revelend (earlier *Rytherfield End*), Bendish, or in the lower stretches of the valleys, with Chalfont and Wheathampstead.

APPENDIX: THE CHILTERNS

The distribution of **lēah** is shown on Fig. 65. Fawley, a DB estate, is still enveloped in woodland. Henley, though not in DB, became important by virtue of its position by the Thames. Dunsley and Pendley are at the scarp foot, so not strictly in the hills. Only north of the river Lea does **lēah** frequently form names of DB estates, now parishes, such as Stopsley, Offley and Lilley, in the gentler, earlier-cleared northern Chilterns. In the central Chilterns the parishes of Lee and Ashley Green are modern creations, of the 16th century and 1875 respectively. In the lower reaches of the rivers Chorleywood is another recently-created parish (1898), but King's and Abbot's Langley appear in DB. Over the greater part of the hills **lēah** is found in the names of minor settlements first recorded in the 13th century. The **lēah** names tend to occur on higher ground. They avoid the Older River Gravels of the southern Chilterns which are so infertile and lacking in water. Many of them represent clearings in the wooded uplands belonging to the scarp foot and valley floor settlements. Peasants concerned with the care of foraging pigs or the cutting of underwood for fuel and tool-fashioning would have made their homes in these clearings, if only seasonally at first, whereas in the Vale settlements with **lēah** names housed more permanent farming communities, many becoming DB estates, such as Haseley, Oakley, Chearsley, Stewkley, Mursley and Steppingley. The terms **feld** and **lēah** both indicate early clearance of woodland in the Chilterns, **feld** indicating large expanses of open land in the southern Chilterns while **lēah** describes small clearings, particularly in the woods of the central Chilterns. Some places with **feld** names were estates by the time of the Domesday Survey, but most of those with **lēah** names, though they may have been named when the word was at its height of popularity as a place-name-forming term between 750 and 950, rarely appear in the records before the 13th century. In the Vale, however, **feld** is less common than **lēah** as an element in estate names.

Other terms which the Anglo-Saxons applied to clearances for new settlement, such as **æcer** and **land**, do not occur with this sense in the Chilterns. When **land** occurs, sparingly, in the 13th and 14th centuries in minor names (there are three Newlands) it usually refers to enlargement of existing arable. In the Chilterns it was not a case of reclaiming land from marsh, heath or moorland, to which **æcer** and **land** usually refer, but from woodland. The rare element **ersc** was used once on the southeast margins of the region in the now lost *Oakhurst* in Shenley HRT. Clearances of the 13th and 14th centuries, made especially by the religious houses in the northern Chilterns, are reflected in the use of ME *breche* (discussed under **brēc**), as in Breach Wood in King's and St Paul's Walden and in Great Gaddesden, and The Brache in Harpenden.

There is an instance of **mǣd** in a cluster of names which vividly illustrates the contrast between the well watered Wye valley and the dry chalk uplands. The Rye (**ēg**), Wycombe Marsh and King's Mead all occur by the river in

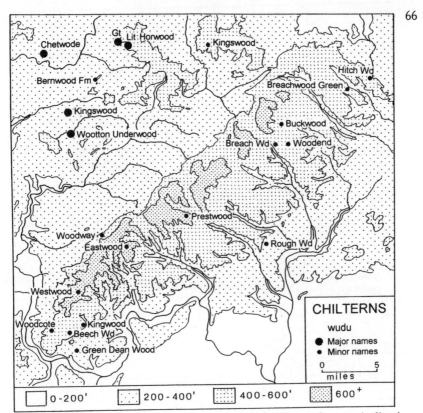

High Wycombe, a name which means 'at the dairy farms', indicating an appropriate use of this land. The term mǣd is, however, generally more appropriate to the Vale of Aylesbury, where it occurs in Meadle and Broadmead.

Some woodland terms other than lēah occur in the Chilterns, though rarely in major names. The district term **wald** is not used, though it would have seemed appropriate to these wooded uplands; this is perhaps because an earlier, probably pre-English name, Chiltern, survived. The term **wudu** occurs in the names of woods and in Woodway and Woodend Farms, but the only settlement of any size with a **wudu** name is the hamlet of Woodcote, first recorded in 1109, lying in woodland on the western edge of the Chiltern plateau above its mother settlement, South Stoke, on the Thames. There are 14th-century records which show the cottagers of Woodcote clearing the waste and selling the timber. Villages such as Kingwood Common and Prestwood are largely modern developments taking their names from the King's and Priest's Woods: Prestwood became a parish in 1849. In the Vale of Aylesbury, by contrast, the parishes of Chetwode, Great and Little Horwood and Wotton Underwood were estates by the time of Domesday (Fig. 66).

APPENDIX: THE CHILTERNS

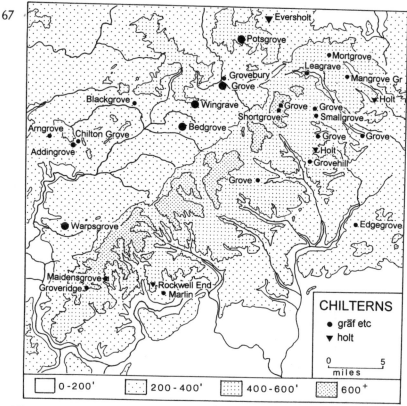

67

The most common of the remaining woodland terms is 'grove', which occurs as a minor name in the south and north Chilterns, but rarely in the central Chilterns. Most of the places are first evidenced in the 13th and 14th centuries, often in 'at the grove' surnames. These groves reflect on the one hand the decreasing woodland resources as the Chiltern woods were assarted and on the other the increase in demand by London for firewood, since both these factors encouraged careful management by coppicing. The 'grove' names occur close to the open **feld** lands of the south, and in the less wooded areas north of the river Gade, where the Roman villas and the important early **lēah** names are. The numerous minor Grove names on the modern map for which no early spellings are available probably also reflect careful management of woodland for London's fuel and later, when the meaning of the term was less exact, for the furniture industry. Major names containing one of the 'grove' words occur in the Vale of Aylesbury where, in fielden country, possession of a coppice was a valuable asset: here, for example, are Warpsgrove, Addingrove, Bedgrove, Grove and Potsgrove (Fig. 67).

The term for a single-species wood, **holt**, is rare in this area of mixed woodland. It occurs in the minor names Rockwell End BUC and The Holt,

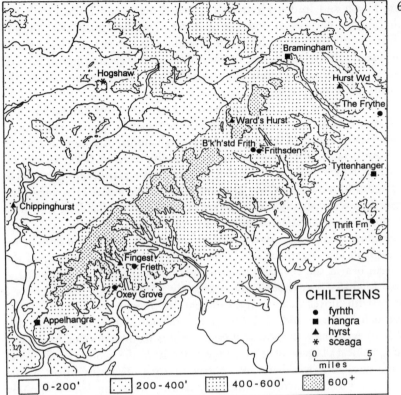

Holt Farm, Holtsmore End and Rucklers Green Wood, all mentioned in 13th- and 14th-century records from Hertfordshire. There is a single major name in **hyrst**, Fingest, which originated as a meeting-place or 'thing' on a wooded hill. Shaw Wood in Aldbury is the only instance of **sceaga**. The older sense of **hangra** is seen in *Appelhangre*, now Beech Farm, near Goring, Tyttenhanger in Ridge and Turlhanger's Wood in Aldbury. Moggerhanger, Polehanger and Bramingham (earlier *Brambelhanger*) are examples from the Vale (Fig. 68).

One of the most informative of the rarer woodland terms is **fyrhth**, which occurs in Frieth, Oxey (earlier *Oxefrid*) Grove in Fawley BUC, and in The Frith and Frithsden in Berkhampstead. Berkhampstead Frith was a wood-pasture common upon which eleven parishes had claim (Hepple and Doggett 1992). Such pressure on a delicately balanced system could tip the balance and lead to degeneration of the tree cover, reducing it to scrub on the edge of woodland. Further pressure here and elsewhere in the Chilterns led to the elimination of trees and the development of heathland, especially in the 13th, 16th and 17th centuries. There is a Little Heath on the edge of Berkhampstead Frith, and two other such heather-clad commons occurred at

APPENDIX: THE CHILTERNS

Goring Heath and Ipsden Heath and spawned the names Little Heath and Heath End for clusters of cottages on them. Turville Heath and Summer Heath are possible examples in Buckinghamshire, as are various Heath Farms in the Hertfordshire Chilterns. Today, without the grazing pressure, such heaths are reverting to scrub and then to woodland, except where the naturalists' trusts intervene.

Discussion of woodland, ploughland and pasture terms has shown up a comparative blank in the central Chilterns. There may be several explanations for this. Firstly, much of the area is in Buckinghamshire, a county for which there is less detailed place-name research. Secondly, there are fewer villages on the uplands which could have become DB estates or parishes with names in this category (those in the valley bottoms usually have names referring to the valley or to a water supply). The blank may, on the other hand, be a genuine effect showing up a different history of colonisation, with the scarp foot villages using the area as outlying pasture: this would fit with the dominance of minor lēah names and the late formation of parishes. The presence of the early term ōra in the central Chilterns indicates an early knowledge of the area by Anglo-Saxons, though not necessarily settlement. It also suggests a landscape with clearer views than the present-day trees permit.

Enough has been said of the Chilterns, the adjacent Thames valley and the Vale of Aylesbury to show that each area has its own distinct assemblage of topographical place-names, linked very closely to the geology, soils, relief and drainage, while woodland, ploughland and pasture terms throw a pale shaft of light on the activities of the early farmers. Over England's infinitely varied countryside no two assemblages of place-names can be the same, and there is endless fascination in studying what they can tell us about our own favourite stamping grounds.

BIBLIOGRAPHY

This bibliography lists only the books and articles to which reference is made in the text; and the chief periodicals in which relevant studies are published are mentioned under a number of items. Full documentation of British place-name studies up to 1989 will be found in *A Reader's Guide to the Place-Names of the United Kingdom. A Bibliography of Publications (1920–89) on the Place-Names of Great Britain and Northern Ireland, the Isle of Man and the Channel Islands*, ed. J. Spittal and J. Field, Stamford, 1990, a *Supplement* to which will be published about the same time as the present book. An excellent general introduction to English place-name studies is provided by the new (1996) version of K. Cameron's *English Place-Names*, London; and a recent collection of papers which deserves mention is *The Uses of Place-Names*, ed. S. Taylor, Edinburgh 1998.

Adkins and Petchey 1984: R. A. Adkins and M. R. Petchey, 'Secklow Hundred Mound and other meeting-place mounds in England', *Archaeological Journal* 141, pp. 243–51.

Atkin 1998: Mary Atkin, 'Places named Anstey', in *Proceedings of the XIXth International Congress of Onomastic Sciences*, vol. 2, ed. W. F. H. Nicolaisen, Aberdeen, pp. 15–23 (see also M. Atkin and V. Watts in EPNS *Journal* 30, pp. 83–98).

Baines 1981: A. H. J. Baines, 'Turville, *Radenore* and the Chiltern *feld*', *Records of Buckinghamshire* 23, pp. 4–22.

Blair 1994: J. Blair, *Anglo-Saxon Oxfordshire*, Stroud.

Brandon 1978: P. Brandon, 'The South Saxon *Andredesweald*', in *The South Saxons*, ed. P. Brandon, Chichester, pp. 138–59.

Charles 1938: B. G. Charles, *Non-Celtic Place-Names in Wales*, London.

Coates 1980: R. Coates, 'Methodological reflexions on Leatherhead', EPNS *Journal* 12, pp. 70–4.

Coates 1981: R. Coates, 'The slighting of Strensall', EPNS *Journal* 13, pp. 50–3.

Coates 1988: R. Coates, *Toponymic Topics*, Younsmere Press, pp. 24–30.

Coates 1989: R. Coates, *The Place-Names of Hampshire*, London.

Coates 1991: R. Coates, 'The name of Lewes: some problems and possibilities', EPNS *Journal* 23, pp. 5–15.

Coates 1995a: R. Coates, 'The place-name *Owermoigne*, Dorset, England', *Indogermanische Forschungen* 100, pp. 244–51.

Coates 1995b: R. Coates, 'English cuckoos, dignity and impudence', EPNS *Journal* 27, pp. 43–9.

Coates 1996, 1997a: R. Coates, 'Clovelly, Devon', 'Clovelly again', EPNS *Journal* 28, pp. 36–44, 29, pp. 63–4.

Coates 1996b: R. Coates, 'Notes', *Locus Focus* (newsletter of the Sussex Place-Names Net), vol. 1, no. 1, 18–20.

Coates 1997b: R. Coates, 'The scriptorium of the Mercian Rushworth gloss: a bilingual perspective', *Notes and Queries* NS 44, no. 4, pp. 453–8.

Cole 1982: Ann Cole, 'Topography, hydrology and place-names in the chalklands of southern England: *cumb* and *denu*', *Nomina* 6, pp. 73–87.

Cole 1985: Ann Cole, 'Topography, hydrology and place-names in the chalklands of southern England: **funta*, *ǣwiell* and *ǣwielm*', *Nomina* 9, pp. 3–19.

Cole 1987: Ann Cole, 'The distribution and usage of the OE place-name *cealc*', EPNS *Journal* 19, pp. 45–55.

Cole 1988: Ann Cole, 'The distribution and usage of the place-name elements *botm*, *bytme* and *botn*', EPNS *Journal* 20, pp. 38–46.

Cole 1990: Ann Cole, 'The origin, distribution and use of the place-name element *ōra* and its relationship to the element *ofer*', EPNS *Journal* 22, pp. 26–41.

Cole 1991: Ann Cole, '*Burna* and *brōc*: problems involved in retrieving the OE usage of these place-name elements', EPNS *Journal* 23, pp. 26–48.

Cole 1992: Ann Cole, 'Distribution and use of the Old English place-name *mere-tūn*', EPNS *Journal* 24, pp. 30–41.

Cole 1993: Ann Cole, 'The distribution and use of *mere* as a generic in place-names', EPNS *Journal* 25, pp. 38–50.

Cole 1994a: Ann Cole, 'Baulking: an Anglo-Saxon industry revealed', EPNS *Journal* 26, pp. 27–31.

Cole 1994b: Ann Cole, 'The Anglo-Saxon traveller', *Nomina* 17, pp. 7–18.

Cole 1997: Ann Cole, '*flēot*: distribution and use of this OE place-name element', EPNS *Journal* 29, pp. 79–87.

BIBLIOGRAPHY

Cole forthcoming: Ann Cole, '*Ersc*: distribution and use of this place-name element'.

Coplestone-Crow 1989: B. Coplestone-Crow, *Herefordshire Place-Names*, British Archaeological Reports, British Series 214.

Cox 1976: B. Cox, 'The place-names of the earliest English records', EPNS *Journal* 8, pp. 12–66.

Cox 1990, 1992: B. Cox, 'Byflete', 'Byflete revisited', EPNS *Journal* 22, 24, pp. 42–6, 49.

Crawford 1953: O. G. S. Crawford, *Archaeology in the Field*, London.

Cronan 1987: D. Cronan, 'Old English *gelad*: "a passage across water"', *Neophilologus* 71, pp. 316–19.

Cullen 1998: P. Cullen, 'Courtup Farm in Nuthurst and the den of *Curtehope* in Lamberhurst', *Locus Focus* (forum of the Sussex Place-Names Net) 2, no. 2, p. 15.

Davis 1962: R. H. C. Davies, 'Brixworth and Clofesho', *Journal of the British Archaeological Association*, Series 3, 25, p. 71.

Dodgson 1973: J. McN. Dodgson, 'Place-names from *hām*, distinguished from *hamm* names, in relation to the settlement of Kent, Surrey and Sussex', *Anglo-Saxon England* 2, pp. 1–50.

Dodgson 1987: J. McN. Dodgson, 'The *-er-* in Hattersley Cheshire and Hothersall Lancashire', *Leeds Studies in English* 18, pp. 135–9.

Duignan 1902: W. H. Duignan, *Notes on Staffordshire Place-Names*, London.

Ekwall 1922: E. Ekwall, *The Place-Names of Lancashire*, Manchester.

Ekwall 1928: E. Ekwall, *English River-Names*, Oxford.

Ekwall 1960: E. Ekwall, *The Concise Oxford Dictionary of English Place-Names*, 4th ed., Oxford.

Everitt 1977: A. Everitt, 'River and wold: reflections on the historical origin of regions and pays', *Journal of Historical Geography* 3, 1, pp. 1–19.

Everitt 1986: A. Everitt, *Continuity and Colonization: The Evolution of Kentish Settlement*, Leicester.

Fellows-Jensen 1972: Gillian Fellows-Jensen, *Scandinavian Settlement Names in Yorkshire*, Copenhagen.

Fellows-Jensen 1978: Gillian Fellows-Jensen, *Scandinavian Settlement Names in the East Midlands*, Copenhagen.

Fellows-Jensen 1981: Gillian Fellows-Jensen, 'Scandinavian Settlement in the Danelaw in the light of the place-names of Denmark', in *Proceedings of the Eighth Viking Congress*, ed. H. Bekkr Nielson *et al*, Odense, pp. 134–45.

Fellows-Jensen 1985: Gillian Fellows-Jensen, *Scandinavian Settlement Names in the North-West*, Copenhagen.

Fellows-Jenson 1989: Gillian Fellows-Jensen, '*Amounderness* and *Holderness*', in *Studia Onomastica: Festskrift till Thorsten Andersson*, ed. L. Peterson, Lund, pp. 87–94.

Forsberg 1950: R. Forsberg, *A Contribution to a Dictionary of Old English Place-Names*, Nomina Germanica 9.

Fox 1981: H. S. A. Fox, 'Approaches to the adoption of the Midland system', in *The Origins of the Open Field Agriculture*, ed. T. Rowley, London, pp. 64–111.

Fox, 1989: H. S. A. Fox, 'The people of the Wolds in English settlement history', in *The Rural Settlements of Medieval England: Studies Dedicated to Maurice Beresford and John Hurst*, ed. M. Aston, D. Austin and C. Dyer, Oxford, pp. 77–101.

Foxall 1980: H. D. G. Foxall, *Shropshire Field-Names*, Shropshire Archaeological Society.

Gelling 1960: Margaret Gelling, 'The element *hamm* in English place-names: a topographical investigation', *Namn och Bygd* 47 (1–4), pp. 140–62.

Gelling 1970: Margaret Gelling, English entries in M. Gelling, W. F. H. Nicolaisen and M. Richards, *The Names of Towns and Cities in Britain*, London.

Gelling 1973: Margaret Gelling, 'Further thoughts on pagan place-names', in *Otium et Negotium: Studies in Onomotology and Library Science Presented to Olof von Feilitzen*, ed. F. Sandgren, Stockholm, pp. 109–28.

Gelling 1974: Margaret Gelling, 'Some notes on Warwickshire place-names', *Transactions of the Birmingham and Warwickshire Arachaeological Society*, 86, pp. 59–79.

Gelling 1977: Margaret Gelling, 'Latin loan-words in Old English place-names', *Anglo-Saxon England* 6, pp. 1–13.

Gelling 1982: Margaret Gelling, 'The *-inghop* names of the Welsh Marches', *Nomina* 6, pp. 31–36.

Gelling 1984: Margaret Gelling, *Place-Names in the Landscape*, London.

Gelling 1987: Margaret Gelling, 'Anglo-Saxon eagles', *Leeds Studies in English* 18, pp. 173–81.

Gelling 1988: Margaret Gelling, *Signposts to the Past: Place-Names and the History of England*, 2nd ed., Chichester.

Gelling 1989: Margaret Gelling, 'Shaw/Shay: the phonological problem', *Nomina* 12, pp. 103–4.

BIBLIOGRAPHY

Gelling 1992: Margaret Gelling, *The West Midlands in the Early Middle Ages*, Leicester.

Gelling 1995: Margaret Gelling, 'Scandinavian settlement in Cheshire: the evidence of place-names', in *Scandinavian Settlement in Northern Britain*, ed. Barbara Crawford, Leicester, pp. 187–205.

Guest 1996: Margery Guest, 'The Frogmere sites of Hertfordshire', EPNS *Journal* 28, pp. 61–70.

Hald 1978: K. Hald, 'A-mutation in Scandinavian words in England', in *The Vikings*, ed. Th. Andersson and K. I. Sandred, Uppsala, pp. 99–106.

Hepple and Doggett 1992: L. W. Hepple and Alison M. Doggett, *The Chilterns*, Chichester.

Higham 1989: Mary C. Higham, 'Shay names – a need for re-appraisal?', *Nomina* 12, pp. 89–102.

Holthausen 1934: F. Holthausen, *Altenglisches Etymologisches Wörterbuch*, Heidelberg.

Hooke 1981: Della Hooke, *Anglo-Saxon Landscapes of the West Midlands: The Charter Evidence*, British Archaeological Reports, British Series 95.

Hough 1995a: Carole Hough, 'OE *græg in place-names', *Neuphilologische Mittelungen* 4, 96, pp. 361–5.

Hough 1995b: Carole Hough, 'The place-names *Bridford*, *Britford* and *Birdforth*', *Nottingham Medieval Studies* 39, pp. 12–18.

Hough 1995c: Carole Hough, 'OE *wearg* in Wanborough and Wreighton', EPNS *Journal* 27, pp. 14–20.

Hough 1996: Carole Hough, 'Old English *rōt* in place-names', *Notes and Queries* 241:2, pp. 128–9.

Huggins 1975: Rhona Huggins, 'The significance of the place-name *Wealdham*', *Medieval Archaeology* 19, pp. 198–201.

Jackson 1953: K. Jackson, *Language and History in Early Britain*, Edinburgh.

Kitson 1990: P. R. Kitson, 'On Old English nouns of more than one gender', *English Studies* 3, pp. 185–221.

Kitson 1994: P. R. Kitson, 'Quantifying qualifiers in Anglo-Saxon charter boundaries', *Folia Linguistica Historica* 14, 1–2, pp. 29–82.

Kitson forthcoming: P. R. Kitson, *Guide to Anglo-Saxon Charter Boundaries*, EPNS.

Kolb 1989: E. Kolb, '*Hamm*: a long-suffering place-name word', *Anglia* 107, pp. 49–51.

Kristensson 1973: G. Kristensson, 'Two Berkshire river-names', *Namn och Bygd* 61: 1–4, pp. 49–54.

Löfvenberg 1942: M. T. Löfvenberg, *Studies on Middle English Local Surnames*, Lund.

Margary 1955: I. D. Margary, *Roman Roads in Britain, Vol. 1: South of the Foss Way – Bristol Channel*, 1957, *Vol. 2: North of the Foss Way – Bristol Channel*, London.

Marwick 1952: H. Marwick, *Orkney Farm-Names*, Kirkwall.

Mills 1991: A. D. Mills, *A Dictionary of English Place-Names*, Oxford.

Nicolaisen 1976: W. F. H. Nicolaisen, *Scottish Place-Names: Their Study and Significance*, London.

Onions 1966: C. T. Onions, *The Oxford Dictionary of English Etymology*, Oxford.

Padel 1981: O. J. Padel, 'Welsh *blwch* "bald, hairless"', *Bulletin of the Board of Celtic Studies* 29 (2), pp. 523–6.

Padel 1985: O. J. Padel, *Cornish Place-Name Elements*, EPNS 56/57.

Padel 1988: O. J. Padel, *A Popular Dictionary of Cornish Place-Names*, Penzance.

Pocock 1968: E. A. Pocock, 'The first fields in an Oxfordshire parish', *The Agriculture History Review* 16 (2), pp. 85–100.

Rackham 1976: O. Rackham, *Trees and Woodland in the British Landscape*, London.

Redmonds 1983: G. Redmonds, *Almondbury: Places and Place-Names*, Huddersfield.

Rivet and Smith 1979: A. L. F. Rivet and C. Smith, *The Place-Names of Roman Britain*, London.

Rumble 1987: A. R. Rumble, 'Old English *Boc-land* as an Anglo-Saxon estate name', *Leeds Studies in English* 18, pp. 219–29.

Rumble 1989: A. R. Rumble, 'A Bedan gloss on Bedfont, Bedwell, etc.', *Nomina* 12, pp. 123–30.

Smith 1956: A. H. Smith, *English Place-Name Elements* Parts 1 and 2, EPNS 25, 26.

Stanford 1980: S. C. Stanford, *The Archaeology of the Welsh Marches*, London.

Stenton 1913: F. M. Stenton, *The Early History of the Abbey of Abingdon*, Oxford; reprinted Stamford, 1989.

Stiles 1997: P. Stiles, "Old English *halh*, 'slightly raised ground isolated by marsh'", in *Names, Places and People: An Onomastic Miscellany in Memory of John McNeal Dodgson*, ed. A. R. Rumble and A. D. Mills, Stamford, pp. 330–44.

Styles 1998: Tania Styles, 'Whitby revisited: Bede's explanation of *Streanæshalch*', *Nomina* 21, pp. 133–48.

Tengstrand 1940: E. Tengstrand, *A Contribution to the Study of Genitival Composition in Old English Place-Names*, Uppsala.

Torvell 1992: D. Torvell, 'The significance of *Here-ford*', EPNS *Journal* 24, pp. 42–48.

Viatores 1967: The Viatores, *Roman Roads in the South-East Midlands*, London.

Watts 1976: V. E., Watts, "Comment on 'The Evidence of Place-Names' by Margaret Gelling", in *Medieval Settlement: Continuity and Change*, ed. P. H. Sawyer, London, pp. 212–22.

Watts 1990: V. E. Watts, 'Shaw/Shay revisited', *Nomina* 13, pp. 109–14.

Watts 1994: 'The place-name Hexham: a mainly philological approach', *Nomina* 17, pp. 119–36.

Watts forthcoming. V. Watts, *The Cambridge Dictionary of English Place-Names*.

Wyn Owen 1988: H. Wyn Owen, 'English place-names and Welsh stress patterns', *Nomina* 11, pp. 99–114.

Wyn Owen 1997: 'Old English place-name elements in Domesday Flintshire', in *Names, Places and People: An Onomastic Miscellany in Memory of John McNeal Dodgson*, ed. A. R. Rumble and A. D. Mills, Stamford, pp. 269–78.

Yorke 1993: Barbara Yorke, 'Lindsey: the lost kingdom found', in *Pre-Viking Lindsey*, ed. A. Vince, Lincoln, pp. 141–50.

INDEX

As throughout the book, all place-names not followed by a language abbreviation are Old English.

Place-names which occure in three or more counties are followed by the abbreviation 'var. cos.' (various counties). The key to the county and other abbreviations appears on pages ix–xi.

á ON 2, 14
Abbefeld BUC 311
Abbotsham DEV 48, 53
Abbot's Langley HRT 312
Abbenhall GLO 128
Aberford YOW 73
Abingdon BRK 168
Abloads Court GLO 81, 82
Abney DRB 39, 41
Abridge ESX 69
Absol ESX 14
Aby LIN 2
Acaster YOW 2
Acklam YOE, YON 240
Acol KNT 233
Acre NFK 264
Acreland YOW 264
Adam's Grave WLT 148
Adber DOR 223
Addingrove BUC 228, 314
Addlestone SUR 115
Adeney SHR 40, 41
Adgardley LNC 185
Adlingfleet YOW 16
Adwell OXF 32, 299
Adzor HRE 202
Agden HNT 115
Agden YOW 117
Agneash IOM 198
Aigburth LNC 151
Aikber YON 151
Ainderby Mires YON 61

Ainsdale LNC 111
Ainstable CMB 186
Ainsty YOW 66, 67
Aisholt SOM 233
Aiskew YON 248
Akefrith LNC 225
Akeld NTB 177
Akeley BUC 240
akr ON 263, 279
Albourn SSX 12
Alburgh NFK 151
Alcumlow CHE 179
Aldbourne WLT 12
Aldbury Nowers HRT 298, 309
Aldcliffe LNC 155
Alderford NFK 75
Alderholt DOR 233
Aldersceugh CMB 248
Aldersey CHE 38, 41
Alderwasley DRB 64
Aldfield YOW 274
Aldford CHE 73
Aldingbourne SSX 12
Aldon SHR 3, 169
Aldreth CAM 85, 86, 87, 88
Alford LIN 76
Alford SOM 73
Algrave DRB 229
Allensmore HRE 59
Allerdale CMB 111
Allerford SOM 75
Allerwash NTB 64

INDEX

Allfleet's Fm ESX 198
Allisland DEV 282
Allithwaite LNC 250
Almeley HRE 240
Almer DOR 26
Almholm YOW 57
Alney GLO 41
Alresford ESX 73
Alresford HMP 75
Alrewas STF 64
Alsager CHE 264, 265
Alsop DRB 139
Alspath WAR 90
Alstonfield STF 274
Altcar LNC 57
Alton var. cos. 3, 299
Amberden ESX 117
Ambleside WML 2
Ambrosden OXF 169
Amounderness LNC 197
Ampleforth YON 75
Ampney GLO 38, 41
Ampthill BDF 193
Amwell HRT 32, 300
Andridge BUC 296
Andwell HMP 32
Anglesey CAM 43
Anmer NFK 26
Anningsley SUR 241
Anser Gallows Fm ESX 67
Ansford SOM 73
Anstey var. cos. 66, 67
Anstie SUR 66, 67
ānstīg/anstig 66–7
Ansty var. cos. 66, 67
Apes Hall CAM 233
Apley var. cos. 240
Apperley var. cos. 240
Appleford BRK, IOW 71, 75
Appleshaw HMP 247
Applethwaite CMB 251
Appley LNC 240
Appuldurcomb IOW 107
Aqualate STF 81

Arborfield BRK 272, 274
Areley WOR 240
Arkendale YOW 118
Arkesden ESX 115
Arksey YOW 38, 43
Arlecdon CMB 118
Arlesey BDF 41
Arley var. cos. 240
Armathwaite CMB 251
Armetridding LNC 244
Arminghall NFK 128
Arneliff(e) YON, YOW 155
Arnecliff YON 155
Arnewas HNT 64
Arnewood HMP 260
Arnford YOW 76
Arnold NTT, YOE 129
Arnside WML 177
Artington SUR 170
Arundel SSX 113
Ashbeach HNT 3
Ashberry Hill YON 149
Ashbourne DRB 12
Ashburnham SSX 12
Ashburton DEV 12
Ashcombe DEV 107
Ashdon ESX 168
Ashdown SSX 168
Ashendon BUC 168
Asheridge BUC 190, 191, 296
Ashford KNT 75
Ashford MDX 34, 76–7
Ashford var. cos. 75
Ashgrove WLT 229
Ashill SOM 193
Ashingdon ESX 169
Ashington NTB 117
Ashley var. cos. 240, 312
Ashness CMB 197
Ashold WAR 233
Ashop DRB 139
Ashour KNT 209
Ashover DRB 202
Ashow WAR 188

325

Ashridge BUC 296
Ashurst KNT, SSX 235
Ashwood STF 260
Askerswell DOR 32
Askrigg YON 214, 215
Askwith YOW 253
Aspall SFK 129
Aspenden HRT 117
Aspenshaw DRB 247
Aspley var. cos. 240
Aspul LNC 193
Assendon OXF 114, 117, 292
Astley var. cos. 241
Ashmore DOR 26
Ashwell var. cos. 32
Astwell NTP 33
Astwood BUC, WOR 259
Athelney SOM 42
Atherfield IOW 276
Attercliffe YOW 156
Atterwith YOW 253
Attington OXF 168
Attleborough WAR 150
Attlebridge NFK 69
Auborn LIN 12
Auburn YOE 12
Audenshaw LNC 247
Aughton WLT xx
Aulden HRE 167, 168
Austcliff WOR 155
Aust Cliff GLO 156
Austerfield YOW 276
Austhwaite CMB 250
Autherlands YOW 243
Aveland LIN 242, 243
Averingdown BUC 296
Avoncliffe WLT 156
Awbridge HMP 191
Awliscombe DEV 107
Axbridge SOM 69
Axford HMP 209
Axford WLT 75
Axholme LIN 57
Aylesbeare DEV 150, 223

Aylesford KNT 73–4
Ayngstree WOR 66, 67
Aynhoe NTP 188
Ayresome YON 2
Ayshford DEV 75
Ayton YON 2, 14

æcer xviii, 262, 263–6, 268, 279, 312
Æscesbyrig BRK xx
ǣwylle 1, 3, 18, 299
ǣwylm 1, 2–3, 18, 299

Babbacombe DEV 107
Babeny DEV 41
Bablock Hythe OXF 20, 87, 88
Bache CHE 3
Backbarrow LNC 151
Backford CHE 74, 144
Backridge YOW 144
Backwell SOM 33, 144
Bacup LNC 136, 139, 144
Badenhall STF 128
Badgemore OXF 296
Badger SHR 202
Badlesmere KNT 27
Badsey WOR 41
Badwell SFK 32
Bagbear(e) DEV 223
Bagborough SOM 150
Bagburrow Wood WOR 223
Bagendon GLO 118
Bageridge DOR, STF 191
Bagley var. cos. 240
Bagnor BRK 209
Bagpath GLO 89, 90
Bagshaw DRB 247
Bagslate Moor LNC 141
Baguley CHE 240
Baildon YOW 167, 170
Bainbridge YON 69
Bainley Bank YON 186
Bakewell DRB 32
Balcombe SSX 108
Baldon OXF 168

INDEX

Baldslow SSX 179
Bale NFK 5
Balkholme YOE 56, 57
Ballidon DRB 117
Ballingdon SFK 168
Ballinger BUC 208, 298, 309
Ballingham HRE 49
Balsall WAR 130
Balscott OXF 130
Balsdean SSX 115
Balsdon CNW 130
Balsham CAM 130
Baltonsborough 150
Bamford DRB, LNC 73
Bampton DEV 5
Bandon SUR 169
Bangrave NTP 229
Bannerdale WML 111
Bannisdale WML 111
Banwell SOM 33
Bapchild KNT 14
Barbon WML 13
Barbourne WOR 12
Bar Bridge GLO 223
Barby NTP 151
Barden YON 118-9
Barden YOW 117, 119
Bardfield ESX 276
Bardney LIN 41
Bardon LEI 167
Bardsea LNC 41
Bardsey YOW 38, 41
Bardwell SFK 34
Bare LNC 223
Barf YON 149
Barford BDF 75
Barford var. cos. 76
Barham var. cos. 150
Barhaugh NTB 130
Barholm LIN 150
Barker's Hill WLT 209
Barkham BRK 52, 54
Barkisland YOW 282
Barkway HRT 95-6

Barley LNC, YOW 241
Barlow DRB 241
Barmoor NTB 60
Barnacle WAR 232
Barney NFK 43
Barnhill YOE 193
Barnsdale RUT 192, 193
Barnshaw CHE 247
Barnskew WML 248
Barnside YOW 219
Barnsley YOW 240
Barnwell CAM 34
Barnwell NTP 34
Barnwood GLO 260
Barpham SSX 150
barr, barrǫg PrW 145
Barr STF 145
Barra, Hebrides 145
Barrasford NTB 74
Barrhead RNF 145
Barrock Fell CMB 145
Barrow LNC 145
Barrow RUT, SOM 149, 150
Barrow var. cos. 221, 223
Barrowby LIN, YOW 151
Barrowden RUT 148, 150
Barrowford LNC 74
Barry GLA 145
Bartley HMP, WOR 240
Bartlow CAM 178, 179
Barugh YON, YOW 149, 150
Barway CAM 42, 150
Barwell LEI 32
Barwise WML 286
Basford CHE 74
Basford STF 74
Bashall YOW 144, 218
Basildon BRK 115
Basildon ESX xix, 165, 168
Bassenthwaite CMB 250
Bassingbourn CAM 12
Bastonford WOR 74
The Batch var. cos. 3-4
Batcombe DOR, SOM 109

Batford HRT 307
Bath SOM 4
Bathampton WLT 5
Bathe Barton DEV 5
Bathgate WLO 224
Bathingbourne IOW 12
Bathley NTT 5
Bathpool CNW 5
Bathurst SSX 235
Batsford GLO 208
Battenhurst SSX 268
Battersea GTL 41
Battisford SFK 74
Battleburn YOE 10
Battlesden BDF 168
Baughurst HMP 235
Baulking BRK 4–5
Bawdeswell NFK 32
Bawdsey SFK 41
Bawsey NFK 42
Baycliff LNC 156
Baycliff WLT 155
Baydon WLT 168
Bayfield NFK 274
Bayford HRT 74
Bayford SOM 75
Baysdale YON 111
bæc 4, 37, 56, 144–5
bæc, bece 3–4, 5, 144–5
bǣr 262, 266
bæth 3–4
Beaconsfield BUC 276
Beadnell NTB 128
Beaford DEV 75
Beal NTB 193
Beal YOW 129
Beamsley YOW 98
Beanacre WLT 265
Beara DEV 221
Beare DEV, SUR 221, 223
Bearl NTB 193
Bearley WAR 241
bearu 221–3, 233, 266
Beausale WAR 129

Beauxfield KNT 274
Beccles SFK 3
Beckermet CMB 6
Beckermonds YOW 6
Beckford GLO 74
Beckhampton WLT 144
Beckney ESX 41
Beckwith YOW 252, 253
Bedale YON 128
Bedburn DRH 12
Beddingham SSX 50, 53
Bedenesteda ESX 100
Bedfield SFK 274
Bedfont MDX 17, 18
Bedford BDF 71, 77, 100
Bedford LNC 74
Bedford SSX 17, 18
Bedgrove BUC 229, 314
Bedingfield SFK 276
Bedlested SUR 100
Bedmond HRT 17, 18, 301
Bednall STF 125, 128
Bedwell HRT 33, 100
Beech Fm OXF 315
Beedon BRK 100
Beeford YOE 75
Beela R. 2
Beeleigh ESX 240–1
Beer(e) var. cos. 221, 223
Beera DEV 221
Beer Crocombe SOM 221
Begbroke OXF 8
Beighton DRB 3
Bekesbourne KNT 10
bekkr ON 5–6
Belchalwell DOR 32
Belchford LIN 77
Belford NTB 77
Belgrave LEI 229
Bellhurst SSX 236
Belsay NTB 188
Bembridge IOW 69
Benacre var. cos. 265
Bendish HRT 269, 311

328

INDEX

Bendysh ESX 269
Benefield NTP 276
Benenden KNT 267
Benfieldside DRH 275
Benfleet ESX 16
Bengeo HRT 189
Bengrove (Sandhurst) GLO 229
Bengrove (Teddington) GLO 229
Benhall SFK 130
Benham BRK 51, 53
Benhilton SUR 193
Bennah DEV 100
Bentfield ESX 275
Benthall SHR 129
Bentham Hill KNT 52
Bentley var. cos. 240
Beobridge SHR 69
Beoley WOR 241
beorg xiii, xv, xvi, 145–51, 152, 192, 221, 296
Berden ESX xii, 118
Bere KNT 266
Bere var. cos. 221
Bere Forest HMP 266
Bere Regis DOR 266
berg ON 143, 145, 149, 151–2
Bergholt ESX, SFK 150, 234
Berkeley GLO 240
Berkeley Harness GLO 197
Berkesden HRT 117
Berkhamstead HRT 150
Berkley SOM 240
Berkshire 145
Berkswell WAR 32
Bermondsey GTL 41
Bernwood BUC 257, 260
Berrier CMB 151
Berrington NTB 170
Berrow SOM, WOR 150
Berry DEV 221
Besford SHR, WOR 74
Beslow SHR 179
Bessacar YOW 264, 265
Bestwood NTT 260

Betchton CHE 3
Betham DEV 100
Bethnal Green GTL 128
Betlow HRT 179
Betteshanger KNT 119, 232
Bettiscombe DOR 107
Bevendean SSX 115
Beversbrook WLT 8
Bexley GTL 240
Bibbern DOR 100
Bickenhill WAR 193
Bicker LNC 57
Bickershaw LNC 247
Bickerstaffe LNC 91, 92
Bickford STF 77
Bickham DEV 109
Bickmarsh WAR 58
Bicknacre ESX 264, 265
Bicknor, DEV, KNT 206, 209
Bicknor GLO/HRE 202
Bidborough KNT 148, 150
Bidden HMP 100
Biddenham BDF 48
Biddesden WLT 115–7
Biddlesden BUC 119
Biddlestone NTB 119
Bideford DEV 77, 100
Bidford WAR 77, 80
Bidna DEV 77, 100
Bidwell BDF, NTP 33, 100, 299
Bigfrith BRK 225
Biggleswade BDF 94, 95
Bigrigg CMB 190
Bilbrook SOM, STF 8
Billesden LEI 168
Billesdon Coplow LEI 159
Billeshurst SUR 268
Billingford NFK 75
Billinghay LIN 42
Billingley YOW 241
Billingshurst SSX 235
Billington BDF 168
Billington LNC 167
Bilney NFK 43

Bilsdale YON 111
Bilsham Fm GLO 50, 53
Binbrook LIN 8
Bincknoll WLT 157
Bincombe DOR 109
Bindon DOR 170
Binegar SOM 230, 232
Binfield BRK 275, 310, 311
Bingfield NTB 276
Bingley YOW 241
Binley WAR 39, 43
Binsey OXF 40, 41
Birchall HRT, WOR 233
Bircham NFK 266
Birchanger ESX, WLT 231
Birchden SSX 267
Birchengrove GLO 229
Bircher HRE 202
Birchgrove SSX 230
Birchitt DRB 177
Bircholt KNT 233
Birchover DRB 202
Birdbrook ESX 8
Birdforth YON 72, 75
Birdsall YOE 130
Birkdale LNC 111
Birkenhead CHE 177
Birkenshaw YOW 247
Birkenside NTB 219
Birkland Barrow LNC 243
Birklands NTT 242, 243
Birkwith YOW 253
Birley HRE 241
Birlingham WOR 48, 49, 53
Birmingham WAR 48
Birstwith YOW 91, 92
Birtenshaw LNC 247
Bisbrooke RUT 8
Bisham BRK 304
Bishopdale YON 111
Bishopsbourne KNT 10
Bishopside YOW 219
Bishopton WAR 169
Bitchfield LIN 274

Bitchfield NTB 275
Bittadon DEV 117
Bitterley SHR 240
Bitteswell LEI 34, 98
Bittiscombe SOM 108
Bixmoor OXF 298, 308
Blackberry Hill LEI 150
Blackborough DEV, NFK 150
Blackburn LNC 11
Blackden CHE 117
Blackdown DOR 167, 168
Blackenhall CHE 129
Blackford SOM 73
Blackfordby LEI 73
Blackgrave WOR 228
Blackgreaves WAR 228
Blackheath var. cos. 279
Blacklache LNC 21
Blackland WLT 282
Blackmoor DOR 60
Blackmoor HMP 27
Blackmore ESX, WLT 60
Blackmore HRT 27
Blacko LNC 174
Blackpool DEV, LNC 29
Blackrod LNC 244
Blackshaw YOW 247
Blackwater R. SUR 30
Blackwell var. cos. 32
Blackwood YOE, YOW 259
Bladon OXF 16
Blagden var. cos. 168
Blagdon NTB 117
Blagrave BRK 228
Blagrove var. cos. 228
blain PrW 152
Blainscough LNC 249
Blaisdon GLO 168
Blakehope NTB 139
Blakeney GLO, NFK 42
Blakeshall WOR 63
Blandford DOR 75
Blankney LIN 40, 42
Blashaw LNC 247

INDEX

Blawith LNC 252
Blaxhall SFK 128
Bleadney SOM 83–5, 88
Bleadon SOM 167, 168
Bleasdale LNC 111
Blechingley SUR 241
Bledlow BUC 179, 296
Blencarn CMB 152
Blencathra CMB 152
Blencogo CMB 152
Blencow CMB 152
Blennerhasset CMB 152
Bletchingdon OXF 168
Bletchley BUC 240
Bletsoe BDF 188
Blindcrake CMB 152
Blithfield STF 275
Blithford STF 76
Blofield NFK 276
Blunham BDF 48
Blunsdon WLT 168
Blunt's Hall ESX 125, 128
Blyford SFK 76
Blymhill STF 194
Blyth NTB 16
Boarhunt HMP 18
Boblow ESX 179
Bodiam SSX 53
Bodney NFK 41
Bognor SSX 205, 208
Bolas SHR 64
Boldon DRH 170
Bolingbroke LIN 9
Bolney OXF 87, 88, 307, 308, 309
Bolney SSX 41
Bolnhurst BDF 236
Bolsover DRB 202
Bomere SHR 26
Bonehill STF 193
Bonhunt ESX 18
Boningale SHR 128
Bonsall DRB 130, 230
Boode DEV 261
Bordesley WAR, WOR 241

Bordley YOW 241
Boresford HRE 75
Borough Fm BRK 305
Borrowby YON 151
Borrowdale CMB, WML 111
Borthwood IOW 260
Boscombe WLT 107
Bosden CHE 168
Bosham SSX 50, 53
Bosmere SFK 27
Bosmore BUC 303
Bossall YON 128
Botesdale SFK 111
Bothall NTB 128
Bothamsall NTT 99, 217, 218
botm, bothm 98–100, 295
botn ON 98–100
Bottesford LEI, LIN xii, 76
Bottom var. cos. 98, 295
Bottoms Fm GLO 98
Botwell GTL 33, 35
Botwood SHR 260
Bouldnor IOW 205, 209
Bouldon SHR 170
Boulmer NTB 26
Bourne LIN 11
Bourn(e) var. cos. 11
Bourne End BUC, HRT 302
Bourne Vale STF 10
Bouthwaite YOW 251
Boveney BUC 305
Boveridge DOR 191
Bovingdon HRT 170, 296
Bowber WML 149
Bowcliffe YOW 155
Bowcombe IOW 108
Bowda DEV 261
Bowden DRB 168, 261
Bowden LEI 167, 168
Bowden WLT 170
Bowderdale CMB 111
Bowdon CHE, DEV 168, 261
Bowland Forest LNC/YOW 283
Bowler's Green SUR 185

Bowmoor GLO 26
Bowness CMB 196, 198
Bowness WML 198
Bowood DEV 261
Bowood DOR, WLT 260
Boxford BRK 209
Boxgrove SSX 229
Boxley KNT 240
Boxwell GLO 32
Boyland NFK 243
Bozen Green HRT 119
Brabourne KNT 11
Bracebridge LIN 69–70
Bracewell YOW 32
The Brache HRT 312
Brackenber WML 151
Brackenborough LIN 151
Brackenthwaite CMB 251
Bracknell BRK 125, 127, 128
Bradavin DEV 46
Bradbourne DRB 11
Bradda Head IOM 176
Bradden NTP 117
Bradfield var. cos. 272, 274
Bradford var. cos. 71, 73
Bradiford DEV 73
Bradle DOR 240
Bradley var. cos. 239, 240
Bradmore NTT 22, 27
Bradney SOM 42
Bradnop STF 139
Bradnor HRE 209
Bradpole DOR 29
Bradshaw var. cos. 247
Bradsole KNT 62
Bradwall CHE 32
Bradway DRB 95
Bradwell var. cos. 32
Brafferton DRH, YON 73
Braffords YOW 250
Brafield NTP 275
Braithwaite CMB, YON 250
Brambridge HMP 69
Bramdean HMP 117

Bramfield HRT 276
Bramfield SFK 275
Bramford SFK 75
Bramhall CHE 129
Bramhope YOW 139
Bramingham BDF 231, 315
Bramley var. cos. 240
Brampford DEV 77
Bramshall STF 216, 218
Bramshaw HMP 247
Bramwith YOW 253
Brancaster Staith NFK 91
Brancepath DRH 89, 90
Brancliffe YOW 155
Brandon LIN 169
Brandon NFK 167
Brandon var. cos. 168
Brandred KNT 244
Brandwood LNC 259
Branscombe DEV 107
Bransdale YON 111
Bransford WOR 74
Bransty CMB 92, 93
Branthwaite CMB 251
Brantwood LNC 259
Brathay R. 2
Bratton SHR, SOM 7
Bratton var. cos. 266 –7
Brauncewell LIN 32
Brawith YON 94
Braxted ESX 266
Braybrooke NTP 8
Brayfield BUC 275
Breach Wood HRT 312
Breadsall DRB 128
Breamore HMP 60
brēc 266–7, 312
Breckles NFK 266
Bredfield SFK 274
Bredhurst KNT 236
Bredon WOR 152, 167, 169
Breedon LEI 152, 167, 169
breʒ PrW 152
Breightmet LNC 285

INDEX

brekka ON 152
Brendon DEV 168
Brendon SOM 169
Brentford BUC 74, 308
Brentford GTL 76, 77
Brent Knoll SOM 77, 157
Brentwood ESX 259
Brereton STF 168
Bretford WAR 73
Bretforton WOR 73
Bretherdale WML 111
Bretton DRB, YOW 266
Brewood STF 152, 259, 260
Brickendon HRT 170
Brickhill BDF, BUC 152, 193
Brickworth WLT 209
Briddlesford IOW 77
Bride Hall HRT 295
Bridford DEV 72, 75
Bridge KNT 68
Bridgend LIN 68
Bridgehampton SOM 68
Bridgemere CHE 26
Bridgerule DEV 68
Bridge Sollers HRE 68
Bridgford NTT, STF 68, 76
Bridgham NFK 67
Bridgnorth SHR 68
Bridgwater SOM 68
Bridmore WLT 27
Briercliffe LNC 156
brig PrW 152
Brigham CMB, YOE 67
Brighouse YOW 68
Brightlingsea ESX 39, 41
Brightwell var. cos. 32, 299
Brightwells Fm HRT 34
Brignall YON 131
Brigsley LIN 68
Brigstock NTP 68
Brig Stones CMB 68
Brill BUC 152, 193
Brimfield HRE 275
Brimpsfield GLO 274

Brimstage CHE 91, 92
Brincliffe YOW 155
Brindle Heath LNC 21, 194
Bringhurst LEI 236
Bringsty HRE 92, 93
Brinkburn NTB 13
Brinklow WAR 179
brinn PrW 153, 260
Brinsford STF 74
Brinshope HRE 136
Brinsop HRE 136, 139, 140
Brinsop LNC 140
Brinsworth YOW 74
Brisco CMB 247
Brisco YON 248
Briscoe CMB 249
Bristol GLO 68
Britford WLT 34, 77
Brithem Bottom DEV 100
Britwell BUC, OXF 34, 299
Brixham DEV 48, 53
Brixworth NTP 101
Broadbottom CHE 100
Broadcar NFK 57
Broadfield HRE, HRT 274
Broadholme NTT 56
Broadmead HRT 313
Broadward HRE 73
Broadwas WOR 64
Broadwater HRT, SSX 30–1
Broadway SOM, WOR 95
Broadwell GLO, OXF 1, 32
Broadwood DEV, SOM 259
brōc xxii, 2, 6–9, 10, 11, 301, 303
Brockdish NFK 8, 269
Brockenhurst HMP 236
Brockford SFK 75
Brockhampton HMP 7
Brockhampton var. cos. 7
Brockhurst var. cos. 235
Brockington DOR 7
Brockley SFK 8
Brockton SHR 7
Brockweir GLO 8

Brockworth GLO 8
Brocton STF 7
Brogden YOW 8, 118
Brokenborough WLT 150
Bromeswold DRB 256
Bromfield CMB 274
Bromfield SHR 275
Bromham BDF 48
Bromholm NFK 57
Bromley var. cos. 240
Bromsberrow GLO xvi, 148, 150
Bromsden Fm OXF 296
Bromsgrove WOR 227, 228
Bromswold NTP/HNT/BDF 254, 256-7
Brook IOW, KNT 7
Brooke NFK, RUT 7
Brookhampton OXF 303
Brookhampton WAR, WOR 7
Brookland KNT 282
Brookthorpe GLO 8
Brookwood SUR 8, 259
Broomfield var. cos. 275
Broomfleet YOE 16
Broomhall BRK 125
Broomhall BRK, CHE 129
Broomhaugh NTB 129
Broomhill KNT 193
Broomhope NTB 139
Broseley SHR 241
Brotton YON 7
Broughton HMP, LIN 148, 150
Broughton var. cos. 7, 303
Brownber WML 149
Brownsea DOR 37, 39, 41
Brownside DRB 219
Broxbourne HRT 12
Broxfield NTB 8, 275
Broxholme LIN 57
Broxted ESX 176
Brundall NFK 129
Brundish NFK 11, 269
Brundon SFK 168
Brun(n)eswold CAM 256

brunnr ON 10, 11
Brunshaw LNC 247
Brunton NTB 11
Brushfield DRB 274
Brushford DEV, SOM 68, 76
brycg 67-70, 308
bryggja ON 68
Brymore SOM 60
Bubbenhall WAR 193
Bubnell DRB 193
Bubwith YOE 253
Buckden HNT 117
Buckden YOW 117
Buckenhill HRE 193
Buckingham BUC 48, 49, 53
Buckhold BRK 233
Buckholt GLO, HMP 233
Buckhurst ESX 235
Buckland var. cos. 284
Buckleigh DEV 155
Buckleshore BDF 298, 309
Buckle Wood GLO 233
Buckley DEV 155
Bucklow CHE 179
Bucknall LIN, STF 131
Bucknell OXF, SHR 194
Bucknowle DOR 157
Buckover GLO 202
Buckshaw DOR 247
Buckwish SSX 286-7
Buda DEV 261
Budbridge IOW 68
Budbrooke WAR 8
Bude DEV 261
Buersill LNC 193
Bugbrook NTP 9
Buglawton CHE 179
Buildwas SHR 64
Bulbridge WLT 69
Bulford WLT 75
Bullaven DEV 46
Bulley GLO 241
Bullingham/-hope HRE 137, 139
Bullington HMP 170

INDEX

Bulman Strands WML 26
Bulmer ESX, YON 26
Bulphan Esx 44, 46
Bulstrode BUC, HRT 63
Bulverhythe SSX 87, 88
Bulwell NTT 33
Bungay SFK 37, 42
Bunny NTT 42
Buntingford HRT 75
Bunwell NFK 32
Burbage var. cos. 144, 145
Burcombe WLT 109
Burden DRH xii, 170
Burdon DRH 118
Burdon YOW 170
Bure Homage HMP 208
Burford OXF, SHR 76
Burgh SUR 150
Burghfield BRK 150, 272, 275
Burghill HRE 194
Burghope HRE 136, 139
Burgh le Marsh LIN 57
Burghley NTP 241
Burland CHE 282
Burland YOE 282
Burlescombe DEV 107
Burley var. cos. 241
Burlingjobb RAD 136, 137, 139
Burlton HRE, SHR 194
Burmarsh KNT 58
burna xxii, 2, 6, 7, 9–14, 121, 301–2
Burnby YOE 11
Burnham var. cos. 11
Burnham LIN 11
Burnham SOM 11, 50, 54
Burnsall YOW 128
Burneside WML 177
Burntwood STP 259
Burradon NTB 170
Burril YON 154
Burrow SOM 149, 150
Burscough LNC 248
Burshill YOE 193
Bursledon HMP 170

Burtholme CMB 57
Burthwaite Forest CMB 250
Burton Stather LIN 91
Burwardsley CHE 241
Burwash SSX 268
Burwell CAM, LIN 33
Busco YON 248
Bush KNT 269
Bushmead BDF 285
Bushwood WAR 259
Buston NTB 170–1
Butcombe SOM 107
Buteland NTB 282
Butterleigh DEV 240
Butterley var. cos. 238, 240
Buttermere CMB, WLT 26
Butterton STF 168
Buxhall SFK 131
byden 17, 18, 77, 100, 107, 301
Byfield NTP 274
Byfleet SUR 16
Byford HRE 74
Bygrave HRT 226
Byker NTB 57
Byland YON 282
bytme, bythme 34, 98–100
Bywell NTB 33

Caber CMB 150
Cabourn LIN 12
Caddington BDF 168, 296
Cadlands HMP 282
Cadmore BUC 27
Cadney LIN 41
Cadwell LIN 111
Cadwell OXF 32
Cainhoe BDF 188
Calbourne IOW 13
Caldbeck CMB 6
Caldbergh YON 150
Caldwell DRB, YON 32
Caldy CHE 42
Calehill KNT 193
Calkeld LNC 18

Callaly NTB 241
Callerton NTB 169
Callingwood STF 260
Callisham DEV 52
Calmsden GLO 117
Calow DRB 129
Calthwaite CMB 251
Caludon WAR 168
Calveley CHE 241
Calver DRB 202
Calverleigh DEV 259
Calverley YOW 241
Calverton BUC 296
camb 153, 159
Camberwell GTL 34
Camblesforth YOW 76
Cambo NTB 153, 188
Cambridge CAM 69
Camerton CMB, YOE 34
Cam Fell YOW 153
Campden GLO 118
Campsall YOW 128
Campsey SFK 43
Cams Head YON 153, 177
Canewdon ESX 171
Canfield ESX 274
Canford DOR 74, 100
Cann DOR 100
canne 100
Canney ESX 100
Cannop Brook GLO 135
Cantsfield LNC 275
Canvey ESX 37, 39, 43
Canwell STF 32, 100
Carbrook NFK 9
Carisbrooke IOW 9
Carlesmoor YOW 60
Carnforth LNC 75
Carrow NFK 188
Carshalton GTL 3
Carswalls GLO 32
Cartington NTB 168
Cascob HRE 136
Cashio HRT 188

Cassop DRH 134, 140
Castleford YOW 74
Castlethwaite WML 251
Castlett Fm GLO 141
Caswell var. cos. 32
Catcliffe YOW 155
Catcomb WLT 107
Catfield NFK 274
Catford GTL 75
Catforth LNC 75
Cathanger SOM, WLT 231
Catley HRE, LIN 241
Catmore BRK 22, 26
Catsfield SSX 275
Catshall SUR 193
Catshaw LNC 247
Cats Head Lodge NTP 175
Catshill WOR 193
Catsley DOR 155
Cattal YOW 129
Cattawade SFK 94, 95
Cattenhall CHE 128
Caudle GLO 32
Caughall CHE 194
Cauldon STF 169
Cauldwell BDF 82
Cavendish SFK 269
Caversfield OXF 274
Caversham BRK 305
Cavil YOE 272, 275
Cawkeld YOE 18
Cawkwell LIN 32
Cawood LNC, YOW 260
cęd PrW 119, 221, 223–4
celde 1, 14, 19
cetel 101, 107
Chackmore BUC 59, 60, 119
Chacombe NTP 107
Chadacre SFK 265
Chaddesden DRB 117
Chaddlehanger DEV 232
Chadlanger WLT 232
Chadnor HRE 202
Chadshunt WAR 18

INDEX

Chadwell ESX, LEI 32
Chaffcombe DEV, SOM 109
Chagford DEV 75
Chaldean's Fm HRT 117
Chalder SSX 209
Chalderbeach CAM 3
Chaldon DOR, SUR 169
Chalfont BUC 18, 301, 311
Chalford GLO, OXF xxii, 73
Chalgrave BDF 226
Chalgrove OXF 226
Chalhanger DEV 232
Chalkdell HRT 295
Chalk Wood OXF 298, 308
Challacombe DEV 108
Challenger DEV 232
Challow BRK 179
Channy Grove BRK 232
Chapel en le Frith DRB 225
Chapmanslade WLT 141
Charborough DOR 151
Charfield GLO 274
Charlcombe SOM 108
Charlinch SOM 182
Charlwood SUR 259
Charndon BUC 169
Charney BRK xvii, 39, 43
Charnwood LEI 257, 259, 260
Charsfield SFK 275
Chartridge BUC 190, 191, 296
Chatford SHR 74
Chatham ESX, KNT 224
Chatmoss LNC 61
Chatteris CAM 214, 215
Chattlehope Burn NTB 101
Chatwall SHR 32
Chatwell STF 32
Chaureth ESX 29
Chawleigh DEV 241
Chaxhill GLO 193
Cheadle CHE, STF 224
Chearsley BUC 312
Chebsey STF 41
Checkendon OXF 119, 293

Checkley HRE, STF 119
Chedburgh SFK 150
Cheddington BUC 168
Cheddleton STF 101
Cheddon SOM 118
Chedgrove NFK 228
Chedzoy SOM 41
Cheesehunger GLO 232
Cheetham LNC 224
Chelborough DOR 151
Cheldon DEV 168
Chelford CHE 77
Chelmarsh SHR 58
Chelmorton DRB 165
Chelmsford ESX 71, 74
Chelsea GTL 87, 88
Chelsfield KNT 274
Cheltenham GLO 50, 54
Chepenhall SFK 128
Chertsey SUR 40, 41
Chesford WAR 79
Chesham BUC 50, 54, 304
Cheshunt HRT 18
Chess R. HRT 302
Chessington GTL 168
Chesterfield DRB, STF 275
Chesterford ESX 74
Chesterhope NTB 139
Chetnole DOR 157
Chettiscombe DEV 101, 107–8
Chettle DOR 101
Chetwode BUC 224, 259, 313
Cheveridge WOR 191
Chickney ESX 42
Chicksgrove WLT 228
Chidden HMP 119, 223
Chideock DOR 224
Chidlow CHE 179
Chignall ESX 129
Chilcombe DOR, HMP 108
Childerley CAM 241
Childrey BRK 29, 285
Childwall LNC 33
Chillenden KNT 117, 267

Chillesford SFK 73
Chillmill KNT 14
Chilmead SUR 285
Chilwell NTT 33
Chimney OXF 39, 40, 41
Chingford ESX 73
Chinnor OXF 204, 208, 297, 308, 309
Chipnall SHR 157
Chippenhall SFK 128
Chippenham WLT 49, 54
Chippinghurst KNT 235
Chiselborough SOM 150
Chishall ESX 193
Chisledon WLT 117
Chislehurst GTL 236
Chisnall LNC 129
Chithurst SSX 224, 236
Chittlehamholt DEV 234
Chittlehampton DEV 101
Chivenor DEV 208–9
Cholsey BRK 39, 42, 305
Cholwell SOM 32
Chopwell DRH 33
Chorley var. cos. 241
Chorleywood HRT 312
Chosen Hill GLO 162
Choulton SHR 218
Chrishall ESX 130
Christon SOM 163
Chunal DRB 128
Churcham GLO 51, 53
Churchdown Hill GLO 162, 165, 170
Church Hanborough OXF 148, 150
Churchill, Church Hill var. cos. 162, 193
Churchill OXF, WOR 194
Churchlawton CHE 179
Churwell YOW 33
Chute WLT/HMP 223
Cinderford GLO 73
Claines WOR 196, 198

Clamhunger CHE 232
Clandon SUR 168
Clanfield HMP, OXF 271, 274
Clannaborough DEV 101, 150
Clanville HMP, SOM 274
Clapdale YOW 111
Clare OXF 209, 308
Clarendon WLT 171
Clatford HMP, WLT 75
Clatterford IOW 73
Clattinger WLT 231
Claverdon WAR 168
Claybrook LEI 8
Claydon var. cos. 168
Clayhanger var. cos. 230, 231, 232
Clayhidon DEV 169
Clayhill Fm GTL 232
Clayhithe CAM 87, 88
Claypole LIN 28, 29
Cleadon DRH 155, 169
Cleave DEV 155
Cleaveanger DEV 232
Cleaving YOE 101
Cleeve var. cos. 153, 155
Cleeve Cloud GLO 156
Clehonger HRE 232
Clennell NTB 193
Clerkenwell GTL 33
Clevancy WLT 155, 297
Cleve DEV, HRE 155
Clevedon SOM 155, 169
Cleveland YON 155, 283
Cleveley LNC, OXF 155
Clevelode WOR 81, 82, 155
Clewer BRK, SOM 155
Cliburn WML 12, 155
Cliddesden HMP 119
clif 144, 153–6, 296–7
Cliff(e) var. cos. 153, 155
Cliffe-at-Hoo KNT 153
Clifford var. cos. 155
Clifton var. cos. 153, 155
Climsland CNW 282
Clinger DOR 232

INDEX

Clingre GLO 232
Clitheroe LNC 174
Clive CHE, SHR 155
Cliveden BUC 118, 153, 155, 297
Cliviger LNC 155, 264, 265
*clof 101, 295
Clofesho 101
Cloffocks CMB 102
Cloford SOM 75
*clōh 102, 295
Closworth SOM 101
Clothall HRT 129
Clotton CHE 102
clough ME 102
Clougha LNC 102, 186, 189
Cloughton YON 102
Clovelly DEV 101
Clumber NTT 152
Clyffe DOR 155
Clyst William DEV 2
cnoll 157
Coalbrookdale SHR 8
Cobhall HRE 34
*cōc(e) 42, 157–8
Cockfield SFK 274
Cockhampstead HRT 158
Cocknowle DOR 157
Codford WLT 74
Codnor DRB 202
Coffinswell DEV 31
Cogdean DOR 117
Cogenhoe NTP 188
Coggeshall ESX 131
Cogload SOM 20
Colburn YON 11
Cold Cam YON 153
Cold Crouch SSX 163
Coldfield WAR 276
Coldmeece STF 61
Coldred KNT 244
Coldrey HMP 29
Coldridge BRK, DEV 191
Coldwell NTB 32
Colebatch SHR 3, 4

Coleford GLO, SOM 76
Coleham SHR 47
Colemore SHR 27
Coleridge BRK 190, 191
Colesden BDF 117
Coleshill BRK 194
Coleshill BUC, FLI 194, 296
Coleshill WAR 193
Colham MDX 51, 53
Collingbourne WLT xx, 12
Colmore HMP 27
Colney HRT 43
Colney NFK 42
Colsterdale YON 111
Coltishall NFK 131
Colwall HRE 32
Colwell DEV, NTB 32
Combe var. cos. 107
Combeinteignhead DEV 107
Comberbach CHE 3, 4
Combermere CHE 22
Combpyne DEV 107
Comhampton WOR 107
Compton DRB 118
Compton var. cos. xii, 97, 105, 107
Combs SFK 153
Combwich SOM 107
Condover SHR 203
Congreve STF 107, 228
Conholt WLT 234
Coniscliffe DRH 156
Conisford NFK 75
Conishead LNC 176, 177
Conock WLT 177
Consall STF 131
Consett DRH 177
Cookham BRK 49, 158
Cook Hill WOR 158, 193
Cookley WOR 156
Cookridge YOW 214, 215–6
Cooksey WOR 42, 158
Cooksland STF 282
Coole CHE 193
Coombe var. cos. 107, 293

339

Coombe's Fm BUC 293
Coombes SSX 107
Copeland CMB, DRH 284
Copford ESX 74
Copgrove YOW 228
Cople BDF 29
Copley Hill CAM 159, 179
Coploe Hill CAM 159, 179
Copnor HMP 205, 209
copp 158–9, 296
Copp LNC 158
Coppenhall CHE, STF 128
Coppingford HNT 75
Coppull LNC 159, 193
Copsegrove GLO 229
Corbridge NTB 69
Coreley SHR 185
corf 102, 295
Corfe SOM 102
Corfe Castle DOR 102
Corfe Mullen DOR 102
Corfham SHR 102
Corfton SHR 102
Cornaa IOM 2
Cornbrook LNC 8
Corney CMB, HRT 39, 42
Cornforth DRH 75
Cornsay DRH 188
Cornwell OXF 32
Cornwood DEV 260
Corpusty NFK 92, 93
corscombe DEV 103
Corscombe DOR 102–3, 109
Corton Denham SOM 102
Corve Dale SHR 102
Cosford WAR 74
Cosgrove NTP 228
Cossall NTT 128
Costessey NFK 42
Cotgrave NTT 228
Cotheridge WOR 191
Cotness YOE 198
Cotswold(s) GLO/OXF/WOR 119, 253, 254, 256, 257

Cottered HRT 30
Cottesbach LEI 3, 4
Cottesbrooke NTP 8
Cottesloe BUC 179
Cottesmore RUT 60
Cottingley YOW 241
Cottisford OXF 74
Coughton HRE, WAR 158
Coundon DRH 169
Coundon WAR 3
Coupland NTB, WML 284
Courns Wood BUC 298, 309
Courteenhall NTP 131
Coveney CAM 42
Covenhope HRE 136, 140
Cowbridge ESX, WLT 69
Cowdale DRB 110, 111
Cowden KNT, SSX 267
Cowden YOE 169
Cowgrove DOR 229
Cowhill GLO 193
Cowley DRB 241
Cowley GLO 241
Cowlinge SFK 53
Cowm LNC 107
Cowpe LNC 136, 139
Cowridge BDF 190, 191
Cowsland NTT 243
Coxford NFK 34, 77
Coxwell BRK 34
Coxwold YON 257
Crabwall CHE 34
Cracoe YOW 174
Craiselound LIN 243
Crakehall YON 129
Crakehill YON 129
Crakemarsh STF 58
Cranage CHE 21
Cranborne DOR 12
Cranbourne HMP 12
Cranbrook KNT 8
Crandon SOM 171
Craneham DEV 52
Cranfield BDF 275

INDEX

Cranford GTL, NTP 75
Cranmere SHR 26
Cranmore SOM 26
Cranoe LEI 188
Cransford SFK 75
Cranshaw LNC 247
Cranwell LIN 32
Cranwich NFK 286, 287
Craswall HRE 32
Cratfield SFK 274
Crawley var. cos. 241
Crawshaw LNC, YOW 247
Cray R. 1
Crayford GTL 76
Creacombe DEV 108
Credenhill HRE 193
Creechbarrow SOM xvi, 148, 159–60, 163
Creech Hill DOR, SOM 163
Creech St Michael SOM 161
Creed SSX 189
Creeksea ESX 88–9
Cregneash IOM 198
Crendon BUC 168
Creskeld YOW 18
Creslow BUC 179
Cresswell DRB, STF 32
Crewood CHE 259
Criccieth CRN 163
Crich DRB 163
Crichel DOR 163
Cricket SOM 162, 163
Crickheath SHR 163
Crickhowell BRE 163
Cricklade WLT 81
Crickley Hill GLO 162, 163
Criddon SHR 168
Criggleston YOW 163
Cringleford NFK 74
Crockernwell DEV 33
Croichlow LNC 163
Cromer NFK 26
Cromford DRB 73
Cromhall GLO 129

Cromwell NTT 33
Crondall HMP 113
Cronkshaw LNC 247
Cronkston DRB 169
Crookbarrow WOR xvi, 148, 159, 163
Crooke DEV 163
Crook Hill DOR, WLT 163
Crooksbury SUR 163
Cropredy OXF 29, 30
Cropwell NTT 30, 193
Croscombe SOM 102–3, 109
Crosland YOW 282
Crossdale CMB 111
Crossens LNC 198
Crossrigg WML 190
Crosthwaite var. cos. 251
Crostwick NFK 251
Crostwight NFK 251
Crouch End GTL 163
Crouch Hill OXF, WLT xvi, 163
Crowborough SSX 150
Crowcombe SOM 108
Crowell OXF 32, 299
Crowfield SFK 275
Crowhurst SSX, SUR 235
Crowland LIN 283
Croxall STF 128
Croxden STF 117
Croxteth LNC 91, 92
Croyde Hoe DEV 189
Croydon CAM 117
Croydon GTL 114, 117
Croydon SOM 169
Croydon Wilds CAM 255, 256
Crucadarn BRE 163
Cruchfield BRK 163, 275
Crudwell WLT 34
crūg PrW xvi, 143, 148, 159–63
Crumacre YOW 264, 265
Crumbacre YON 265
Crutch WOR 161, 163
Cryfield WAR 275
Cucket Nook YON 257

Cuckfield SSX 275
Cuckmere R. SSX 25–6
Cuckney NTT 42
Cuddesdon OXF 168
Cuerdale LNC 111
Culcheth LNC 223–4
Culford SFK 74
Culgaith CMB 223–4
Culham Fm BRK 304
Culham OXF 49, 53
Culphoe SFK 189
cumb xiii, xv, xxii, 97, 103–9, 114, 125, 292, 293, 294
Cumberland 283
Cumberlow HRT 180
cumbo- British 106–7
Cumcatch CMB 106
Cumcrook CMB 106, 163
Cumdivock CMB 106
Cummersdale CMB 111
Cumnor BRK 209
Cumrew CMB 106
Cumwhinton CMB 106
Cumwhitton CMB 106
Cundall YON 110, 112, 113
Cunscough LNC 249
Curbridge OXF 69
Curdridge HMP 191
Curland SOM 282
Curload SOM 20
The Curraghs IOM 61
Curridge BRK 191
Curthwaite CMB 251
Cusop HRE 136
Cutsdean GLO 119, 257
Cutteridge WLT 191
Cutteslowe OXF 178, 179
Cuxham OXF 51, 53
Cuxwold LIN 257

Daccombe DEV 109
Dadford BUC 74
Dagnall BUC 128, 294
Dalby var. cos. 110

dale ME 110
Dale Abbey DRB 110, 111
Daleham SSX 112
Dale Park SSX 112
Dalestorth NTT 63
Dale Town YON 110
Dalham KNT, SFK 112
Dalkeith MLO 224
Dallinghoo SFK 189
dalr ON 97–89 110
Dalton var. cos. 110, 113
Dalwood DEV 112
Damerham HMP 51, 53
Danaway KNT 112
Danehill SSX 267
Darfield YOW 275
Darley DRB 241
Darnall YOW 129
Darnford SFK 73
Darnhall CHE 129
Dartford KNT 71, 76
Dartmoor DEV 58, 59, 60
Darwen R. 1
Dassett WAR 224
Datchet BUC 224
Dauntsey WLT 40, 42
Dawdon DRH 110
Dawley GTL 113
Daylesford WOR 74
dæl 110, 111–12
Deadwin Clough LNC 102
Deal KNT 112
Dean var. cos. xxi, 115, 293
Deane HMP, LIN 115
Deanham NTP 115
Deanshanger NTP 232
Dearnbrook YOW 8
Debach SFK 144
Debdale Lodge NTP 110
Debden ESX 117
Deene NTP 115
Deepdale LNC, YON 111
Deerhurst GLO 235
Defford WOR 73

INDEX

Deirewald YOW 254
dell 113, 295
Denes NFK 115
Denford BRK, NTP 115
Dengie ESX 42
Denham BUC 115
Denholme YOW 115
Denmead HMP 285
denn 114, 262, 267
Denner BUC 209, 298, 309
Denny CAM 42
Denton var. cos. xii, 115
denu xiii, xv, xviii, xxi, xxii, 97, 98, 103, 112, 113–22, 125, 267, 292, 293, 303
Denver NFK 70–1
Denwick NTB 115
Depden SFK 117
Deptford GTL, WLT 73
Desborough BUC 150, 296
Desford LEI 75
Deuxhill SHR 194
Dewsall HRE 33
Dibden HMP, KNT 117
Dickley ESX 241
Didbrook GLO 8
Didham DEV 52
Didmarton GLO 22, 27
Digswell HRT 34
Dillicar WML, YOW 264, 265
Dillington HNT, SOM 209
Dinchope SHR 136, 137, 139
Dinsdale YON 129
Dippenhall SUR 129
Diptford DEV 73
Dipton DRH 117
Dishforth YON 74
Ditchburn NTB 12
Ditchford NTP, WAR 74
Ditchley OXF 241
Ditteridge WLT 191
Dittisham DEV 52, 53
Dixton GLO 165, 171
Dockenfield HMP 275

Dodderhill WOR 193
Doddershall BUC 193
Doddinghurst ESX 236
Dodford NTP, WOR 74
Dodwell WAR 32
Doepath NTB 90
Dogmersfield HMP 26, 275
Dolton DEV 283
Donhead WLT 167, 175, 177
Doniford SOM 77
Donyland ESX 282
Dordon WAR 169
Dorney BUC 42, 305
Dornford OXF 73
Dosthill WAR 193
Dotland NTB 282
Dovercourt ESX 131
Doverdale WOR 112, 113
Doveridge DRB 69
Dowdeswell GLO 34
Dowland DEV 283
Down(e) DEV, KNT xxi, 167
Downham LNC, NTB 167
Downham var. cos. 167
Downham Hithe CAM 86, 88
Downhead SOM 167, 175, 177
Downholme YON 167
Downstow DEV 13
Downton var. cos. 167
Downwood HRE 167, 259
Doxey STF 42
Drakelow var. cos. 178
Drakenage WAR 173
Drascombe DEV 108
Draycott Moor BRK 58
Drellingore KNT 209
Driffield GLO, YOE 272, 274
Dronfield DRB 275
Droxford HMP 78
Droylsden LNC 119
Druridge NTB 191
Dudbridge GLO 69
Duddo NTB 188
Duddoe NTB 117

343

Duddon CHE 168
Dudley WOR 240
Dudswell HRT 300
Duffield DRB, YOE 272, 275
Duffield Frith DRB 225
Dukinfield CHE 275
Dulwich GTL 287
Dumbleton GLO 283
Dummer HMP 27, 167
dūn xiii, xv, xvi, xvii, xviii, xix, xxi, xxii, 37, 38, 114, 125, 143, 164–71, 192, 296
Dunbridge HMP 69, 115
Dunchideock DEV 224
Dundon SOM 118, 167
Dundry SOM 167
Dungrove DOR 228
Dunhampton WOR 167-8
Dunhamstead WOR 167
Dunkenhalgh LNC 128
Dunkenshaw LNC 247
Dunkeswell DEV 32
Dunnett SOM 167, 175, 177
Dunnockshaw LNC 247
Dunnose IOW 197
dūno- British 164
Dunsden OXF 117, 292
Dunsfold SUR 13
Dunsford DEV 74
Dunsforth YOW 74
Dunsley HRT 312
Dunsmore WAR 58, 59, 60
Dunstan NTB 167
Duntisbourne GLO 13
Duntish DOR 167
Dunton var. cos. 167
Dunwood HMP 167, 259
Dupath CNW 90
Durborough SOM 150
Durford SSX 75
Durham DRH 55, 56
Durleigh SOM 241
Durley HMP 241
Durnford WLT 73

Duxford BRK 71, 74
Dyrham GLO 50, 54

ēa 2, 6, 14–16, 120
Eaglesfield CMB 276
Ealand LIN 15, 282
Eamont WML 15
Earith HNT 86, 88
Earle NTB 194
Earley BRK 240
Earnley SSX 240
Earnshaw YOW 247
Earnshill SOM 193–4
Earnwood SHR 260
Earsdon NTB 168
Easebourne SSX 12
Easinghope WOR 137, 139
Easingwold YON 256
Eastbourne SSX 10, 12
Eastbridge KNT 69
Eastburn YOE 10
Eastburn YOW 12
Eastham WOR 50
Easthope SHR 136, 139
Easthwaite CMB 251
Eastleach GLO 21
Eastlound LIN 243
Eastness ESX 198
Eastnor HRE 203
Eastrea CAM 43
Eastridge WLT 191
Eastwell KNT, LEI 33
Eastwood ESX 259
Eastwood NTT 251
Eaton var. cos. 14–15
Eaton var. cos. 39, 41
Eau R. LIN 14
Ebbesborne WLT 12
Ebbsfleet KNT 16
Ebury GTL 41
Ecchinswell HMP 34
Ecclesall YOW 130
Ecclesfield YOW 276
Eccleshall STP 130

INDEX

Eccup YOW 139
ecg 173–4
Eckland NTP 243
Edale DRB 39, 41, 110, 111
Edder Acres DRH 265
Edenbridge KNT 69
Edenfield LNC 275
Edenhall CMB 129
Edenhope SHR 139
Edensor DRB 202
Edgbaston WAR 174
Edge var. cos. 173–4
Edgefield NFK 276
Edgerley SHR 174
Edgeworth GLO 174
Edgefield NFK 174, 269
Edgeley CHE, SHR 174, 269
Edginswell DEV 31
Edgton SHR 168
Edgware GTL 174
Edingale STF 128
Edington WLT 168
edisc 269
Edlesborough BUC 148, 150, 296
Edvin HRE 46
Efford var. cos. 73
ēg xviii, xix, xxi, xxii, 36, 37–44, 55, 110, 125, 165, 305
Egdean SSX 117
Egford SOM 74
Eggardon DOR 167, 168
Eggbear DEV 223
Eggbeer DEV 223
Eggesford DEV 74
Eggington BDF 168
Egglescliffe DRH 156
Egmere NFK 27
Egrove BRK 228
Eleanor Fm SSX 205, 209
Elbridge KNT 68
Elcombe WLT 107
Eldersfield WOR 276
Eldmire YON 26
Eldon DRH 168

Elford NTB, STF 74
Elham KNT 54
Elland YOW 15, 282
Ellel LNC 128
Ellenbrook LNC 8
Ellenhall STF 130
Ellerbeck YON 6
Ellerburn YON 12
Ellerker YOE 57
Ellershaw CMB 248
Ellesborough BUC 150, 296
Ellesmere SHR 1, 22, 27
Ellisfield HMP 274
Ellishaw NTB 247
Elmbridge WOR 191
Elmdon ESX, WAR 168
Elmesworth IOW 205–6, 209
Elmley KNT, WOR 237, 239, 240
Elmore GLO 200, 202
Elmsall YOW 129
Elmswell SFK 32
Elmswell YOE 33
Elphaborough YOW 17
Elsdon NTB 117
Elsfield OXF 274
Elslack YOW 142
Elveden SFK 117
Elvendon OXF 293, 296
Elwell DOR 33
Emanuel Wood ESX 131
Emblehope NTB 139
Embleton DRH 117
Emborough SOM 150
Embsay YOW 38, 42
Emlett DEV 233
Emmott LNC 15
Emstrey SHR 41
Enborne BRK 12
Endon STF 169
Enfield MDX 276
Enford WLT 75
Englebourne DEV 12
Englefield BRK 272, 276
Englefield SUR 274

345

Enmore SOM 26
Ennerdale CMB 111
Enville STP 274
Epney GLO 42
Eppleworth YOE 253
Epwell OXF 32
Eridge SSX 191
Eriswell SFK 34
Erith GTL 87, 88
Erringden YOW 117
ersc 267–9, 312
Escombe DRH 269
Escrick YOE 214, 215
Esholt YOW 233
Espley NTB 240
Essendine RUT 117, 120
Essendon HRT 119–20
Etal NTB 131
Etchingham SSX 53
Eton var. cos. 14–15, 303
Ettingshall STF 131
Etwall DRB 32
Eudon SHR 168
Evedon LIN 168
Evelith SHR 182, 185
Evenjobb RAD 136, 137, 139
Evenlode WOR 81
Evenwood DRH 259
Evercreech SOM 162–3
Everden NTP 169
Everley WLT, YON 241
Eversden CAM 169
Evershaw BUC 247
Eversheds Fm SUR 175
Eversley HMP 241
Eversholt BDF 234
Evesbatch HRE 3, 4
Evesham WOR 48, 49, 53
Ewart NTB 15
Ewell GTL 3
Ewelme OXF 2, 299
Ewen GLO 2
Ewhurst var. cos. 235
Ewood LNC 15, 259

Exbourne DEV 12
Exceat SSX 26
Exford SOM 76
Exhall WAR 130
Exlade Street OXF 295
Exmoor DEV/SOM 58, 59, 60
ey ON 55
Eyam DRB 39, 41
Eydon NTP 168
Eye var. cos. xxi, 39, 41
Eyecote GLO xii, 41
Eyeworth BDF 41
Eyford GLO 76
Eynsford KNT 74
Eynsham OXF 50, 54
Eyton SHR 41

Faccombe HMP 107
Facit LNC 219
Fairbourne KNT 12
Fairburn YON 12
Fairfield DRB, KNT 274
Fairfield WOR 276
Fairford GLO 73
Fairlight SSX 240
Fairwood YON 256
Falfield GLO 274
Fallowdon NTP 168
Fallowfield LNC, NTB 274
Falmer SSX 27
Falsgrave YON 192
Fambridge ESX 44, 46, 69
Fancott BDF 46
Farcet HNT 176
Farewell STF 32
Farforth LIN 78
Faringdon var. cos. xvii, xx, 168
Farleigh var. cos. 240
Farley var. cos. 240
Farlow SHR 178
Farmborough SOM 150
Farmington GLO 22, 27
Farnacres NTB 265
Farnborough var. cos. 150

INDEX

Farncombe SUR 107
Farndale YON 111
Farndon var. cos. 169
Farnham SUR 51, 54
Farnhill YOW 193
Farnley YOW 240
Farnsfield NTT 275
Farringdon DEV 169
Farthingdale SUR 113
Farthinghoe NTP 189
Farway DEV 70, 96
Faulkbourne ESX 12
Faulkland SOM 284
Fawcett Forest WML 219
Fawcettlees CMB 219
Fawdon NTB 168
Fawley BUC 312
Faws Grove OXF 228
Fawside DRH 219
Faxfleet YOE 16
Fazakerley LNC 265
fær 65, 70–1
Fearnhead LNC 177
Feckenham WOR 52, 53
Felbridge SUR 69, 274
Felbrigg NFK 68
feld xviii, xxi, xxii, 262, 268, 269–78, 279, 280, 310–11, 312, 314
Feldom YON 274
Feldon WAR 274
Fell Beck Bridge YOW 68–9
Fell Briggs YON 69
Felliscliffe YOW 155
Felmersham BDF 48
Felsted ESX 274
Feltham MDX 274
Felton var. cos. 274
Feltwell NFK 32
Fenacre DEV 265
Fenby LIN 46
Fenchurch GTL 44, 46
Fencote HRE, YON 46
Fencott OXF 46
Fen Ditton CAM 44

Fenemere SHR 27
Fenham NTB 44, 46
Fenhampton HRE 46
fenn 37, 44–6, 83
Fenny Drayton LEI 44
fenten Cornish 18
Fenton var. cos. xii, 46
Fenwick NTB, YON 46
ferja ON 71
Fernham BRK, DEV 52, 54
Fernhurst SSX 235
Ferriby LIN, YOE 71
Ferrybridge YOW 71
Ferry Fryston YOW 71
Fersfield NFK 275
Fiddleford DOR 74
Field STF 274
Figheldean WLT 117
Filey Brigg YOE 68
Filgrave BUC 228
Fillongley WAR 241
Fimber YOE 22, 27
Finborough SFK 150
Finburgh WAR 150
Finchale DRH 129
Finchfield STF 275
Finchingfield ESX 276
Finchley GTL 241
Findon SSX 169
Finedon NTP 118
Fineshade NTP 177
Fingest BUC 236, 315
Finghall YON 128
Fingringhoe ESX 189
Fingrith ESX 29
Finkley HMP 241
Finmere OXF 26
Finnamore Fm BUC 306
Finningley NTT 241
Finsthwaite LNC 250
Firbank WML 225
Firbeck YOW 225
Firths YOE 225
Fishacre DEV 265

Fishbourne SSX 12
Fishburn DRH 12
Fishlake YOW 20
Fishmere LIN 27
Fitz SHR 188
Fitzworth DOR 206, 209
Flashbrook STF 9, 16
Flaska CMB 249
Flaskew WML 249
Flaunden HRT 117
Flawborough NTT 148, 150
Flawforth NTT 73
Flawith YON 94
Flaxley GLO, YOW 241
Fleckney LEI 43, 189
Flecknoe WAR 189
Fleet var. cos. 16
Fleetham NTB, YON 16
Fleetwood LNC 16
flēot 16–17
Fletching SSX 189
Flete DEV 16
Fletton LIN 16
Flixborough Stather LIN 91
Flordon NFK 171
Flotterton NTB 96
Fluskew CMB 249
Flyford WOR 225–6
Foleshill WAR 194
Fonthill WLT 18
Fontmell DOR 18
Forcett YON 72
ford xxii, 65, 66, 71–80, 91
Ford var. cos 72
Fordham var. cos. xii, 72
Fordingbridge HMP 69
Fordley SFK 72
Fordon YOE 170
Fordwich KNT 72
Foreland KNT 284
Forsbrook STF 8
Forton var. cos. 72
Fotheringhay NTP 42
Foulden NFK 169

Fouldray LNC 197–8
Foulness ESX 196, 198
Foulridge LNC 191
Founthill SSX 18
Fovant WLT 18
Fowlmere CAM 22, 26
Fownhope HRE 136, 139
Foxhanger's Fms WLT 231
Foxley var. cos. 241
Foxton DRH, NTB 117
Fradswell STF 32
Framfield SSX 276–7
Framilode GLO 81, 82
Framland LEI 242, 243
Framsden SFK 117
Framwellgate DRH 32
Fredley SUR 225
Freeford STF 78
Frenchay GLO 247
Freshford SOM 73
Freshwater IOW 30
Frieth BUC 225, 315
Fright Fm SSX 225
Frilford BRK 71, 74
Friskney LIN 15
Frith var. cos. 225, 315
Fritham HMP 225
Frithsden HRT 118, 225, 292, 315
Frithville LIN 225
Fritwell OXF 34
Frizenham DEV 52
Frizinghall YOW 128
Frodsham CHE 47, 53
Frogmore var. cos. 24–5, 26
Frome R. var. cos. 16
Frontridge SSX 18
Frostenden SFK 120
Froxfield HMP, WLT 275
The Frythe HRT 225
Fryup YON 136–7, 140
Fulbeck LIN 6
Fulbourn CAM 7, 12
Fulbrook var. cos. 7, 8
Fulford var. cos. 73

INDEX

Fulham GTL 48, 49, 53
Fulmer BUC 22, 26, 303
Fulready WAR 29
Fulshaw CHE, LNC 247
Fulwell DRH, OXF 32
Fulwith YOW 253
Fundenhall NFK 131
funta 1, 17–18, 300
Funthams CAM 18
Funtington SSX 18
Funtley HMP 18
Furness LNC 197–8
Furtho NTB 72, 189
fyrhth(e) 221, 224–6, 315

Gabwell DEV 32
Gad Bridge BRK 69
Gaddesden HRT xvii, 120, 292
Gade R. HRT 120
Gadshill KNT 194
Gagingwell OXF 33
Gainford DRH 73
Gaisgill WML 123
Galsworthy DEV 209
Gangsdown OXF 117, 293
Gappah DEV 90
Garendon LEI 167, 168
Garford BRK 71, 74
Garforth YOW 74
Gargrave YOW 228
Garmondsway DRH 95
Garsdale YOW 111
Garsden WLT 168
Garsington OXF 168
Garswood LNC 261
Gartside LNC 219
Gascow LNC 248
Gatacre SHR 264, 265
Gatcombe IOW 108
Gatecliff YOW 186
Gateford NTT 76
Gateforth YOW 76
Gateley NFK 241
Gatesbury HRT 120

Gateshead DRH 175, 176, 212
Gathurst LNC 236
Gatley CHE 155
Gatley HRE 183, 184, 185
Gattertop HRE 136, 140
Gauxholme LNC 56
Gawber YOW 151
Gaydon WAR 171, 261
Gayhurst BUC 236
Gaywood NFK 171, 261
Gazegill YOW 123
Gedgrave SFK 229
Gedney LIN 42
gelād 20, 21, 65, 81–3, 184
Georgeham DEV 53
gewæd 65, 94–5
Gibsmere NTT 22, 27
gil ON 123
Gilcrux CMB 163
Gillcambon CMB 123
Gilsland CMB 282
Gipsey Race R. ERY 11
Gisburn YOW 13
Gisleham SFK 13
Gislingham SFK 13
Givendale YOE 111
Givendale YON 110, 111, 113
Glaisdale YON 111
Glandford NFK 75
Glandford Brigg LIN 68, 75
Glapwell DRB 34
Gledholt YOW 234
Glemsford SFK 75
Glen LEI 123
Glencoyne CMB/WML 123
Glendon NTP 168
Glendue NTB 123
Glenfield LEI 274
glennos British 123
Glenridding WML 123
Glossop DRB 139
Glusburn YOW 11
Glydwish SSX 286
Glynde Reach SSX 215

Gnatham DEV 52
Gnosall STF 131
Goatacre WLT 265
Goathill DOR 193
Goathland YON 282
Goathorn DOR 206, 209
Goathurst SOM 236
Gobsore DEV 210
Godington OXF 168
Godney SOM 42
Godshill HMP, IOW 194
Golborne CHE, LNC 12
Golder OXF 209, 308
Goldhanger ESX 232
Gomeldon WLT 171
Gomershal YOW 128
Gomshall SUR 218
Goodacre DEV 265
Goodwood SSX 260
Goosewell DEV 33
Goosey BRK 39, 42
Gopshall LEI 194
Gordano SOM xvi, 117
Gornal STF 130
Gosfield ESX 276
Gosford var. cos. 76
Gosforth CMB, NTB 76
Goudhurst KNT 235
Gowbarrow CMB 151
Gowdall YOW 129
Graffham SSX 228
Grafham HNT 228, 258
Grafton var. cos. 227, 228, 258
Grandborough BUC 150
Gransden CAM, HNT 117
Gransmoor YOE 60
Grasmere WML 26
Graston DOR 228
Gratton DEV 168
Graveley CAM 228
Graveney KNT 15
Gravenhill OXF 228
Gravenhunger SHR 228, 230, 232
Gravenhurst BDF, SSX 228, 236

Gravesend KNT 226, 228
Grayrigg WML 190
Grayshott HMP 228
Graythwaite LNC 250
Grazeley BRK 62
græfe, grāf(a) 221, 226–30, 233, 314
Greatford LIN xxii, 73
Great Ormes Head CRN 176
Greenacre KNT 265
Greenfield OXF 311
Greenford GTL 73
Greenhaugh NTB 129
Greenhill WOR 194
Greenhithe GTL 87, 88
Greenscoe LNC 249
Greenway DEV 95
Greetland YOW 282
Greetwell LIN 32
Grendon HRE 117
Grendon var. cos. 168
Gressenhall NFK 129
Gresty CHE 92
Greta R. 2
Greywell HMP 32
Grimswolds Fm WAR 257
Grindalythe YOE 185
Grindle SHR 193
Grindon var. cos. 168
Gringley-on-the-Hill NTT 241
Grinshill SHR 195
Grisedale CMB, WML 111
Grizedale LNC 111
Groombridge KNT 69
Grove var. cos. 228, 314
Grovebury BDF 228
Grovely Wood WLT 228
Groveridge OXF 296
Guestwick NFK 251
Guildford SUR 71, 73
Gunthwaite YOW 250
Gurnard IOW 206, 209
Gurnard's Head CNW 175
Gusford SFK 74
Guy's Cliff WAR 155

INDEX

Gwash R. 64

Haccombe DEV 108
Hackford NFK 74
Hackforth YON 74
Hackness YON 197
Hackney GTL 43
Hackpen Hill DEV, WLT 212
Hackwood HMP 260
Haddiscoe NFK 248
Haddlesey YOW 30
Haddon HNT 168
Haddon var. cos. 169
Hadfield DRB 274
Hadleigh ESX, SFK 240
Hadley HRT, SHR 240
Hadlow CHE 179
Hadnall SHR 128
Hadzor WOR 200, 202
Hagbourne BRK 12
Hagley var. cos. 240
Haighton LNC 128
Hainault Forest ESX 234
Hainford NFK 78
Haithwaite CMB 251
Halam NTT 128
Hale var. cos. 128, 294
Hale LIN 126
Halefield NTP 128
Hales var. cos. 128
Halewood LNC 259
Halford var. cos. 74, 128
halh xxii, 37, 97, 123–33, 192, 294
Halliford MDX 73
Hallam DRB 128
Hallaton LEI 128
Hal(l)ikeld YON 18–19
Halliwell LNC 33
Halloughton NTT, WAR 128
Hallow WOR 126, 128
Halnaker SSX 264
Halsall LNC 131
Hallsenna CMB 174
Hallshot DEV 233

Halshanger DEV 232
Halsinger DEV 232
Halse NTB 189
Halston SHR 128
Halsway SOM 95
Haltcliff CMB 102
Haltham LIN 228, 233
Halton var. cos. 128, 294
Halwell DEV 33
Halwill DEV 33
Ham var. cos. 47–53
Hambleden BUC 117, 293
Hambledon var. cos. 168
Hameldon LNC 168
hamm 36, 37, 39, 46–55, 56, 133, 196, 304
Hammerden SSX 267
Hammill KNT 254, 256
Hammoon DOR 49, 52
Hamp SOM 50, 53
Hampden BUC 120, 292, 294
Hampen GLO 212
Hampole YOW 29
Hampreston DOR 49, 52
Hampsthwaite YOW 250
Hampton var. cos. 49, 53
Hampton Loade WOR 82
Hamsey SSX 53
Hamworthy DOR 50, 53
Handbridge CHE 68
Handforth CHE 76
Handley var. cos. 241
Handsacre STF 264, 266
Hanford STF 76
Hang YON 231
Hanger var. cos. 231
Hangerberry GLO 231
hangra 230–4, 315
Hankelow CHE 179
Hankford DEV 74
Hankham SSX 53
Hanley STF, WOR 239, 241
Hanlith YOW 186
Hanmer CHE 22

Hannah LIN 42
Hanney BRK 39, 42
Hanslope BUC 61
Hanwell GTL 33
Hanwell OXF 95
Hanwood SHR 261
Hapsford CHE 74
Harberton DEV 12
Harbledown KNT 167, 168
Harborne STF 10, 11
Harborough LEI 150
Harborough WOR 150
Harbridge HMP 69
Harding WLT 117
Hardisty YOW 93
Hardley HMP, IOW 240
Hardmead BUC 285
Hardwell BRK 33
Harefield MDX 277
Harehope NTB 139
Harepath DEV 90
Haresceugh CMB 249
Harescombe GLO 107
Haresfield GLO 274
Harewood var. cos. 261
Harford GLO 75
Hargrave var. cos. 228
Hargrove DOR, GLO 228
Harlington BDF 169
Harlow ESX 179, 180
Harlsey YON 42
Harnage SHR 173
Harnham Bridge WLT 94
Harnhill GLO 195
Harpenden HRT xvii, 120, 307
Harpford DEV, SOM 74
Harpole NTP 28, 29
Harpsden OXF 120, 292, 307
Harpswell LIN 35
Harragrove DEV 228
Harraton DEV 261
Harrietsham KNT 52, 53
Harrold BDF 254, 256
Harrop CHE, YOW 139

Harrowbridge CNW 69, 70
Harrowden BDF, NTP 170
Harswell YOE 32
Hartanger KNT 231
Hartburn DRH, NTB 12
Hartest SFK 235
Hartford CHE, NTB 75
Hartford HNT 78
Hartforth YON 75
Hartgrove DEV, DOR 228
Harthill var. cos. 193
Hartington DRH 170
Hartington NTB 96
Hartland DEV 281, 283
Hartlepool DRH 28, 39, 42
Hartley var. cos. 241
Harton DRH 169
Hartridge BRK 190, 191
Hartshead LNC, YOW 175, 176
Hartshill WAR 192, 193
Hartshurst SUR 235
Hartside CMB 176
Hartsop WML 139
Hartwell var. cos. 32
Hartwith YOW 253
Harty KNT 42
Harvington (near Evesham) WOR 78
Harwell BRK 33
Harwell NTT 32
Harwood var. cos. 261
Hascombe SUR 108
Haselbech NTP 3, 4
Haselbury DOR, SOM 223
Haselden SSX 267
Haseley var. cos. 240, 312
Haselor WAR, WOR 200, 202
Haselour STF 202
Hasfield GLO 275
Hasilacre SUR 264, 265
Hasland DRB 243
Haslemere SUR 26
Haslingden LNC 117
Haslingfield CAM 276

INDEX

Hassall CHE 130
Hassop DRB 140
Haswell DRH, SOM 32
Hatcliffe LIN 155
Hatfield var. cos. 274
Hatford BRK 71, 74, 175, 176
Hatherden HMP 117
Hatherleigh DEV 240
Hatherley GLO 240
Hathersage DRB 173
Hathershaw LNC 247
Hathershelf YOW 218
Hatton var. cos. 279
Haughstrother NTB 63
Haughton NTT 188
Haughton (halh) var. cos. xii, 128
haugr ON 174, 188
Hauxwell YON 32
Haulgh LNC 128
Hautbois NFK 286, 287
Havant HMP 18
Haverbrack WML 152
Haverhill SFK 193
Haverholme LIN 57
Haveringland NFK 282
Haversham BDF 48
Haverthwaite LNC 251
Hawkedon SFK 169
Hawkes End WAR 92
Hawkesley WOR 179
Hawkhill NTB 193
Hawkhurst KNT 235
Hawkridge KNT 235
Hawkwell ESX 35
Hawkwell NTB, SOM 32
Hawne WOR 128
Hawold YOE 257
Hawridge BUC 190, 191, 296
Haxey LIN 40, 42, 57
Haycrust SHR 235
Hayden Fm OXF 296
Haydon CAM 118
Haydon var. cos. 169
Haydon Bridge NTB 117

Hayfield DRB 277
Haynes BDF 197, 198
Haythwayt CMB 251
Haywood HRE, STF 260
Hazard DEV 209
Hazel Hall SUR 233
Hazel Hanger GLO 231
Hazelholt SSX 234
Hazelhurst SSX 269
Hazeland WLT 233–4
Hazelmere BUC 26
Hazelslack WML 142
Hazelwood GLO, WAR 234
Hazlebadge DRB 3, 4
Hazlebarrow DRB 223
hǣth 87, 262, 279
Headingley YOW 241
Headlam DRH 240
Headley var. cos. 240
hēafod 114, 175–7
Heage DRB 173
Healaugh YON, YOW 241
Healey var. cos. 241
Heanor DRB 202
Hearthcote Fm DRB 279
Heath DRB 243
Heath var. cos. 279, 315–6
Heathcote DRB, WAR 279
Heathfield var. cos. 272, 274
Heathwaite LNC, YON 251
Heatley CHE 240
Hebden YOW 117
Heckfield HMP 272, 274–5
Heddon NTB 169
Hedley var. cos. 240
Hednesford STF 74
Hedon YOE 37, 165, 169
Hedsor BUC 209, 309
Heeley YOW 241
helde 144, 153, 177
Hele DEV, SOM 128
Hellesdon NFK 168
Hellidon NTP 120
Hellifield YOW 274

Hellingly SSX 242
Helmdon NTP 114, 117
Helmsley YON 42
Helstone DEV 281
Helton WML 177
Helwith YON 94
Hemerdon DEV 169
Hemingfield YOW 274
Hempholm YOE 57
Hempnall NFK 128
Hemswell LIN 33
Henbury CHE 286
Hendal SSX 112
Hendall SSX 112
Hendon GTL 168
Hendon DRH 117
Hendred BRK 29
Henegar DEV 232
Henfield SSX 277
Henhull CHE 193
Henhurst KNT 235
Henley var. cos. 239, 241, 312
Henlow BDF 179
Henmarsh GLO 58
Hennett CNW 141
Henney ESX 42
Hennor HRE 202
Henon CNW 141
Hensborough WAR 150
Hen's Grove OXF 229
Henshall YOW 128
Henshaw NTB 128
Henstridge SOM 191
Henwood WAR 259
Hereford HRE 71, 78
herepæth 90
Hernhill KNT 193
Heronbridge CHE 69
Heronshead Fm SUR 175
Herringfleet SFK 16–17
Hemingford HNT 75
Herringswell SFK 35
Herstmonceux SSX 235
Hertford HRT 71, 75

Hertfordlythe YOE 185
Hescombe SOM 108
Hesket CMB 175, 177
Hesket var. cos. 176
Hesledon DRH 117
Hessay YOW 38, 42
Hessleskew YOE 248
Hestercombe SOM 108
Heswall CHE 32
Hethe OXF 279
Hethel NFK 193
Hethfelton DOR 274
Heugh DRH 188
Hewelsfield GLO 274
Hexgreave NTT 229
Hexham DRH 39
Heybridge ESX 70
Heydon NFK 169
Heyford NTP, OXF 76
Heyrod LNC 244
Heyshott SSX 279
Heywood LNC 259
Heywood SHR, WLT 260
Hezzlegreave YOW 229
Hidden BRK 87, 88, 118
Hiendley YOW 241
Higford SHR 74
Higginbotham CHE 100
High and Over SSX 199, 202
High Bray DEV 152
Highbridge SOM 70
Highleigh SSX 241
Highnam GLO 51, 53
Highover HRT 202
Highway WLT 96
High Wycombe BUC 313
Hilbre CHE 40, 42
Hilfield DOR 192, 275
Hilgay NFK 42
Hill var. cos. 192, 296
Hillam YOW 192
Hillbeck WML 6
Hillerton DEV 171
Hillesden BUC 168

INDEX

Hillhampton WOR 192
Hillingdon GTL 168
Hilsea HMP 43
Hilton DOR, WML 177
Hilton var. cos. 192
Hinchingbrook HNT 9
Hinckley LEI 240
Hinderskelf YON 218
Hinderwell YON 35
Hindley LNC 241
Hindon WLT 169
Hinksey BRK 39, 40, 42
Hinnegar GLO 231
Hinxhill KNT 193
Hippenscombe WLT 108
Hipswell YON 35
Hirst NTB, YOW 235
Hive YOE 87, 88
Hixon STF 168
hlāw 178–80, 296
*hlenc 180
hlinc 180–2
hlith 182–5, 296
hlíth ON 143, 182, 185–6
Hoath KNT 279
Hoathly SSX 240
Hoborough KNT 150
Hoby LEI 188
Hockcliffe BDF 156
Hockenhull CHE 193
Hockerill HRT 93, 259
Hockerwood NTT 259
Hockwold NFK 256
Hoddesdon HRT 168
Hoddlesden LNC 117
Hoddnant Wales 141
Hodnell WAR 193
Hodnet SHR 141
Hodore SSX 209
Hoe HMP, NFK 188
Hofflet LIN 17
hofuth ON 175
Hogshaw BUC 247

hōh xv, 101, 102, 144, 174, 186–90, 192, 199, 296, 297, 309
Holbeach LIN 3, 37, 144, 145
Holbeche STF 3
Holbeck var. cos. 6
Holborn GLO 11
Holbrook DRB, SFK 7, 8
Holcombe var. cos. 108
Holdelythe YOE 185
Holden YOW 117
Holdenhurst HMP 235
Holderness YOE 197
Holebrook DOR 7, 8
Holford SOM 73
Holker LNC 57
Holland HNT 243
Holland LIN 283
Holland var. cos. 188, 282, 283
Holleth LNC 176
Hollinfare LNC 70, 71
Hollingbourne KNT 12
Holloway GTL 95
Hollowell NTP 32
Holme var. cos. 56
Holme on the Wolds YOE 174
Holme Lacey HRE 49, 52
Holmer BUC, HRE 27
Holmes ORK 56
Holmes Chapel CHE 56
Holmesfield DRB 275
holmr ON 36, 55–6, 133
Holmwood SUR 259
holt 221, 222, 227, 233–4, 248, 314–5
Holt var. cos. 233, 314–5
Holton LIN 188
Holton OXF, SOM 128
Holtsmore End HRT 315
Holway SOM 95
Holwell DOR, OXF 33
Holwell var. cos. 32
Holybourne HMP 12
Holywell LIN 33

Holywell var. cos. 33
The Home SHR 47
Homer OXF 303
Homersfield SFK 272, 274
Honeybourne GLO/WOR 11
Honeychild KNT 14
Honor SSX 209
Honor End BUC 209, 298, 309
Hoo var. cos. xiii, 188, 297
Hooe DEV, SSX 188
Hoofield CHE 276
Hoon DRB 174
Hooton CHE, YOW 188
hop xviii, 97, 114, 133–40, 295
hóp ON 133, 135
Hope var. cos. 133–9
Hope Dale SHR xviii, 136, 139
Hopleys Green HRE 136
Hoppen NTB 134
Hopton var. cos. 133–9
Hopwas STF 64, 139
Hopwell DRB 33, 139
Hopwood LNC, WOR 136, 139, 259
Horcum YON 108
Horden DRH 117
Hordle HMP 194
Hordlow var. cos. 178
Horelaw LNC 179
Horfield GLO 275
Horgrove GLA 228
Hormead HRT 285
Horncliffe NTB 153, 155
Horndean HMP 117
Horndon ESX xix, 169
Horndon on the Hill ESX 169
Horner DEV 209
Horninghold LEI 256
Horningsea CAM 42
Hornsea YOE 30, 197
Horrabridge DEV 69, 70
Horringford IOW 75
Horrington SOM 169
Horsenden BUC 169, 296

Horsendon GTL 169
Horsey NFK, SOM 42
Horse Eye SSX 42
Horseheath CAM 279
Horseway CAM 86, 88
Horsford NFK 76
Horsforth YOW 76
Horsley var. cos. 241
Horspath OXF 90
Horwood BUC, DEV 259, 313
Hose LEI 188
Hoseley FLI 183, 184–5, 185
Hothersall LNC 131, 277
Hothfield KNT 274
Hough CHE, DRB 188
Houghall DRH 129, 188
Houghton DEV 121
Houghton LNC, YOW 128
Houghton var. cos. 188, 297
Houndean SSX 114, 117
Houndsfield WOR 276
Hounslow GTL 179
Howcans YOW 100
Howden NTB 117
Howden YOE 114, 117, 118, 175, 176
The Howe OXF 309
Howell LIN 35
Howfield KNT 121, 277
Howgill YOW 123
Howgrave YON 228–9
Howler's Heath GLO 185
Howsell WOR 195
Howtel NTB 129, 233
Hoyland YOW 188, 282
hrīsbrycg 68, 69
hrycg 190–2, 214, 296
hryggr ON 190, 214
Hubbridge ESX 69, 188
Hucknall DRB, NTT 128
Huddersfield YOW 277
Huddlesceugh CMB 248
Hudspeth NTB 89, 90
Hudswell YON 32

INDEX

Hughenden BUC 120-1, 277
Huglith SHR 183, 184, 185
Hulgrave CHE 229
Hull SOM 192
Hulland DRB 188, 282
Hulm(e) var. cos. 55
Hulton LNC, STF 192
Humbledon DRH, NTB 168
Humblescough LNC 249
Humshaugh NTB 128
Hundersfield LNC 275
Hunderthwaite YON 251
Hundon SFK 117
Hundon LIN 168
Hundridge BUC 190, 191
Hungerford BRK, SHR 78
Hunsdon HRT 168
Hunshelf YOW 218
Hunsingore YOW 200, 202
Hunslet YOW 16, 17
Huntercombe OXF 108, 293
Huntingdon HNT 169
Huntingfield SFK 276
Huntingford DOR, GLO 75
Huntington CHE 169
Huntington STF, YON 169
Huntroyde LNC 244
Huntshaw DEV 247
Huntspill SOM 28
Hurdlow var. cos. 178
Hurdsfield CHE 276
Hurgrove BRK 229
Hurley BRK, WAR 241
Hursley HMP 241
Hurst var. cos. 235
Hurstbourne HMP xxii, 11, 12
Hurst Castle HMP 235
Hurstingstone Hundred HNT 235
Hurstpierpoint SSX 235
Hurtmore SUR 26
Husborne BDF xxii, 11, 12
Husthwaite YON 251
Huthnance CNW 141
Huthwaite NTT, YON 188, 251

Huttoft LIN 188
Hutton var. cos. 188
Huyton LNC 87, 88, 92
hváll ON 192
***hwæl** 192
Hyde DEV 83, 88
hyll 192-5, 296
hylte 233
Hylton DRH 177
hyrst 221, 222, 227, 234-6, 269, 315
hȳth 66, 83-9, 308, 309
Hythe var. cos. 87, 88, 308
Hyton CMB 87, 88

Iburndale YON 12
Ickford BUC 74
Ickleford HRT 78
Icklesham SSX 50, 53, 78
Ickelton CAM 78
Icknield Way 78, 95, 307
Icknor KNT 210
Ickornshaw YOW 247
Ickwell BDF 35
Icomb GLO 107
Icornshaw WML 247
Iddinshall CHE 128
Ideford DEV 74
Iden SSX 54, 267
Ifield KNT, SSX 275
Ightenhill LNC 195
Ightfield SHR 275
Iham SSX 50, 54
Ilford GTL 76
Ilford SOM 76
Ilfracombe DEV 107
Ilketshall SFK 128
Ilmer BUC 26, 303
Ilmington WAR 169
Iltney ESX 42
Imber WLT 27
Impney WOR 42
Inglesham WLT 49, 53
Inglewood CMB 257, 259
Ingoe NTB xv, 186, 188

Inkberrow WOR 148, 150
Inkersall DRB 195
Inkpen BRK 212
Innsacre DOR 264
Intwood NFK 260
Ipplepen DEV 212
Ipsden OXF 118, 293
Ireleth LNC 186
Iridge SSX 191
Irmingland NFK 282
Ironbridge ESX 69
Isel CMB 129
Isfield SSX 274
Islip NTP, OXF 61, 62
Ismere WOR 27
Isombridge SHR 69
Itchenor SSX 205, 209
Itchingfield SSX 276
Ivegill CMB 123
Ivinghoe BUC xv, 186, 189, 297, 309
Ivonbrook DRB 8
Ivybridge DEV 69
Iwade KNT 94, 95
Iwode HMP 260

Kaber WML 150
Katesgrove BRK 228
Keele STF 193
Keighley YOW 240
Kelbarrow CMB, WML 18
Kelber YOW 18
Kelbrooke YOW 9
kelda ON 1, 18–19
Keldbottom YON 100
Keldholme YON 19, 57
Kelfield LIN, YOE 272, 275
Kellet LNC 18, 186
Kelleth WML 18, 186
Kelloe DRH 179
Kelmarsh NTP 58
Kelsale SFK 128
Kelsall CHE 128
Kelsall HRT 193

Kelsey LIN 43
Kelshall HRT 128
Kelvedon ESX 117
Kempsey WOR 42
Kempsford GLO 74
Kendal WML 111
Kendale YOE 18
Kenninghall NFK 128
Kensal Green GTL 234
Kentford SFK 76
Kentisbeare DEV 223
Kentmere WML 27
Kentwood BRK 259
Keresforth YOW 78
Kerridge CHE 191
Kersal LNC 129
Kersall NTT 128
Kersey SFK 37, 39, 42
Kershope CMB 139
Kersoe WOR 188
Kerswell DEV 32
Kesgrave SFK 230
Kessingland SFK 282
Kesteven LIN 224
Kettleshulme CHE 56
Kettlewell YOW 33, 101
Kew SUR 188
Kexmoor YOW 60
Keymer SSX 26
Keynor SSX 205, 209
Keynsham SOM 49, 54
Keysoe BDF 188
Kidall YOW 129
Kidbrooke KNT, SSX 8
Kiddal YOW 111
Kidland DEV 282, 283
Kidland Forest NTB 283
Kigbeare DEV 223
Kilburn var. cos. 13
Kildale YON 111
Kildridge OXF 296
Killamarsh DRB 58
Killinghall YON 128
Killinghall YOW 128

INDEX

Kilmersdon SOM 167, 168
Kilnhurst YOW 236
Kilnsea YOE 30
Kilnsey YOE 30
Kilpin YOE 212
Kilsall SHR 128
Kimmeridge DOR 214, 216
Kinard's Ferry LIN 71
Kingcombe DOR 107
Kingrove GLO 228
Kingsbridge DEV 69
King's Cliffe NTP 155
Kingsdon SOM 169
Kingsdown KNT 169
Kingsey BUC 41, 305
Kingsford WOR 75
Kingsford WAR 75
Kingsholme GLO 51, 53
Kingsley var. cos. 241
King's Mead BUC 312
Kingsterndale DRB 110
Kingswood var. cos. 259
Kingwood Common OXF 313
Kinnersley SHR 42
Kinsham HRE 285–6
Kinver STF 152
Kirby Knowle YON 157
Kirdford SSX 74
Kirkber WML 151
Kirkburn YOE 10
Kirkdale LNC, YON 111
Kirkhaugh NTB 130
Kirkland CMB 281
Kirkland LNC 243
Kirkleatham YON 186
Kitchenham SSX 50, 55, 210
Kitchingham SSX 55, 210
Kitchenour SSX 210
Kitnore SOM 209
Kittisford SOM 74
kjarr ON 57
Knaith LIN 87, 88
Knapwell CAM 35
Knaresdale NTB 111

Kneesall NTT 131
Kneesworth CAM 131
Knettishall SFK 129
Knightley STF 241
Knightlow Hundred WAR 180
Knightsbridge GTL, OXF 69, 308
Knill RAD 157
Knockhall KNT 233
Knockholt KNT 233
Knoddishall SFK 129
Knole KNT, SOM 157
Knoll var. cos. 157
Knottingley YOW 242
Knowl(e) var. cos. 157
Knowlton DOR 157
Knutsford CHE 74
Kyo DRH 188

Lach Dennis CHE 21
Lackford SFK 75
Lacon SHR 20
lacu 2, 19–20, 21, 118, 303
lād 20–1, 81, 82, 89
Ladbroke WAR 8
Lagness SSX 269
Laindon ESX xix, 169
Lake WLT 20
Lakenheath SFK 86, 88
Lamarsh ESX 58
Lamas NFK 58
Lamberhurst KNT 236
Lambeth GTL 87, 88
Lambley NTB, NTT 241
Lambourn BRK 12
Lambourne ESX 12
Lambrigg WML 190
Lambrook SOM 9, 281
Lambsceugh CMB 249
Lambside DEV 219
Lamesley DRH 241
Lamplugh CMB 141
land xviii, 262, 268, 272, 279–84, 312
Landbeach CAM 3, 144, 145

Landermere Hall ESX 281
Landmoth YON 281
Landford WLT 78
Landwade CAM 95
Langbaurgh YON 149
Langcliffe YOW 155
Langdale WML 117
Langdon var. cos. 168
Langford NTT 74
Langford var. cos. 73
Langhale NFK 129
Langham LIN 57
Langho LNC 129
Langley var. cos. 239, 240
Langness IOM 198
Langney SSX 42
Langsett YOW 219
Langthwaite var. cos. 250
Langstrothdale YOW 63
Langwathby CMB 94
Langwith var. cos. 94
Langworth LIN 94
Lansdown SOM 169
Lanshaw WML 247
Lapal WOR 78
Lapford DEV 78
Lapley STF 78
Larkbeare DEV 223
Lashbrook DEV, OXF 21, 303
Lastingham YON 39
Latchford CHE, OXF 21, 74
Latchingdon ESX 171
Lattiford SOM 75
Launceston CNW 167, 175, 177
Launder's Fm OXF 298, 308
Lavendon BUC 117
Laver ESX 70–71
The Lawley SHR 183
Lawford ESX, WAR 74
Laxey IOM 2
Laxfield SFK 274
Layham SFK 13
læcc, læce 21, 303
Lea var. cos. 240

Leach R. GLO 21
Leagrave BDF 230
Leafield OXF 274
lēah xvii, xx, 220–1, 237–42, 257, 262, 287, 312, 313, 314
Leake var. cos. 21
Lealholm YON 56
Leam DRH, NTB 240
Learchild NTB 13, 177
Leath Ward CMB 185
Leatherhead SUR 91
Leathley YOW 185
Leaveland KNT 282
Lechlade GLO 21, 81
Leck LNC 21
Leckford HMP 74
Leconfield YOE 277
Lee var. cos. 240, 312
Leece LNC 240
Leeford DEV 73
Leek STF 21
Lees LNC 240
Leese CHE 240
Legbourne LIN 13
Leigh var. cos. 240
Leighton NTB 168
Leinthall HRE 129
Leith HRT 185, 296
Leith Hill SUR 184, 185
Lenacre KNT 265
Lenborough BUC 151, 185
Lench LNC 180
Lench WOR 180
Lendrick Hills YOW
Lenwade NFK 94, 95
Letcombe BRK 107
Letwell YOW 33
Levens WML 198
Levenshulme LNC 56
Lew OXF 179
Lewell DOR 33
Lewes SSX 179
Lewknor OXF 308, 309
Lewsey Fm BDF 305

INDEX

Lexden ESX 117
Leybourne KNT 13
Leyburn YON 13
Leyland LNC 282
Leysdown KNT 171-2
Lichfield STF 223, 261, 275
Lidcott CNW 223
Lidcutt CNW 223
Lightcliffe YOW 155
Lightshaw LNC 247
Lilbourne NTP 10, 13
Lilburn NTB 13
Lilford NTP 74
Lilleshall SHR 193
Lilley HRT 312
Limber LIN 150
Limehurst LNC 236
Limpenhoe NFK 189
Limpsfield SUR 275
Linacre DRB, LNC 265
Linacres NTB, WOR 265
Linch SSX 182
Linchmere SSX 27
Lindal(e) LNC 111
Lindeth LNC, WML 177
Lindfield SSX 275
Lindhurst NTT 235
Lindley var. cos. 241
Lindrick NTT, YOW 215
Lindridge DRB 215
Lindridge KNT 191
Lindsey LIN 39
Lindsey SFK 42
The Linegar GLO 265
Lineholt WOR 234
Lineover GLO 202
Linersh SUR 269
Linethwaite CMB 251
Linford BUC 75
Lingfield SUR 277
Linley SHR 241
Lingwood NFK 182, 259
Linkenholt HMP 180, 182, 234
Linslade BUC 81, 182

Linthwaite YOW 251
Lintz DRH 182
Lintzford NTB 76
Linwood HMP, LIN 260
Lipwood NTB 261
Liscombe SOM 108
Litchborough NTP 151
Litchfield HMP 219
Litherland LNC 186, 282
Litherskew YON 186
Littlebourne KNT 10
Littleham DEV 52, 53
Littlehope HRE 136
Littlemore OXF 60
Littleover DRB 202
Litton DRB, YOW 185
Livermere SFK 27
Liverpool LNC 29
Liversedge YOW 173
Liza R. CMB 2
Llinegar FLI 265
Llwytgoed DEN, GLA 223
Load SOM 20
Load Hill YOW 89
Loatland Wood NTP 242, 243
Lockeridge WLT 191
Locksash SSX
Lockwood YON 253
Loddiswell DEV 32
Lode CAM 20
Lode GLO 82
Londonthorpe LIN 242, 243
Longborough GLO 150
Longbottom YOW 100
Longden var. cos. 168
Longdendale DRB 110, 117
Longdon var. cos. xiii, 168
Longford var. cos. 73
Longhirst NTB 236
Longhold NTP 256
Longhope GLO 136, 139
Longlands var. cos. 281
Longload Lane DRB 89
Long Mynd SHR 195

Longney GLO 42
Longnor STF 202
Longridge LNC, STF 191
Longsdon STF 169
Longslow SHR 179
Longstone DRB 169
Long Strath CMB 63
Longthwaite CMB 250
Longville SHR 275
Lonsdale LNC/WML 111
Loscoe var. cos. 248, 251
Loscombe DOR 108
Loskay YON 248, 251
Lostford SHR 75
Lostrigg Beck CMB 214
Lothersdale YOW 118
Loud R. LNC 172
Lound var. cos. 242, 243
Louth LIN 172
Loversall YOW 128
Low var. cos. 179
Lowbury Hill BRK 178
Lowthwaite CMB 251
Loxbeare DEV 223
loekr ON 21
Lubenham LEI 188
Luccombe IOW, SOM 109
Ludbrook DEV 8
Ludeott CNW 223
Luddenden YOW 118
Luddesdown KNT 172
Ludford LIN 76
Ludford SHR 74
Ludgershall BUC, WLT 130
Ludlow SHR 178
Ludstone SHR 172
Ludwell var. cos. 1, 32
Luffenhall HRT 128
Luffield BUC 274
Luggershall GLO 130
Lugharness HRE 197
Lulham HRE 49
Lulsley WOR 42
Lumby YOW 242, 243

Lund var. cos. 242, 243
Lund Forest YON 242
lundr 221, 242–3
Lund Wood YOE 243
Lunt LNC 243
Lupridge DEV 177, 191
Lupsett YOW 177, 191
Lupton DEV 177, 191
Lurgashall SSX 130
Lurkenhope SHR 140
Lushill WLT 193
Luton Hoo BDF 297
Luxborough SOM 150
Lydcott CNW 223
Lydden KNT 121
The Lyde BUC 184, 185, 296
Lydford DEV 76
Lydlinch DOR 182
Lydney GLO 42
Lye var. cos. 240
Lyegrove GLO 229
Lynch HRE, SOM xiii, 182
Lyndhurst HMP 235
Lyndon RUT 169
Lyneal SHR 131
Lynford NFK 75
Lyng NFK 182
Lynsore KNT 209
Lyonshall HRE 129–30
Lyscombe DOR 107
Lytchett DOR 223
Lyth SHR 182, 184, 185
Lyth WML 185
Lytham LNC 184, 185
Lythe var. cos. xiii, 185
Lythe Beck YON 186
Lythe Brow LNC 184
Lytheside WML 185
Lythwood SHR 259
Lyveden NTP 117

Macclesfield CHE 274
Macknade KNT 177
Mackney BRK 39, 42, 305

INDEX

Madehurst SSX 236, 285
Madingley CAM 242
Madresfield WOR 276
Maghull LNC 129
Maidenhead BRK 87, 88, 308, 309
Maidensgrove OXF 229
Maidenwell LIN 33
Maidford NTP 75
Maids Moreton BUC 59
Mainsforth DRH 74
Makeney DRB 42
Malborough DEV 150
Malden SUR 170
Maldon ESX 170
Mallerstang WML 152
Malpas var. cos. 71
Malvern WOR 153, 283
Malvern Link WOR 181–2
Mamhead DEV 177
Manea CAM 43
Maney WAR 39, 43
Manfield YON 274
Mangotsfield GLO 274
Mangrove HRT 229
Manningford WLT 75
Mansell HRE 193
Mansfield NTT 275
Manshead Hundred BDF 175
Manuden ESX 118, 170
Maplebeck NTT 6
Mapledurwell HMP 32
Maplesden SSX 267
Mappleborough WAR 150
Marbury CHE 26
Marcham BRK 51, 54
Marche SHR 57
Marchwood HMP 260
Mardale WML 26, 111
Marden SSX 170
Marefield LEI 275
Mareham-le-Fen LIN 26
Maresfield SSX 26
Marfleet YOE 16, 17, 26
Marford HRT 307

Marham NFK 26
Marholm NTP 26
Marishes YON 57
Marker SSX 209
Markfield LEI 276
Markingfield YOW 276
Markshall NFK 132
Marland DEV, LNC 26, 282
Marlborough WLT 150
Marlbrook SHR 9
Marlcliff WAR 156
Marldon DEV 169
Marlesford SFK 78–9
Marley var. cos. 26
Marley DEV, DRH 241
Marlingford NFK 79
Marlow BUC 303
Marlston CHE 78–9
Marnhull DOR 195
Marrick YON 214, 216
Marsden LNC, YOW 118
Marshacre GLO 265
Marsham NFK 58
Marshaw LNC 247
Marsh Baldon OXF 57
Marshborough KNT 150
Marsh Gibbon BUC 57
Marshfield GLO 276
Marshwood DOR 58, 61, 259
Marske YON 57
Marston var. cos. xii, 58
Marten WLT 22, 26
Martin var. cos. 22, 26
Martindale WML 111
Martinhoe DEV 189
Martley SFK, WOR 241
Marton var. cos. 22, 26
Marvel IOW 275
Marwell HMP 33
Marwood DEV 259
Marwood DRH 261
Mason NTB 36
Matfen NTB 46
Matfield KNT 274

363

Matson GLO 157, 165, 172
Matterdale CMB 111
Mattersea NTT 42
Mattishall NFK 128
Maulden BDF 170
Mawfield HRE 275
Maxey NTP 40, 42
Maydencroft HRT 285
Mayfield STF 275
Mayford SUR 75
mǣd 262, 284–6, 312–3
Meaburn WML 11, 285
Meadle BUC 313
Meaford STF 79
Meare SOM 26
Mearley LNC 26, 241
Mease R. 61
Meathop WML 134, 140
Meaux YOE 30
Medbourne LEI, WLT 11, 285
Meddon DEV 169, 285
Medhurst KNT 236, 285
Medley OXF 40, 43
Medlock R. LNC 20
Medmenham BUC 304
Meend GLO 196
Meerbrook STF 9
Meersbrook DRB 9
Meertown STF 26
Meesden HRT 61, 169
Meese Brook STF/SHR 61
Melbourn CAM 12
Melbourne YOE 12
Melbourne DRB 9, 12
Melchbourne BDF 11
Melchett HMP/WLT 223
Melcombe DOR 108
Meldon DEV 168
Meldon NTB 170
Meldreth CAM 29
Melford SFK 76
Mellersh SUR 269
Melliker KNT 265
Mellor CHE, LNC 152

Mellor Knoll YOW 252
Melwood LIN 259
Membland DEV 282
Mendip Hills SOM 196
Mentmore BUC 60
Menutton SHR 196
Menwith YOW 253
Meon Hill GLO 196
mēos 61, 121, 300
Mepal CAM 125–6, 132
Meppershall BDF 129
Merdegrave LEI 229
mere xv, 2, 21–7, 58, 303
Mere var. cos. 26
Meriden WAR 117
Merrifield DEV 275
Merrington SHR 168
Merrivale DEV 275
Merryfield DEV 272, 275
mersc 36–7, 57–8, 305
Mersea ESX 22, 26, 37, 39
Merston var. cos. 58
Merton var. cos. xii, 22, 26
Metcombe DEV 285
Metfield SFK 275, 285
Methersham SSX 50, 53
Methley YOW 38
Methwold NFK 256
Methwold Hythe NFK 86, 88
Mewith Forest YOW 252
Michel Grove SSX 228
Michelmarsh HMP 58
Michen Hall SUR 125
Micklefield YOW 275
Mickleover DRB 202
Micklethwaite var. cos. 250
Middle Hope SOM 133
Middlehope SHR 136, 139
Middleney SOM 43
Middlesceugh CMB 248
Middlewood HRE 259
Middlezoy SOM 43
Middop YOW 140
Mid(d)ridge DRH, SOM 140

INDEX

Midford SOM 74
Midge Hall LNC 129
Midgehall WLT 129
Midgley YOW 241
Midhope YOW 140
Midhurst SSX 140, 236
Midloe HNT 189
Midsyke YON 140
Migley DRH 241
Milbo(u)rne var. cos. 9, 12
Milburn WML 9, 12
Milcombe OXF 108
Mildenhall SFK 125, 132
Mildenhall WLT 128, 132
Miles Hope HRE 136
Milford var. cos. 76
Millbrook BDF, HMP 9
Millichope SHR 136, 137, 139
Millmeece STF 61
Millow BDF 188
Milverton SOM, WAR 76
Mindrum NTB 196
Minehead SOM 177, 196
Minety WLT 42
Minn CHE 195
Minsden HRT 117
Minshull CHE 218
Minsmere SFK 27
Minsteracres DRH 265
Minton SHR 196
Mirfield YOW 275
Misbourne R. BUC 121, 301, 302
Misbrooke SUR 61
Missenden BUC 121, 292, 301
Miswell HRT 32, 299, 300
Mitford NTB 74
Mixbrooks SSX 61
Mixnams SUR 49, 55
Mobwell BUC 300
Moddershall STF 128
Modney NFK 43
Moelfre, Wales 152
Moggerhanger BDF 232, 315
Molesey SUR 40, 42

Molland DEV 283
Molton DEV 283
Monewden SFK 118, 171
Mongewell OXF 32, 299
mönith PrW 143, 195–6
Monkspath WAR 90
Monmore STF 27
Monsal DRB 128
Montford SHR 79
Moor BUC, WOR 305–6
Moorbath DOR 5, 60
Moorby LIN 60
Moore CHE 59
Mooresbarrow CHE 223
Moorlinch SOM 180–1, 182
Moorshall WLT 245, 247
Moorsley DRH 179
Moorton OXF 59
mōr 36–7, 44, 58–60, 305–6
mór ON 59
Morborne HNT 10, 11, 12, 60
Morchard DEV 224
Morcombelake DOR 20, 109
Morcott RUT 60
Morden var. cos. 60, 169
Mordiford HRE 79
Mordon DRH 60, 169
More SHR 59
Morebath DEV 5, 60
Moreby YOE 60
Moredon WLT 60
Moreleigh DEV 60, 241
Morestead HMP 60
Moreton var. cos. 37, 59
Moretonhampstead DEV 59
Morgrove WAR 229
Morland WML 60, 243
Morley var. cos. 60, 241
Morpeth NTB 89, 90–1
Morphany CHE 249
Morston NFK 58
Mortgrove HRT 229
Morthoe DEV 20, 189
Mortlake GTL 20

Mortlock's Fm ESX 20
Morton var. cos. 59
Morville SHR 277
Morwick NTB 60
mos 60–1
Mosborough DRB 60
Mose SHR 61
Mosedale CMB 61, 111
Moseley YOW 61
Moseley WOR 241
mosi ON 60–1
Moss YOW 61
Mosser CMB 61
Moston var. cos. 61
Mostyn FLI 61
Motcombe DOR 108
Mottisfont HMP 18
Mouldridge Grange DRB 214, 216
Moulsecoombe SSX 108
Moulsford BRK 74
Moulsoe BUC 188
Mountfield SSX 274
Mousen NTB 46
Mow Copp CHE 158
Mowsley LEI 241
Moxhull WAR 219
Moze ESX 61
Muchelney SOM 36, 42
Mudford SOM 73
Muker YON 264, 265
Mulfra CNW 152
Mullingar DEV 231
Mulwith YOW 94
Munden HRT 121, 171
Mundford NFK 74
Mundon ESX 37, 172
Mundon HRT 117
Munslow SHR 179
Murcot OXF 59, 60
Murcott WLT 60
Mursley BUC 312
Murton var. cos. 59
Musgrave WML 229
Muswell BUC, GTL 32, 61, 300

Mutford SFK 75
Muxbere DEV 223
Muxworthy DEV 55
Muze DEV 61
Muzwell Fm BUC 61, 300
Myerscough LNC 61, 249
Mynde HRE 195
Mynd Town SHR 196
Myne SOM 196
mýrr ON 61
Mytholmroyd YOW 244
Mythop LNC 134, 140

Naburn YOE 13
Nackington KNT 168
Nail Bourn R. KNT 11
Nailsbourne SOM 12
Nailsea SOM 40, 42
Nanplough CNW 141
nant Cornish, Welsh 140–1
Nass GLO 198
Nassaborough Hundred NTB 198
Nassington NTP 198
Natland WML 243
Navant SSX 233
Navestook ESX 198
The Naze ESX 196, 198
Nazeing ESX 198
Naze Wick ESX 198
næss, ness 196–9
Neadon DEV 170
Neasden GTL 168, 197
Neasham DRH 197
Needwood STF 260
Nepicar KNT 264, 265
nes ON 143, 197–8
Nesfield YOW 277
Ness var. cos. 196, 197, 198
Neston WLT 197
Neswick YOE 196, 198
Netherpool CHE 28
Netteswell ESX 32
Nettlecombe var. cos. 107
Nettleden HRT 117, 292

INDEX

Nevenden ESX 117
Newbourn SFK 11
Newburn NTB 11
New Hythe KNT 87
Newland(s) var. cos. 279, 281, 282, 312
Nibthwaite LNC 252
Ninfield SSX 275
Ningwood IOW 260
Norbreck LNC 152
Nordley SHR 241
Nore var. cos. 206, 208
Norgrove WOR 229
Norland YOW 282
Norley CHE 241
Norridge WLT 191
Northbourne KNT 12
Northfield WOR 276
Northiam SSX 53, 54
Northleach GLO 21
Northload SOM 20
Northmoor OXF 59
Northolt GTL 130
Northover SOM 200, 203
Northowram YOW 199, 202
Northsceugh CMB 248
North Sunderland NTB 282
Northumberland 283
Northwold NFK 256
Northwood GTL, IOW 259
Norwell NTT 33
Norwood GTL, SUR 259
Noska YOW 248
Noss Mayo DEV 197, 198
Nosterfield CAM, YON 277
Notgrove GLO 228
Notley BUC, ESX 240
Nower var. cos. 206, 208, 309
Nowhurst Fm SSX 208
Nuffield OXF 275, 310, 311
Nunburnholme YOE 11, 56
Nuneaton WAR 15
Nunney SOM 42
Nunwell IOW 33

Nutbourne (in Pulborough) SSX 12
Nutbourne (in Westbourne) SSX 12
Nutfield SUR 275
Nutford DOR 75
Nuthall NTT 129
Nuthurst SSX, WAR 236
Nutshaw LNC 247
Nutwell DEV 32
Nutwith YON 253
Nympsfield GLO 275

Oakenbottom LNC 100
Oakenshaw YOW 247
Oakford DEV 75
Oakhall GLO, WOR 233
Oakhanger HMP 231
Oakhunger GLO xxii, 231
Oakhurst HRT 268, 312
Oakle GLO 240
Oakleigh KNT 240
Oakley var. cos. 240, 312
Oakmere CHE 26
Oaksey WLT 42
Oar(e) var. cos. 206, 208
Oatersh GLO 269
Oborne DOR 11
Occleshaw LNC 247
Occold SFK 233
Ockbrook DRB 8
Ockendon ESX xix, 168
Ockeridge KNT 191
Ockham SSX 53
Ockold GLO 233
Ockwells BRK 233
Ocle HRE 240
Odcombe SOM 107
Oddingley WOR 242
Oddington OXF 168
Odell BDF 193
Odiham HMP 27
ofer xv, xvi, 144, 186, 199–203, 206, 208, 308
Offley HRT 312
Offord HNT 73

Offwell DEV 32
Ogbear DEV 223
Ogbourne WLT 12
Ogwell DEV 35
Okeford DOR 75
Okeover STP 202
Old NTP 256
Oldacres DRH 265
Oldberrow WAR 150
Oldham LNC 55, 56, 57
Old Hurst HNT 235
Oldland GLO 283
Old Weston HNT 254
Ollerenshaw DRB 247
Olney BUC 42
Onecliffe YOW 155
Onesacre YOW 264
Onesmoor YOW 264
Openshaw LNC 247
ōra xvi, 79, 144, 186, 199, 203–10, 296, 297, 298, 308–9, 316
Orchard DOR 224
Orcop HRE 158, 206, 208
Ordsall LNC, NTT 132
Ore SSX 206, 208
Oreham SSX 208
Orford LIN 75
Orford LNC 79
Orford SFK 79, 208
Orgrave LNC 208, 226
Orgreave STF 228
Orgreave YOW 226
Ormerod LNC 244
Orsett ESX 208
Orwell CAM 33
Orznash SSX 269
Osborne IOW 12
Osea ESX 39, 42
Osland NTT 243
Osney OXF 40, 42
Oswaldslow WOR 178
Otford KNT 74
Othery SOM 43
Otley YOW 240

Otmoor OXF 37, 58, 59, 60
Otterbourne HMP 12
Otterburn NTB, YOW 12
Otterpool KNT 29
Ottershaw SUR 247
Oughtershaw YOW 247
Oughtibridge YOW 69
Oughton Head HRT 299
Ousden SFK 117
Ousefleet YOW 16, 17
Outerness Wd HNT 196, 198
Outhwaite LNC 251
Outwell CAM 31
Ovenden YOW 117
Over var. cos. 199, 200, 202
Overacres NTB 265
Ormside WML 176
Overpool CHE 28
Overs SHR 202
Overton var. cos. 200-2
Overy Staithe NFK 91
Ovingdean SSX 118
Ower DEV, HMP 206, 208
Owermoigne DOR 208
The Owers SSX 206
Owlands YON 243
Owlpen GLO 212
Oxcombe LIN 106
Oxenhall GLO 129
Oxenhope YOW 139
Oxenton GLO 165, 169
Oxey Grove OXF 315
Oxford OXF xx, 71-2, 76
Oxhill WAR 218
Oxley STF 241
Oxney KNT, NTP 42
Oxnop YON 139

Packington var. cos. 261
Packmores WAR 261
Packwood WAR 261, 277
Paddington SUR 117
Padfield DRB 275
Padgham SSX 53

INDEX

Padnells Wd OXF 294
Padside YOW 219
Painley YOW 89, 90, 129
Pakefield SFK 261, 277
Pakenham SFK 261
Palgrave NFK 228
Palgrave SFK 229
Palstre KNT 43
Pamber HMP 151, 213
Pamphill DOR 193
Panborough SOM 90
Panfield ESX 275
Pangbourne BRK 12
Pangdean SSX 118
Pannal YOW 129
Panshanger HRT 232
Panxworth NFK 79
Parford DEV 74, 89, 90
Parndon ESX 169
Parracombe DEV 108
Partney LIN 42
Passenham BUC 48, 49, 53
Pateley Bridge YOW 89, 90
Pathe SOM xxi, 89, 90
Pathlow WAR 90, 179
Patmore HRT 27
Patney WLT 42
Patrixbourne KNT 10
Patshull STF 193
Patterdale WML 111
Pattingham STF 48
Pattishall NTP 193
Pave Lane SHR 89
Pavenham BDF 48
Paxhill SSX 218
Paygrove GLO 228
pæth xxi, 65, 89–91, 92, 184, 307
pēac 210
Peak Cavern DRB 210
Peakdean SSX 210
Peamore DEV 26
Peasemore BRK 26
Peasenhall SFK 129
Peasmarsh SSX 58

Pebmarsh ESX 268
Peckforton CHE 74, 210
Peckham KNT, SUR 210
Pedmore WOR 60
Pedn Kei CNW 212
Pedn Tenjack CNW 212
Pednor BUC 298, 309
Pedwell SOM 32
Peek DEV 210
Pegsden BDF 118, 210
Peldon ESX 172
Pelsall STF 132
Pemboa CNW 212
Pembridge HRE 70
Pembroke PEM 211
Penbough CNW 212
Pencaitland ELO 224
Pencarrow CNW 212
Pencombe HRE 106, 108, 212
Pencoyd HRE 211
Pencraig HRE 211
Pendeford STF 74
Pendle LNC 193, 211
Pendlebury LNC 211
Pendley HRT 312
Pendock HRE 213
Pendoggett CNW 224
Pendomer SOM 211
Penfranc CNW 212
Penge GTL 210, 211
Pengethly HRE 211
Penhill YOW 193, 212
Penhurst SSX 236
Peniarth MTG 211
Peniston YOW 212
Penketh LNC 211
Penkevil CNW 212
Penkhull STF 194, 211
Penkridge STF 159, 163, 212
penn PrW xvi, 143, 210–3, 223
Penn BUC, STF 211
Pennance CNW 141
Pennant CNW 141
Pennard SOM 211

Pennersax DMF 212
Penquit DEV 211
Penquite CNW 211
Penrith CMB 91, 211
Penryn CNW 211
Pensax WOR 212
Penselwood SOM 213
Pensford SOM 79
Pensham WOR 48, 49, 53
Penshurst KNT 236
Pensnett STF 212, 213
Pentney NFK 43
Pentrich DRB 212
Pentridge DOR 212
Pentyrch GLA 212
Penvith CNW 212
Penwith CNW 211
Penyard HRE 211
Penyghent YOW 211
Penzance CNW 210, 211
Perdiswell WOR 32
Pershore WOR 206, 209
Pertenhall BDF 128
Petersfield HMP 277
Petham KNT 54
Pethill DEV 128
Petsoe BUC 188
Petton SHR 210
Pevensey SSX 15, 91
Pewsey WLT 42
Pexall CHE 193, 210
Phoside DRB 219
Phynis YOW 197, 198
pīc 213
Pickburn YOW 12, 213
Pickeringlythe 185
Pickhill ESX 213
Pickhill YON 129
Pick Hill KNT 213
Pickhurst KNT 213
Pickmere CHE 22, 26, 213
Pickthorne SHR 213
Pickup Bank LNC 158, 213
Pickwell LEI 33, 213

Pickwick WLT 213
Picton PEM 213
Piddinghoe SSX 186, 189
Piercebridge DRH 69
Pike Law LNC 179
Pilland DEV 28, 282
Pilleth RAD 183, 185
Pilley HMP, YOW 214
Pilsdon DOR 172
Pilsdon Pen DOR 212
Pilton DEV, RUT, SOM 28
Pinchbeck LIN 3, 144, 145
Pinden KNT 118, 267
Pinhoe DEV 189, 212
Pinner GTL 189, 206, 209, 309
Pinnik CNW 212
Pinsgrove BRK 228
Pinvin WOR 46
Piper's Grove GLO 229
Pipewell NTP 33
Pirbright SUR 225
Pirzwell DEV 32
Pishill OXF 193, 296, 311
Pishobury HRT 188
Pitchcombe GLO 109
Pitch Font SUR 18
Pitchford SHR 74
Pitcombe SOM 108
Pitney SOM 42
Pitsea ESX 40, 42
Pitsford NTP 74
Pitshanger GTL 231
Plaitford HMP 75
Playden KNT 267
Playford SFK 75
Pledgdon ESX 117
Plenmeller NTB 152
Ploughley OXF 179
Plowden SHR 118
Plumbland CMB 242, 243
Podmore STF 60
pol Cornish 28
pōl 22, 26, 28-9
Polden SOM 172

Polebrook NTP 9
Polehanger BDF 29, 232, 315
Polesden SLTR 121
Poling SSX 121
Polperro CNW 28
Polruan CNW 28
Polsham SOM 53
Polsloe DEV 29, 62
Polstead SFK 28, 29
Poltimore DEV 44, 60
Ponteland NTB 282
Pontesford SHR 74
Pontesford Hill SHR 183
Pool DEV, YOW 28
Poole var. cos. 26, 28
Poolham LIN 28
Poolhampton HMP 29
Poringland NFK 282
Portisham DOR 50, 54
Portishead SOM 177
Portland DOR 283
Portnall SUR 125, 128
Portsea HMP 37, 43
Portslade SSX 81
Portswood HMP 259
Posenhall SHR 129
Postlip GLO 61
Potheridge DEV 191
Potlock DRB 20
Potsgrove BDF 230, 314
Poughill DEV 193
Poulner HMP 210
Poulshot WLT 234
Poultney LEI 44
Poulton var. cos. 28, 29
Poundisford SOM 75
Poundon BUC 172
Pownall CHE 129
Prattshide DEV 83, 88
Prawle DEV 195
Preesall LNC 176
Presthope SHR 136, 140
Pressen NTB 46
Presteigne RAD 285–6

Prestwold LEI 256
Prestwood BUC 313
Prickshaw LNC 247
Priestcliffe DRB 156
Priest Grove OXF 229
Prittlewell ESX 32
Prudhoe NTB 188
Puckeridge HRT 191, 214
Puddlestone HRE 169
Pudsey ESX 88
Pudsey YOW 38
Puesdown GLO 172
Pulborough SSX 29, 151
Puleston SHR 167, 172
Pulford CHE 29, 74
Pulham var. cos. 29, 54
pull 28–9
Pulloxhill BDF 195
Pulverbatch SHR 3, 4
Puncknowle DOR 157
Purbrook HMP 8
Purfleet ESX 17
Purleigh ESX 241
Purley BRK 241
Purslow SHR 179
Pusey BRK xvii, 39, 42
Putford DEV 74
Putney GTL 87, 88
Putteridge HRT 190, 191
Pyecombe SSX 108, 213
pyll 28–9
Pylle SOM 28
Pyrford SUR 75

Quantoxhead SOM 177
Quarlton LNC 169
Quarnford STF 76
Quarrington DRH 169
Quatford SHR 79
Quatt SHR 79
Queenhill WOR 194
Queenhithe GTL 87, 88
Quemerford WLT 79
Quendon ESX 118

Quernmore LNC 60
Quethiock CNW 224
Quixhill STF 193
Quorndon LEI 167, 169
Quy CAM 42

Rackenford DEV 76, 79
Rackheath NFK 87, 88
Radbourn WAR 12
Radbourne DRB 12
Radcliffe LNC, NTT 155
Radclive BUC 155
Raddon DEV 167, 168
Radford OXF 73, 76
Radford var. cos. 73
Radge DEV 269
Radipole DOR 28, 29
Radley BRK 240
Radmanthwaite NTT 26-7, 251
Radmore STF 60
Radnor (Pyrton) OXF 308
Radnor CHE, RAD 202
Radnor SUR 209
Radway WAR 95
Radwell BDF, HRT 32
Ragdale LEI 111
Ragleth SHR 183
Ragnall NTT 193
Raincliff(e) YOE, YON 155
Rainford LNC 74
Rainhill LNC 193
Rainow CHE 186, 188
Rainworth NTT 94
Raleigh DEV 240
Ramacre Wd ESX 264
Rampside LNC 176
Ramridge BDF 190, 191
Rams Cliff WLT 155
Ramsbottom LNC 100
Ramsdale YON 111
Ramsdell HMP 113
Ramsden ESX, OXF 117
Ramsey ESX, HNT 42
Ramsgreave LNC 229

Ramsholt SFK 234
Ramshorn STF 202
Ranskill NTT 219
Rastrick YOW 214, 216
Ratchwood NTB 259
Ratcliff(e) var. cos. 155
Ratfyn WLT 45, 46
Rathmoss LNC 61
Ratlinghope SHR 136, 137, 139
Rattlesden SFK 121
Ravendale LIN 111
Ravenfield YOW 275
Ravenscliff DRB 155
Ravensdale DRB 110, 111
Ravensden BDP 117
Ravens' Hall CAM 234
Ravenshead Wd NTT 175
Ravenstonedale WML 111
Ravensty LNC 93
Ravensworth YON 94
Rawcliffe var. cos. 155
Rawden YOW 168
Rawdon YOW 168
Rawerholt HNT 234
Rawmarsh YOW 58
Rawreth ESX 29
Rawridge DEV 192
Rawthey R. 2
Ray, Rea R. 14
Ray Court BRK 305
Raydon SFK 169
*ræc 86, 213, 214
Reach BDF, CAM 86, 213
The Reaches NTP 213
Read LNC 176
Reagill WML 123
Reathwaite CMB 251
Redbourn HRT 12, 302
Redbourne LIN 12
Redbridge HMP 67, 69
Redcar YON 57
Redcliff GLO, SOM 155
Reddaway DEV 95
Redenhall NFK 129

INDEX

Redfern WAR 46
Redgrave SFK 228
Redhill SUR 177
Redland GLO 282
Redmarley WOR 27
Redmarshall DRH 27, 194
Redmire YON 27
Rednal WOR 132
Reedness YOE 198
Reeth YON 29
Reighton YOE 214, 215
Remenham BRK 49, 304
Renacres LNC 264, 265
Rendcombe GLO 108
Renhold BDF 129
Renishaw DRB 247
Renscombe DOR 108
Repton DRB 129
Retford NTT 73
Rettenden ESX 169
Revelend HRT 311
Reydon SFK 169
Ribbesford WOR 75
Ribby LNC 190
*ric 214
Riccal YOE 128
Ricebridge SUR 68
Rice Bridge ESX, SSX 68
Riche LIN 215
Rickinghall SFK 128
Riddlecombe DEV 109
Riddlesden YOW 117
Ridge HRT, STF 190
Ridgeacre WOR 264, 265
Ridgebridge SUR 68
Ridgewardine SHR 191
Ridgewell ESX 35
Ridware STF 91
Rigsby LIN 190
Ringland NFK 282
Ringmer SFK 27
Ringmore DEV 60
Ringsfield SFK 272, 277
Ringshall Fm BUC 294

Ringwood HMP 259
Ringwould KNT 254, 257
Ripley var. cos. 240
Ripon YOW 129
Rippingale LIN 128–9
Ripponden YOW 118
Risborough BUC 150, 296
Risebridge DRH, ESX 68
Risebrigg YOW 68
Riseley BDF, BRK 240
Rishangles SFK 231
Risingbridge NTP 68
Risley DRB, LNC 240
Risplith YOW 185
Rissington GLO 165, 169
rīth, rīthig 2, 29–30
Ritton NTB 29
ritu- British 91
Rivacre CHE 264, 265
Rivenhall ESX 125, 129
Riverhead KNT 87, 88
Road SOM 244
Roade NTP 244
Roadway DEV 95
Robin Wood's Hill GLO 157
Roborough DEV 150
Rochdale LNC 111
Rochford ESX, WOR 76
Rockbear(e) DEV 223
Rockbourne HMP 12
Rockcliff CMB 155
Rockford HMP 75
Rockland NFK 243
Rockwell BUC 234, 314
Rodbourne WLT 12
Rodd HRE 244
Roddam NTB 244
Rodden SOM 117
Rode CHE 244
Roden SHR 16
Rodmarton GLO 22, 27
rodu 243–4
Rodway SOM 95
Roe HRT 244

Roeburn LNC 12
Roecliffe YOW 155
Roel GLO 32
Rofford OXF 74
Romaldesham SHR 47, 53
Romford GTL 73
Romney Marsh KNT 14
Romsey HMP 42
Rookabear DEV 223
Rookhope DRB 139
Rookwith YON 253
Roppa YON 89, 90
Rora DEV 208
Roscoe LNC 248, 249
Roseacre LNC 265
Roseberry YON 151, 152
Rosedale YON 111
Rosgill WML 123
Rossall LNC, SHR 129
Rossett YOW 236
Rossway HRT 306
*roth 244
Rothay R. 2
Rothend ESX 244
Rotherbridge SSX 69
Rotherfield var. cos. 87–8, 276, 307, 311
Rotherhithe GTL 87, 88
Rotherwas HRE 64
Rothley LEI, NTB 244
Rothwell var. cos. 33, 244
Rotsea YOE 30
Rottingdean SSX 118
Rounden SSX 267
Roundthwaite WML 252
Roundway WLT 95
Rousdon DEV 167
Routh YOE 244
Rowberrow SOM 150
Rowborough IOW 150
Rowland DRB 243
Rowley var. cos. 240
Rowley Hill Fm STF 159
Rowner HMP, SSX 205, 209

Rowridge IOW 191
Roxhill BDF 193
Roxton BDF 169
Roxwell ESX 32
Roy's Bridge NTT 68
Royd LNC, YOW 243
Roydon var. cos. 169
Ruckholt ESX 234
Ruckland LIN 243
Rucklers Green HRT 234, 315
Rudford GLO 75
Rudge SHR, WLT 190
Rudgwick SSX 191
Rudloe WLT 191
Rudway DEV 95
Rufford LNC, NTT 73
Rufforth YOW 73
Rugeley STF 191
Ruislip GTL 61
Rumbelow STF 178
Rumbridge HMP 69
Rumsden SSX 267
Rumwell SOM 33
Runhall NFK 132
Runwell ESX 33
Rushcliffe NTT 156
Rushall NFK 132
Rushall STF 129
Rushall WLT 128
Rushbrooke SFK 8
Rushden HRT, NTP 117
Rushford NFK 75
Rushmere SFK 27
Rusland LNC 282
Russellhead ESX 198
Ruthall SHR 132
Ruthwaite CMB 250
Rutland 283
Rutleigh DEV 155
Ryal NTB 193
Ryarsh KNT 268, 269
*rȳd 244
Rydal WML 111
Ryde IOW 29

INDEX

*ryden 244
*ryding 244
Rye var. cos. 41, 305, 312
Rye Fm SSX 286-7
Ryehurst BRK 269
Ryeish BRK 269
Ryersh SUR 269
Ryhall RUT 130
Ryhill YOE, YOW 193
Ryhope DRH 134, 139
Ryle NTB 193

Sabden LNC 117
Sadberge DRH 151
Saham NFK 30
St Ives HNT 61
St Lawrence IOW 94
St Mary le More BUC 305
St Nicholas at Wade KNT 94, 95
Salcombe DEV 108
Salden BUC 117
Salehurst SSX 235
Salford BDF, LNC 75
Salford OXF, WAR 76
Salkeld CMB 177, 233
Sall NFK 240
Salph BDF 188
Saltburn YON 11
Salterford NTT 75
Salterforth YOW 75
Saltfleet LIN 17
Saltford SOM 76
Saltmarsh(e) HRE, YOE 58
Saltwood KNT 260
Sambourn WAR 11
Sambrook SHR 8
Sampford var. cos. 73
Sandal(l) YOW 129
Sandbach CHE 3, 4
Sandford var. cos. 73
Sandhoe NTB 188
Sandhurst var. cos. 235, 236
Sandiacre DRB 265
Sandleford BRK 74

Sandlin WOR 182
Sandon var. cos. 168
Sandown SUR 168
Sandridge var. cos. 191
Sandwith CMB 94
Sandy BDF 42
Saniger GLO 230-1
Sansaw SHR 247
Saredon STF 172-3
Sarnesfield HRE 275
Sarscow LNC 249
Sarsden OXF 121
Satterthwaite LNC 251
Saughall CHE 129
Saul GLO 240
Saunderton BUC 168, 296
Sawbridge WAR 69
Sawdon YOW 117
Sawley YOW 240
Sawtry HNT 29
Saxondale NTT 110, 111, 118
sǣ, sā 30
Scackleton YON 121
Scaitcliffe LNC 155
Scalderskew CMB 249
Scaldwell NTP 32
Scaleber LNC 151
Scalescough CMB 248
Scalford LEI 73
Scammonden YOW 117
Scarcliff DRB 155
Scargill YON 123
Scarisbrick LNC 152
Scarthingwell YOW 33
Scawdale CMB 111
sceaga 245-7, 248, 315
scelf 144, 216-9
Sceugh var. cos. 248
Scorborough YOE 248
Scotforth LNC 75
Scottow NFK 188
Scottsquar GLO 203
Scow var. cos. 248
Scrafield LIN 272, 275

Scrainwood NTB 261
Scugdale YON 111
Seaborough DOR 151
Seabrook BUC 303
Seacombe CHE 30, 108
Seacroft YOW 30
Seaford SSX 30, 74
Seagrave LEI 230
Seagry WLT 30
Seaham DRH 30
Seamer YON 27, 30
Seanor DRB 202
Seascale CMB 30
Seathwaite CMB 251
Seathwaite LNC 30, 251
Seaton var. cos. 30
Seawell NTP 33
Seckington WAR 168
Secklow BUC 178
Sedberg YOW 151
Sedborough DEV 151
Sedbury YON 151
Seddlescombe SSX 108
Selborne HMP 12
*sele 63
Selgrove KNT 63
Selly Oak WOR 218, 241
Selsdon SUR 173
Selsey SSX 42
Selstead KNT 63
Selwood SOM 257, 260
Semer var. cos. 27, 30
Setchey NFK 87, 89
Sewell BDF 33, 299
Sewsterne LEI 278
Shadforth DRH 73
Shadoxhurst KNT 236
Shadwell MDX 32
Shadwell NFK, YOW 33
Shafford Fm HRT 307–8
Shafford's Knoll HRT 308
Shalbourne WLT 11
Shalcombe IOW 108
Shalden HMP 117

Shalfleet IOW 17
Shambridge OXF 296
Shanklin IOW 182
Shardlow DRB 179
Shareshill STF 219
Sharnbrook BDF 8
Sharncliffe GLO 155
Sharnford LEI 73
Sharow YOW 189
Sharpenhoe BDF xxii, 188, 297
Sharpness GLO 196, 198
Shaugh DEV 247
Shave Cross DOR 247
Shaw var. cos. 247, 315
Shawbury SHR 247
Shawdon NTB 247
Shawell LEI 33
Shawford HMP 73
Shay var. cos. 245–6
The Shays SHR 245
Shebbear DEV 223
Shebdon STF 168
Shedfield HMP 278
Sheepy LEI 42
Sheerness KNT 196, 198
Sheffield BRK 276
Sheffield SSX 276
Sheffield YOW 275
Shefford BDF, BRK 76
Sheldon DEV 114, 118
Sheldon WAR 168
Shelf DRB, YOW 217, 218
Shelfanger NFK 218
Shelfield STF, WAR 193, 218
Shelford CAM, NTT 73
Shell WOR 218
Shelland SFK 216, 218, 282
Shelley var. cos. 218, 241
Shellingford BRK 71, 75
Shelswell OXF 35
Shelton var. cos. 218
Shelton under Harley STF 216
Shelve SHR 217, 218
Shenfield ESX 275

INDEX

Shenington OXF 168
Shephall HRT 129
Sheppey KNT 42
Shepreth CAM 29
Shepshed LEI 176
Sherborne var. cos. 7, 11
Sherburn var. cos. 7, 11
Shereford NFK 73
Sherfield HMP 272, 275
Sherford DEV, SOM 73
Shernborne NFK 11
Sherwood NTT 257, 259
Shifford OXF 71, 76
Shifnal SHR 125, 127, 128
Shildon DRH 168, 218
Shilford NTB 73
Shillingbottom LNC 100
Shillingford DEV, OXF 80
Shillington BDF 173
Shilstone DEV 218
Shilton var. cos. 216, 218
Shilvinghampton DOR 218
Shincliffe DRH 156
Shinfield BRK 272, 276
Shingay CAM 42
Shingleswell KNT 33
Shipbourne KNT 12
Shipden NFK 117
Shiplake OXF 20, 303
Shiplate SOM 81
Shipley var. cos. 241
Shipmeadow SFK 285
Shipperbottom LNC 100
Shirburn OXF 10, 11, 299, 302
Shirebrook DRB 7, 9
Shirecliffe YOW 155
Shirland DRB 243
Shirley HRE 183, 184, 185
Shirley var. cos. 240
Shirwell DEV 32
Shobdon HRE 168
Shobrooke DEV 8
Shocklach CHE 21
Shoddesdon HMP 117

Shogmoor BUC 303
Sholden KNT 168
Shooter's Hill GTL 177
Shootersway HRT 306
Shopland ESX 282
Shoppenhangers BRK 232
Shortengrove WLT 228
Shortgrove var. cos. 228
Shorwell IOW 33
Shotford NFK/SFK 75
Shotover OXF xv, 200, 203, 308
Shottery WAR 29
Shottesbrooke BRK 9
Shotteswell WAR 32
Shottle DRB 193
Shovelstrode SSX 216, 218
Showell OXF 33
Shrivenham BRK 51, 55
Shuckburgh WAR 151
Shulbrede SSX 218
Shurlach CHE 21
Shurlacres LNC 264, 265
Shutford OXF 75
Shuthanger DEV 232-3
Shutlanger NTP 232
Sibertswold KNT 254, 256
Sibford OXF 74
Sibsey LIN 42
Sibson LEI 167, 168
sīc 2
Sicklinghall YOW 128
Sidcup GTL 159
sīde 219
Sidebottom CHE 100
Sidford DEV 76
Sidgreaves LNC 228
Sigford DEV 74
Silkmore STF 60
Silloth CMB 30
Silpho YON 188, 218
Silsden YOW 117
Silsoe BDF 188
Silton YON 218
Silverdale LNC 111

Simonburn NTB 12
Simonswood LNC 260
Sinderland CHE 284
Sinfin DRB 46
Singleborough BUC 150
Singlesole NTP 234
Sisland NFK 282
Sissinghurst KNT 236
Sittingbourne KNT 12
Sizewell SFK 32
Skeaf var. cos. 248
Skegness LIN 197
Skelbrooke YOW 9
Skelderskew YON 249
Skelding YOW 118
Skellingthorpe LIN 137, 139
Skellow YOW 130
Skelton var. cos. 216, 218
Skewfe var. cos. 248
Skewkirk YOW 248
Skews var. cos. 248
Skewsby YON 248
Skidbrook LIN 8
Skiddaw CMB 174
Skinburness CMB 198
Skippool LNC 28
Skirbeck LIN 6
Skirpenbeck YOE 6
Skirwith CMB 253
Skitterick var. cos. 214
skjalf ON 143, 218-9
skógr ON 221, 245, 248-9
Skuff var. cos. 248
Skutterskelfe YON 218, 219
Slad Valley GLO 142
Slaggyford NTB 73
Slaidburn YOW 12
Slaithwaite YOW 252
slakki ON 142
Slapton var. cos. 61, 62
Slaughter GLO 62
slæd 141-2, 295
slǣp 37, 61-2
Sleaford LIN 76

Sleagill WML 123
Sleap SHR 61
Sleddale WML 141
Sledmere YOE 27, 141
Sledwich DRH 286
Sleekburn NTB 11
Sleightholme var. cos. 56, 57
Sleningford YON 75
Slethat DMF 252
Slimbridge GLO 68, 69
Slindon SSX, STF 173
slōh 37, 62
Slough BUC 62
Slow Grove BRK 230
Smallacomb(e) DEV 106, 108
Smallburgh NFK 151
Smallcombe DEV 106, 108
Smallhanger BRK, DEV 232
Smallicombe DEV 106, 108
Smallridge DEV 191
Smardale WML 112
Smarden KNT 267
Smawith YOW 252
Smerrill DRB 193
Smithdown LNC 168
Smithills LNC 193
Smytham DEV 52
Smythapark DEV 89, 90
Snailham SSX 54
Snailwell CAM 32
Snarehill NFK 193
Snarford LIN 74
Snaygill YOW 129
Snelland LIN 243
Snelshall BUC 128
Sneismore BRK 58, 59, 60
Snilesworth YON 94
Snitterfield WAR 275
Snodhill HRE 193
Snodland KNT 282
Snorscomb NTP 107
Snowford WAR 80
Snowshill GLO 193
Snydale YOW 129

INDEX

Socknersh SSX 269
Softley DRH, NTB 240
Soham CAM 30
Soham SFK 30
Solberg YON 151
Solcum WOR 63
Soldridge HMP 191
Soles KNT 62
sol(h) 37, 62–3
Solihull WAR 193
Solinger BUC 231
Sollers Hope HRE 136
Solway Firth CMB 94
Solway Moss CMB 63
Somborne HMP 12
Somerford var. cos. 73
Somersal DRB 132
Songar WAR 232
Sonning Eye OXF 305
Sookholme NTT 57
Soothill YOW 193
Soppit NTB 90
Sopwell HRT 32
Sotwell BRK 35, 109
Soudley GLO 241
Southall GTL 130
Southam Holt WAR 256
Southampton HMP 49, 50, 53
South Hams DEV 52
Southburgh NFK 150
Southburn YOE 10
Southery NFK 43
Southfleet KNT 17
Southgrove WLT 229
Southolt SFK 234
Southmere NFK 27
Southmoor BRK 58
Southoe HNT 189
Southover DOR, SSX 200, 203
Southowram YOW 199, 202
Southrey LIN 43
Southwaite WML 250
Southwell NTT 33
Southwold SFK 254, 255

Southwood NFK 259
Soyland YOW 282
Sparcell's Fm WLT 234
Sparhanger DEV 232
Sparket CMB 177
Sparkford SOM 73
Sparkwell DEV 32
Sparsholt BRK, HMP 234
Speldhurst KNT 236
Spelhoe Hundred NTP 189
Spernall WAR 203
Spexhall SFK 129
Spofforth YOW 80
Spondon DRB 169
Spotland LNC 282
Springfield ESX 278
Spruisty YOW 93
Spurway DEV 96
Staddiscombe DEV 108
Staddon DEV 169
Staffield CMB 192
Stafford DEV, DOR 73
Stafford STF 71–2, 76, 91, 92
Stagenhoe HRT 188
Stagsden BDF 118
Stainburn CMB, YOW 10, 11
Staincliffe YOW 155
Staindrop DRH 139
Stainfield LIN 250, 275
Stainforth YOW 73
Stainland YOW 282
Stainmore WML/YOW 58, 59, 60
Stainsacre YON 265
Staithes YON 91
Stalbridge DOR 69
Stalisfield KNT 276
Stallenge DEV 182
Stambourne ESX 12
Stambridge ESX 69
Stamford var. cos. 73
Stamfordham NTB 73
Stanbridge BDF, HMP 69
Stancliff DRB 155
Standean SSX 117

Standedge YOW 173
Standen var. cos. 117
Standen IOW 168
Standish LNC 269
Standlake OXF 20
Standlynch WLT 182
Standon HRT, SFK 168
Stanedge DRB 173
Stanfield NFK 275
Stanford var. cos. 73
Stanford le Hope ESX 133
Stanhoe NFK 188
Stanhope DRB 139
Stanley var. cos. 240
Stanlow CHE 179
Stanmer SSX 22, 27
Stanmore GTL 27
Stanners Hill SUR 209
Stanney CHE 42
Stanninghall NFK 129
Stansfield SFK, YOW 278
Stanshope STF 140
Stantway GLO 95
Stanway var. cos. xxii, 95
Stanwell MDX 18, 32
Stapeley CHE 241
Stapely HMP 241
Stapenhill STF 193
Stapleford var. cos. 73
Staplegrove SOM 229
Staplehurst KNT 236
Staploe BDF 188
Starbotton YOW 99, 100
Startforth YON 74, 93
Statenborough KNT 149, 150
Statham CHE 91
Stathe SOM 91
Staveley var. cos. 241
Staverton DEV 73
Stavordale SOM 112, 118
Stawell SOM 32
stæth 66, 91-2
Stechford WAR 80
Stennerley LNC 186
Stenwith LIN 94
Stepney GTL 87, 88
Steppingley BDF 242, 312
Sterndale DRB 111
Sternfield SFK 278
Stewkley BUC 312
Stibbard NFK 276
Sticelett IOW 17
Stickford LIN 44, 80
Sticklepath DEV 90
Sticklinch SOM 181, 182
Stickney LIN 44
Stiffkey NFK 42
Stifford ESX 75
stīg 65, 90, 92-3, 184
stígr ON 92-3
Stildon WOR 173
Stillingfleet YOE 16, 17
Stinchcombe GLO 108
Stinsford DOR 75
Stirchley SHR 241
Stirzacre LNC 265
Stitchbrook STF 9
Stitchcombe WLT 108
Stivichall WAR 129
Stixwould LIN 256
Stoberry SOM 150
Stoborough DOR 150
Stobswood NTB 259
Stockbridge HMP 69
Stockeld YOW 177
Stockholt BUC 234
Stockingford WAR 73
Stockland DEV, SOM 282
Stockland YON 243
Stocklinch SOM 180
Stockwith 87, 88
Stockwood DOR 259
Stodmarsh KNT 58
Stoford SOM, WLT 73
Stoke Ferry NFK 71
Stonar KNT 206, 209
Stondon BDF, ESX 168
Stonea CAM 42

INDEX

Stoneacre KNT 265
Stoneleigh WAR 239, 240
Stonelink SSX 182
Stonely HNT 240
Stonesfield OXF 278
Stonnal STF 129
Stonor OXF 209, 298, 308
Stonyhurst LNC 236
Stoodleigh DEV 241
Stopsley BDF 312
Storiths YOW 63
Storrs LNC 63
Stortford HRT 74
storth ON 63
Storth CMB 63
stoth ON 91-2
Stottesdon SHR xvii, 169
Stowey SOM 95
Stowell var. cos. xxii, 1, 32
Stowford var. cos. 73
Stowood OXF 259
Stramshall STF 193
Stratfield BRK/HMP 93, 272, 275
Stratford var. cos. 71, 74, 93
Stratton var. cos. 65, 93
strǣt 65, 93-4, 306
Streat SSX 93
Streatham GTL 93
Streatlam DRH 93, 241
Streatley BDF, BRK 93, 241, 306
Street var. cos. 93
Streetley ESX 241
Streetly CAM 93, 241
Strefford SHR 74, 93
Strelley NTT 93, 241
Strensall YON 130
Strete DEV 93
Stretford HRE, LIN 74, 93
Strethall ESX 93, 129
Stretham CAM 93
Stretley ESX 93
Stretton var. cos. 65, 93
Strickland WML 282
Stroat GLO 93

strōd 63
Strood KNT 63
strōther 63
Stroud GLO 63
Strumpshaw NFK 247
Strutforth WML 253
Studdal KNT 254, 256
Studdridge BUC 311
Studham BDF 311
Studland DOR 282
Studley var. cos. 241
Stuntney CAM 42
Sturmer ESX 22, 27
Sturton var. cos. 93
Styal CHE 92, 129
Styford NTB 74, 92
Subberthwaite LNC 251
Sudbourne SFK 12
Sudbrooke (in Ancaster) LIN 8
Sudbrooke (near Lincoln) LIN 9
Sudeley GLO 241
Sudgrove GLO 229
Suffield NFK, YON 276
Sugarswell OXF, WAR 33
Sugnall STF 195
Sugwas HRE 64
Sulber YOW 151
Sulgrave NTP 229
Summersdale SSX 112
Sunderland var. cos. 284
Sunderlandwick YOE 284
Sundon BDF 168
Sundridge KNT 268, 269
Sunninghill BRK 194
Sunningwell BRK 33
Surfleet LIN 16, 17
Susacres YOW 265
Susscarrs YOE 248
Sutcliff YOW 156
Sutcombe DEV 35, 109
Sutgrove GLO 229
Sutherland YON 243
Swafield NFK 278
Swalcliff OXF 155

Swalecliffe KNT 155
Swaledale YON 111
Swallowcliffe WLT 155, 297
Swallowfield BRK 272, 275
Swalwell DRH 32
Swanborough SSX 151
Swanbourne BUC 12
Swangrove GLO 229
Swanland YOE 243
Swanmore HMP 26
Swanside YOW 219
Swarland NTB 282
Swarling KNT 182
Swarthmoor LNC 60
Swavesey CAM 86, 88
Swayfield LIN 278
Sweethope NTB 139
Swilland SFK 282
Swimbridge DEV 68
Swinbrook OXF 8
Swinburn NTB 12
Swinchurch STF 176
Swincliffe YOW 155
Swinden YOW 117
Swindon var. cos. 165, 169
Swinefleet YOW 16, 17
Swinehead Hundred GLO 175
Swineshead var. cos. 175-6, 212
Swinesherd WOR 175
Swinfen STF 46
Swinford var. cos. 76
Swinhoe NTB 188
Swinhope LIN 137, 139
Swinithwaite YON 250
Swinscoe STF 249
Swinside Hall ROX 175, 176
Swintley RUT 155
Swithland LEI 243
Swyncombe OXF 108, 293, 311
Syde GLO 219
Sydenham var. cos. 53-4
Sydenham GTL 54
Sydling DOR 182
Symondsbury DOR 150

Symond's Hall GLO 128
Syndercombe SOM 108
Sywell NTP 33

Tachbrook WAR 9
Taddiford HMP 75
Tadlow CAM 179
Tadmarton OXF 22-24, 26
Tandridge SUR 192
Tanfield DRH 275
Tanfield YON 278
Tangmere SSX 27
Tanholt NTP 234
Tanshelf YOW 218
Tansor NTP 203, 278
Tansterne YOE 278
Taplow BUC 178, 179, 296
Tarbock LNC 8
Tarlscough LNC 249
Tarnacre LNC 265
Tatenhill STF 193
Tathwell LIN 32
Tatsfield SUR 274
Tattenhall CHE 128
Tattenhoe BUC 190
Tatterford NFK 74
Tattershall LIN 128
Taxal CHE 132
Tebay WML 39, 42
Tedburn DEV 12
Teddington WOR 133
Tedsmore SHR 27
Teffont WLT 18
Teiscombe SSX 107
Tempsford BDF 76
Tenterden KNT 267
Tessal WOR 132
Tetford LIN 75
Tetney LIN 42
Tetsill SHR 193
Tettenhall STF 133
Thackthwaite CMB 250
Thame OXF 16
Thame R. 1

INDEX

Thatto LNC 33
Thedden HMP 118
Thenford NTP 75
Therfield HRT 275
Thetford CAM, NFK 75
Theydon ESX 117
Thicket YOE 177
Thiefside CMB 176
Thinghill HRE 194
Thixendale YOE 111
Thoresway LIN 95
Thorfinsty LNC 93
Thorley HRT, IOW 240
Thornage NFK 269
Thornborough BUC 148, 150
Thornbrough YON 150
Thorncombe DOR 107
Thornden SSX 267
Thorndon SFK 169
Thorner YOW 202
Thorney var. cos. 42
Thornford DOR 75
Thorngrove var. cos. 229
Thornhill var. cos. 193
Thorniethwaite DMF 251
Thornley DRH, LNC 240
Thornthwaite var. cos. 251
Thorpland NFK 282
Thorverton DEV 75
Threapland CMB 283
Threapwood CHE 260
Threlkeld CMB 18
Thremhall ESX 128
Threshfield YOW 276
Thrift var. cos. 225
Thriplow CAM 179
Throckenholt CAM 234
Throckmorton WOR 27
Throphill NTB 194
Thrybergh YOW 151
Thunderlow ESX 180
Thurgoland YOW 282
Thurland LNC 282
Thurlbear SOM 150

Thurnscoe YOW 248
Thurrock ESX xix
Thursford NFK 75
Thurstonland YOW 282
thveit ON 221, 249–52
Thwaite var. cos. 250
Tibshelf DRB 218
Ticehurst SSX 236
Tichborne HMP 12
Tickenhurst KNT 236
Tickford BUC 76
Tickham KNT 52
Tickhill YOW 193
Ticknall DRB 129
Tidcombe WLT 107
Tideford CNW 76
Tidenham GLO 52, 53
Tideswell DRB 32
Tidmarsh BRK 58
Tidnor HRE 202
Tidover YOW 199
Tiffield NTP 278
Tilberthwaite LNC 251
Tilbrook HNT 9
Tilehurst BRK 236
Tilford SUR 80
Tillingdown SUR 169
Tiln NTT 44
Tilney LIN, NFK 38, 44, 125
Timbercombe SOM 108
Timberhanger WOR 232
Timberland LIN 242, 243
Timbridge WLT 191
Timsbury SOM 223
Tinacre DEV 264
Tincleton DOR 118
Tingreave LNC 228
Tingrith BDF 29
Tinhay DEV 16
Tinney DEV 16
Tintinhull SOM 195
Tinwell RUT 35
Titchfield HMP 276
Titchmarsh NTP 58

Titchwell NFK 33
Titsey SUR 42
Tittensor STF 202
Tittleshall NFK 128
Tiverton DEV 73
Tivetshall NFK 129
Tixall STF 129
Tixover RUT 202
Tockwith YOW 253
Todber DOR 151
Todber YOW 150
Toddington BDF 169
Todenham GLO 50, 53
Todlith MON 185
Todmorden YOW 118
Togstone NTB 117
Tolland SOM 283
Toller Whelme DOR 2
Tolleshunt ESX 18
Tolpits HRT 90, 91
Tonbridge KNT 70
Topcliffe YON 153, 156
Topley DRB 156
Toppesfield ESX 275
Topsham DEV 50, 54
Torkington CHE 44
Torksey LIN 44
Tormarton GLO 22, 27
Torrisholme LNC 56
Toseland HNT 243
Tosside YOW 219
Tothill LIN, MDX 194
Totland IOW 282
Totnes DEV 197, 198
Tottenham Court GTL 128
Tottenhill NFK 193
Towersey BUC 41, 305
Trafford LNC 74, 93
Trafford NTP 75
Tranwell NTB 32
Trawden LNC 117, 142
Treborough SOM 150
Trenance CNW 141
Trenant CNW 141

Trenholm YON 57
Trentishoe DEV 188
Tretire HRE 91
Treyford SSX 73
Tring HRT 233
Trobridge DEV 69
trog 107, 142
Trottiscliffe KNT 153, 155
Trough var. cos. 142
Troughfoot CMB 142
Troughton LNC 142
Trow WLT 142
Trowbridge WLT 69
Trowell NTT 33
Trumfleet YOW 17
Trussel Bridge CNW 69
Tubney BRK 39, 42
Tudhoe DRH 188
Tugford SHR 74
Tughall NTB 128
Tungrove GLO 229
Tupholme LIN 56
Turlhanger's Wd HRT 315
Turn Bridge YOW 69
Turtley DEV 155
Turvey BDF 42
Turville BUC 275, 311
Turzes SSX 269
Tusmore OXF 27
Tutnall WOR 193
Tuxford NTT 35, 80
Tuxwell SOM 35
Twemlow CHE 178
Twickenham GTL 48, 49, 54
Twigmore LIN 60
Twineham SSX 16
Twinyeo DEV 16
Twyford var. cos. 73
Twywell NTP 32
Tyburn GTL 12
Tylerhill KNT 177
Tyringham BUC 48
Tysoe WAR 144, 186, 189
Tyttenhanger HRT 232, 315

INDEX

Uckfield SSX 274
Udimore SSX 27
Ugborough DEV 150
Ugford WLT 74
Uggeshall SFK 128
Ughill YOW 193
Ulcombe KNT 108
Uldale CMB 111
Ullenhall WAR 129
Ulleskelf YOW 218
Ullthwaite WML 251
Ulpha CMB 174
Underley HRE 183, 185
Underwood KNT 260
Unhill Wd BRK 234
Uphill SOM 28
Upleatham YON 186
Upnor KNT 210
Upsland YON 243
Upwell CAM 31
Upwood HNT 259
Urchfont WLT 18
Urpeth DRH 90
Uxbridge GTL 69
Uxmore OXF 303, 304
Uzzicar CMB 265

Vange ESX 44, 46
vath ON 94
Verwood DOR 259
Vexour KNT 209
View Edge SHR 144
vithr ON 221, 252-3
Viveham DEV 52

Waberthwaite CMB 251
Wackland IOW 283
Wadborough WOR 151
Wadden Hall KNT 130
Waddesdon BUC 168
Waddicar LNC 266
Waddington SUR 169
Waddon DOR, SUR 169
Wade HMP, SFK 94

Wadebridge CNW 94, 95
Wadenhoe NTP 188
Wadhurst SSX 235
Wadshelf DRB 218
Wainfleet LIN 17
Wainhill OXF 133, 294
Wainlode GLO 81
Waithe LIN 94
Waithwith YON 252
Wakebridge DRB 69
Wakefield NTP, YON 276
Walbrook GTL 8
Walburn YON 12
wald 246, 253-7, 313
Walden var. cos. 118
Walderchain KNT 254, 255, 256
Waldersea CAM 42
Waldershare KNT 254, 255, 256
Waldershelf YOW 218
Walderslade KNT 141, 256
Waldingfield SFK 255, 256, 276
Waldridge BUC 191
Waldridge DRH 191
Waldringfield SFK 276
Waldron SSX 256
Walfield CHE 275
Walford DOR 80
Walford HRE, SHR 32, 74
Walford (near Ross) HRE 75
Walgrave NTP 229
Walham GLO 50, 53
Walkden LNC 121
Walker NTB 57
Walkerith LIN 87, 88
Walkerslow SHR 179
Walkham R. DEV 121
Walkwood WOR 260
Wall SHR 31
Wallasey CHE 42
Walleybourne SHR 12
Wallhope GLO 32, 135, 139
Wallingford BRK 71, 75
Wallingwells NTT 32
Wallop HMP, SHR 32, 135, 139

385

Walltown SHR 31
Walmer KNT, WOR 27
Walpole NFK, SFK 28, 29
Walsall STF 128
Walsden LNC 121
Walshaw LNC 247
Walshford YOW 75
Walter Hall ESX 70-7
Waltham var. cos. 255, 256
Wambarrow SOM 150
Wambrook SOM 9
Wanborough SUR, WLT 150
Wandon End HRT 117
Wanford SFK 86
Wangford SFK 80
Wansford YOE 75
Wansunt GTL 18
Wantage BRK xix, xx
Wantisden SFK 117
Wapley HRE 183-4, 185
Wappenshall SHR 128
Wapsbourn SSX 96
Warborough OXF 148, 151
Warbreck LNC 152
Warcop WML 158-9
Warden BDF 169
Wardle CHE, LNC 194
Wardour WLT 209
Warenford NTB 76
Warfield BRK 276
Warford CHE 76
Wargrave BRK 226
Warmfield YOW 278
Warminghurst SSX 236
Warmscombe OXF 293
Warmwell DOR 32
Warnborough HMP 13
Warndon WOR 168
Warne DEV 46
Warnford HMP 80
Warninglid SSX 184, 185
Warningore SSX 210
Warsop NTT 137, 139
Warpsgrove OXF 96, 229, 314

Warthill YON 194
Wasdale CMB, WML 111
Washbourne DEV 13
Washbourne GLO 12, 64
Washbrook SFK 9
Washfield DEV 275
Washford DEV 73
Washford SOM 76
Washingley HNT 242
Wasperton WAR 64
Wassand YOE 94
Watchet SOM 224
Watchfield BRK 271, 274
Watchingwell IOW 33
Watcombe DEV 108
Watcombe OXF 108, 293
Waterbeach CAM 3, 30, 144, 145
Watercombe DOR 30, 108
Waterdale Fm HRT 295
Waterden NFK 30, 117
Waterdine SHR 30, 117
Waterperry OXF 30
Watership HMP 31
Waterslade CAM 141
Waterstock OXF 30
Watford HRT, NTP 75
Wath YON, YOW 94
Watnall NTT 190
Wattisfield SFK 274
Watton YOE 169
Wavendon BUC 168
Way DEV 95
Wayford SOM 95
*wæsse xiii, 26, 37, 63-4
wæter 30-1
Weald var. cos. 254, 255, 256
The Weald KNT/SSX 220, 253-4, 257
Weald/Wild CAM 254
The Weald Moors SHR 58, 59, 60
Weathergrove DOR 228
Weddicar CMB 265
Wedmore SOM 60
Wednesfield STF 276

INDEX

Weedon BUC, NTP 170
Weeford STF 76
Weeley ESX 240
Weetslade NTB 141
Weetwood NTB 259
weg 65, 95–6, 192, 306
Welbatch SHR 3, 4
Welborne NFK 12, 32
Welbourn LIN 12, 32
Welburn YON 12, 32
Welbury YON 32, 151
Welby LIN 31
Welcombe DEV, WAR 32, 48
Weldon NTP 32, 169
Welford BRK 75
Welford GLO, NTP 32, 74
Welham NTT, YOE 31
well xxi, xxii, 1, 18, 31–5, 299
Well var. cos. xxi, 31
Welland WOR 282
Wellesbourne WAR 13
Wellingham SSX 53
Wellingore LIN 202
Wellow NTT 32
Wellow LIN 32
Wells NFK, SOM 31
Wellsborough LEI 151
Welney NFK 15, 31
Welwick YOE xii, 31
Welshpool MTG 28
Welton var. cos. 31
Wembdon SOM 169
Wendens ESX 117
Wendover BUC 294, 299, 302
Wendy CAM 42
Wenslow BDF 180
Wentford SFK 80
Wentnor SHR xvi, 200, 202
Weo SHR 186, 189
Wepre FLI 121
***werpels** 96
Werrar IOW 209
Wervin CHE 46
Wescoehill YOW 248

Westbourne SSX 10, 12
Westbourne GTL 12
Westbrook BRK 9
Westbrook IOW, KNT 9
Westerdale YON 111
Westerfield SFK 272, 276
Westfield NFK 276
Westhall SFK 130
Westhope HRE, SHR 136, 139
West Leith HRT 184
Westley CAM, SFK 241
Westley Fm DEV 287
Westley SHR 64
Westmorland 284
Westoe DRH 190
Westonzoyland SOM 43
Westover var. cos. 200, 203
Westridge BRK 190, 191
Westwell KNT, OXF 33
Westwood KNT 260
Westwood var. cos. 259
Westworth YON 253
Wetheral CMB 129
Wetherden SFK 117
Wethersfield ESX 274
Wetmoor STF 27
Wettenhall CHE 129
Wetwood STF 259
Wexcombe WLT 108
Weybourne NFK 13
Weybread SFK 95
Weybridge SUR 69
Weycock Hill BRK 158
Weyhill HMP 95
Whaddon var. cos. 165, 169
Whaddon WLT 117
Whale WML 192
Whaley CHE 95
Whalley LNC 192
Whalton NTB 192
Whaplode LIN 20
Wharncliffe YOW 156
Whatborough LEI 151
Whatcombe BRK, DOR 108

Whatfield SFK 276
Wheatacre NFK 265
Wheatenhurst GLO 236
Wheatfield OXF 275
Wheathampstead HRT 311
Wheathill SHR, SOM 193
Wheatley var. cos. 241
Wheatshaw LNC 247
Wheddon SOM 117
Wheldale YOE 111
Wheldrake YOE 214, 216
Whelpo CMB 174
Whernside YOW 219
Wherwell HMP 32, 101
Whessoe DRH 190
Whichford WAR 75
Whinburg NFK 150
Whissendine RUT 118
Whistley BRK, NTP, WLT 287
Whitacre WAR 265
Whitaker LNC 265
Whitbarrow WML 150
Whitbourne HRE 12
Whitcliff(e) var. cos. 155
Whitcombe DOR, IOW, WLT 108
Whiteacre KNT 265
Whitecliff DOR 155
Whitefield var. cos. 272, 275
Whitehaven CMB 176
Whitehill DRH 193
Whitehill OXF 194
Whiteoxmead SOM 285
Whiteway DEV 95
Whitfield var. cos. 275
Whitford DEV 73
Whitgreave STF 228
Whitleigh DEV 240
Whitley var. cos. 240
Whitlinge WOR 182
Whitmore STF 60
Whitney HRE 44
Whitnole DEV 157
Whitsun Brook WOR 8
Whittingslow SHR 179

Whittington Tump WOR 161
Whittle LNC, NTB 193
Whittles Fm OXF 296
Whittlesey CAM 42
Whittlesford CAM 74
Whittlewood NTP 257, 260
Whitton LIN 44
Whitwell var. cos. 32
Whitwood YOW 259
Whixall SHR 128
Wibsey YOW 38, 40, 42
Wichenford WOR 75
Wickford ESX 76
Wicklewood NFK 259
Wickmere NFK 27
Widcombe SOM 108
Widdale YON 252
Widdecombe DEV 106, 108
Widdicombe DEV 107
Widdop YOW 139
Widefield DEV 275
Widford var. cos. 75
Widley HMP 240
Widmer BUC 303
Widmere BUC 303
Widmerpool NTT 27, 29
Widmore BUC, HRT, OXF 303
Widnell BUC 294
Widnes LNC 196, 198
Wield HMP 256
Wigborough ESX 150
Wigford LIN 76
Wiggenhall LIN 125, 128
Wiggenholt SSX 234
Wiggold GLO 256
Wighill YOW 130
Wightfield GLO 275
Wigland CHE 283
Wigley DRB, HMP 241
Wigmore HRE, SHR 59, 60
Wigton YOW 170
Wilcot WLT 31
Wild var. cos. 255, 256
Wildboarclough CHE 102

INDEX

Wilden BDF 121
Wilden WOR 169
Wilderhope SHR 136, 139
Wildmore LIN 60
Wildon YON 168
Wilford NTT, SFK 75
Willand DEV 282
Willenhall STF, WAR 133
Willersey GLO 42
Willesborough KNT 150
Willey Fm CAM 86, 88
Willey DEV 240
Willey var. cos. 240
Willingale ESX 129
Willinghurst SUR 269
Wilmslow CHE 179
Wilsden YOW 117
Wilsford LIN, WLT 75
Wilsill YOW 129
Wilton SOM, WLT 31
Wimbledon GTL 168
Wimborne DOR 12
Wimpole CAM 28, 29
Winchcombe GLO 105, 108
Winchelsea SSX 42
Winchendon BUC 173
Winchfield HMP 275
Wincle CHE 195
Windermere WML 1, 22, 27
Winderwath WML 94
Windhill YOW 193
Windle LNC 193
Windsor var. cos. 208
Winestead YOE 190
Wingfield BDF 278
Wingfield DRB 275
Wingfield SFK 276
Wingfield WLT 274
Wingrave BUC 227, 228
Winkburn NTT 10
Winkfield BRK 274
Winnal HRE 133
Winnersh BRK 268, 269
Winscombe SOM 107

Winsford CHE 74
Winsham DEV 52
Winshill STF 193
Winslow BRK 156
Winslow BUC 179
Winsor DEV, HMP 208
Winswell DEV 32
Winterbottom var. cos. 100
Winterbo(u)rne var. cos. 11, 302
Winterbrook OXF 303
Winterslow WLT 179
Wirples Moss LNC 61, 96
Wirrall CHE 129
Wirswall CHE 32
Wisbech CAM 3, 144, 155, 287
Wisborough Green SSX 151, 286
wisc 286–7
Wiscombe Park DEV 287
The Wish SSX 286
Wish Tower SSX 286
Wishanger GLO 232, 287
Wishaw WAR 247
Wishford WLT 75
Wisley SUR 287
***wisse** 286–7
Wistlers Wd SUR 287
Wistley Hill GLO 287
Wiswell LNC 35
Witcham CAM 50, 54
Witcham Hithe CAM 86, 88
Witchford CAM 75
Witcombe GLO 108
Witcombe SOM 107
Withcote LEI 158
Witheridge DEV 191
Witheridge OXF 296
Withernsea YOE 30
Withersdale SFK 111
Withersfield SFK 276
Witherslack WML 142, 252
Withiel SOM 240
Withnell LNC 193
Withybrook WAR 8
Withycombe DEV, SOM 107

389

Withyham SSX 50, 54
Witney OXF 39–40, 42
Wittenham BRK 49, 53
Wittersham KNT 50, 53
Witton var. cos. 258
Wivenhoe ESX 190
Wixford WAR 74
Wixhill SHR 193
Wixoe SFK 188
Woburn BDF, SUR 12
Wokefield BRK 272, 274
Wolborough DEV 150
Wold BDF 254, 256
The Wold CAM 254
Woldingham SUR 255, 256
The Wolds var. cos. 256
Wolferlow HRE 179
Wolf Hall WLT 129
Wolmersty LIN 93
Wolsty CMB 92, 93
Wolvershill WAR 192, 193
Wombourn STF 10, 12
Wombridge SHR 70
Wombwell YOW 32, 70
Womenswold KNT 254, 256, 257
Wonersh SUR 268, 269
Wonford DEV 74
Wooburn BUC 12, 302
Woodacre LNC 265
Woodale YON 111
Woodbarrow SOM 150, 259
Woodbeer DEV 223
Woodborough NTT 259
Woodborough WLT 150, 259
Woodbridge SFK 70, 258
Woodburn NTB 12, 259
Woodbury DEV 259
Woodchester GLO 259
Woodchurch CHE, KNT 258
Woodcote var. cos. 258, 313
Woodcott CHE, HMP 258
Woodcroft NTP 259
Wooden NTB 169
Woodford var. cos. 74, 259

Woodgarston HMP 259
Woodgreaves YOW 229
Woodhall var. cos. 248, 258
Woodham BUC 51, 53
Woodham DRH 258
Woodham var. cos. 258
Woodhay BRK, HMP 259
Woodhorn NTB 259
Woodhouse var. cos. 248, 258, 259
Woodhurst HNT 235
Woodkirk YOW 258
Woodland DEV 259
Woodlands var. cos. 259, 281, 282
Woodlesford YOW 74
Woodleigh DEV 259
Woodloes WAR 179
Woodnesborough KNT 148–9, 151
Woodrow WLT, WOR 259
Woodsetts YOW 259
Woodsfield WOR 278
Woodsford DOR 74
Woodsome YOW 258, 259
Woodstock OXF 259
Woodthorpe DRB, LIN 259
Woodton NFK 258
Wool DOR 31
Woolacombe DEV 32, 108
Woolcombe DOR 32, 108
Wooldale YOW 111
Wooler NTB xv, 199, 203
Woolhanger DEV 230, 232
Woolhope HRE 136, 139
Woolland DOR 282
Woolley SOM 32
Woolley var. cos. 241
Woolmer HMP 26
Woolsgrove DEV 228
Wools Grove WLT 228
Woolstanwood CHE 260
Woolstone BRK xx
Woolvers Barn BRK 209
Wooperton NTB 122
Wootton var. cos. 258
Wordwell SFK 32

INDEX

Worfield SHR 275
Worksop NTT 137, 140
Wormdale KNT 112
Wormegay NFK 42
Wormhill DRB 193
Wormingford ESX 74
Worminghall BUC 129
Wormley HRT 241
Worms Head GLA 176
Worms Heath SUR 175
Wormsley HRE 241
Wormwood Scrubs GTL 234
Wornish CHE 269
Worplesdon SUR 96
Worral YOW 129
Worsall YON 128
Worthele DEV 130
Worton OXF 208
Wothersome YOW 258, 259
Wotton var. cos. 258, 313
Wrabness ESX 196, 198
Wraxall var. cos. 129
Wreighill NTB 194
Wrentnall SHR 128
Wribbenhall WOR 128
Wrinehill CHE 194
Wroxall IOW, WAR 125, 129
Wroxhills Wd OXF 294
wudu xxi, 221, 228, 252, 257–61, 313
Wuerdle LNC 195
Wychnor STF 200, 202
Wychwood OXF 257, 260
Wycliffe YON 156
Wycombe Marsh BUC 305, 312
Wydale YON 252
Wyddial HRT 129
Wyford Grange OXF 311
Wymeswold LEI 256
Wysall NTT 190
Wythburn CMB 98–9, 100
Wytheford SHR 75
Wythenshawe CHE 247
Wythop CMB 139

Wyville LIN 33

Yafford IOW 76
Yafforth YON 80
Yagdon SHR 169
Yantlet Creek KNT 21
Yanwath WML 252
Yarcombe DEV 108
Yardgrove DOR 229
Yardley var. cos. 241
Yarley SOM 241
Yarner DEV 202
Yarnfield STF 275
Yarnfield WLT 275
Yarnscombe DEV 108
Yarsop HRE 139
Yarwell NTP 33
Yattendon BRK 118
Yaverland IOW 282
Yazor HRE 202
Yeadon YOW 173
Yealand LNC 282
Yearsley YON 241
Yeaton SHR 14
Yeavering NTB 153
Yelden BDF 118
Yelford OXF 71, 74
Yell Bank SHR 177
Yelverton DEV 74
Yen Hall CAM 129
Yeo R. DEV, SOM 14
Yeoveney MDX 42
Yewdale LNC 111
Yewden BUC 293
Yockenthwaite YOW 250
Yokefleet YOW 16, 17
Youlgreave DRB 228
Yoxall STF 133
Yoxford SFK 76